Polymer Surfaces in Motion

Juan Rodríguez-Hernández • Carlos Drummond
Editors

Polymer Surfaces in Motion

Unconventional Patterning Methods

Springer

Editors
Juan Rodríguez-Hernández
Institute of Polymer
Science and Technology
Madrid, Spain

Carlos Drummond
Centre de Recherche Paul-Pascal
CNRS, Université de Bordeaux
Pessac, France

ISBN 978-3-319-17430-3 ISBN 978-3-319-17431-0 (eBook)
DOI 10.1007/978-3-319-17431-0

Library of Congress Control Number: 2015943852

Springer Cham Heidelberg New York Dordrecht London

Printed on acid-free paper

Springer International Publishing AG Switzerland is part of Springer Science+Business Media (www.springer.com)

Contents

Chapter 1
Nonconventional Methods for Patterning Polymer Surfaces

Carlos Drummond and Juan Rodríguez-Hernández

1.1 Introduction

Structured surfaces are present in Nature for many different purposes. Self-cleaning leaves with microscopic wax patterns, antireflective properties of moth eyes, superior adhesion of gecko feet via the enhanced adhesive interaction of nano-hairs, reduced friction and wear of the skin of sandfish via nanostructured scales, or turbulence reduction near shark' scales by micro-grooves are few examples of enhanced surface functionality by microstructuring or nanostructuring. Commercial materials with enhanced properties strongly inspired in some of these examples of Nature are already in the market. Dirt-repellent coatings and fabrics, nanostructured finish to reduce the need for car-waxing, self-cleaning vehicle windshields and mirrors, or antireflection and anti-UV coatings are some examples of surface properties enhancement by microstructuring and nanostructuring. A vast number of other applications may be anticipated for structured surfaces. Magnetic, electric, or chemical properties can be exploited to build information storage devices, displays, lighting and energy systems, actuators, and sensors. The impact of nanostructured surfaces will extend all the way from materials science and microelectronics to bioengineering. However, there is a great deal of research that needs to be conducted in this field, in order to achieve technical control of surface structure and functionality at the nanometric scale. This includes studying the structuring

C. Drummond
Centre de Recherche Paul Pascal, CNRS-Université Bordeaux 1,
Avenue Albert Schweitzer, 33600 Pessac, France
e-mail: drummond@crpp-bordeaux.cnrs.fr

J. Rodríguez-Hernández (✉)
Instituto de Ciencia y Tecnología de Polímeros, CSIC, Juan de la Cierva 3,
28006 Madrid, Spain
e-mail: jrodriguez@ictp.csic.es

J. Rodríguez-Hernández, C. Drummond (eds.), *Polymer Surfaces in Motion*,
DOI 10.1007/978-3-319-17431-0_1

possibilities of new materials, achieving new structures and functionalities for old materials, as well as developing new patterning techniques.

1.2 Classic Approaches for Patterning of Polymer Surfaces

During the last few decades numerous strategies of surface structuration have been reported. They are commonly classified as top-down or bottom-up methods. The first group involves shaping the microstructure of a large piece of material to the desired structure. It includes techniques as photolithography, sputtering, inkjet printing, or molecular beam epitaxy. Although the invention of photolithography can be traced back to the nineteenth century, the microelectronics industry has been the main driving force for its continuous development. These techniques have been greatly improved as a result of growing efforts to increase the density of data storage or the number of components in circuit boards. State-of-the-art photolithography facilities can be used to produce structures of few tens of nm in size. To achieve these very small features extremely costly clean room facilities of limited access are required, which involves high operational costs. However, the minimum feature size that can be obtained by conventional photopatterning is restricted by the diffraction limit of the light. In addition, in many cases these techniques are not adequate for biological or organic materials, or non-planar surfaces. These limitations may be surpassed by using near-field or contact mask techniques.

Soft lithography techniques, based on the use of polymer stamps, have been proposed as easier and cheaper alternatives for surface patterning. The use of soft stamps allows patterning of planar or non-planar substrates, transferring the pattern from the master to the substrate by techniques like Replica Molding (by polymerizing, reticulating or solidifying a liquid material against the soft mold) or Microcontact Printing (by transferring an ink to a surface using a patterned stamp). Typically sub-100 nm patterns can be produced by these methods; however, as the stamps readily deform under compression as a consequence of their low elastic modulus, producing smaller features with these techniques represents a challenging task.

Bottom up strategies are based on molecular self-assembly or self-recognition to promote matter organization in a desired configuration, arranging the elementary bricks in more complex arrangements by building structures atom-by-atom, or molecule-by-molecule. Techniques based on molecular or atomic manipulation (e.g., dip pen nanolithography) can also be included in this category. Recent advances in macromolecular chemistry make possible the production of complex block-copolymers; a large variety of self-assembly structures can be rationally produced based on these materials. Some of those arrangements are discussed in Chap. 4. Self-assembly techniques may provide exquisite control of the final material under reasonably manageable conditions, but they face obvious scale-up challenges.

The need for easier and more versatile strategies of surface patterning has motivated the development of many research programs in this area. A number of new patterning techniques, some of which are difficult to catalog in the classification described above, have been reported during the last few decades. The purpose of this volume is to bring together some of the different strategies of control of surface morphology of polymers. We focus on methods based on controlling the outcome of "pushing" a particular system close or beyond an unstable condition, or dynamically manipulating the process of material preparation. Many of these techniques are low-cost and relatively easy to implement, and may prove to be useful for patterning of polymer films and surfaces at small scales or for prototyping.

1.3 Surface Patterning Approaches Based on Surface Instabilities or Dynamic Process Control

Unstable conditions are often undesirable and are associated to lack of control of a given process. For instance, film dewetting may produce imperfect coatings resulting in improper functionality or bad appearance. However, understanding the dynamic behavior of a specific unstable condition may open pathways to valid patterning techniques, if the outcome of the surface reconstruction due to the instability can be mastered. The interest of using surface instabilities to pattern polymer surfaces relies on the rich and complex patterns obtained as a result of spontaneous processes. These patterns are rather difficult, if not impossible, to fabricate by using traditional patterning techniques. In addition, most of these techniques do not require the use expensive equipment; fast and inexpensive modification of polymer surfaces can be performed in regular laboratory environment by these means. Similar considerations apply to methods based on the dynamic control of process of preparation and shaping of polymer materials. A fine control of reaction or drying conditions may allow the tuning of surface morphology. For these reasons many research groups worldwide are currently working on understanding and developing patterning strategies based on controlling surface instabilities or dynamic processes of polymer materials. We can classify the techniques described in this book in three broad categories, based on the methods used to guided polymer patterning. Figure 1.1 shows illustrative images of the surface patterns obtained by using the approaches depicted below.

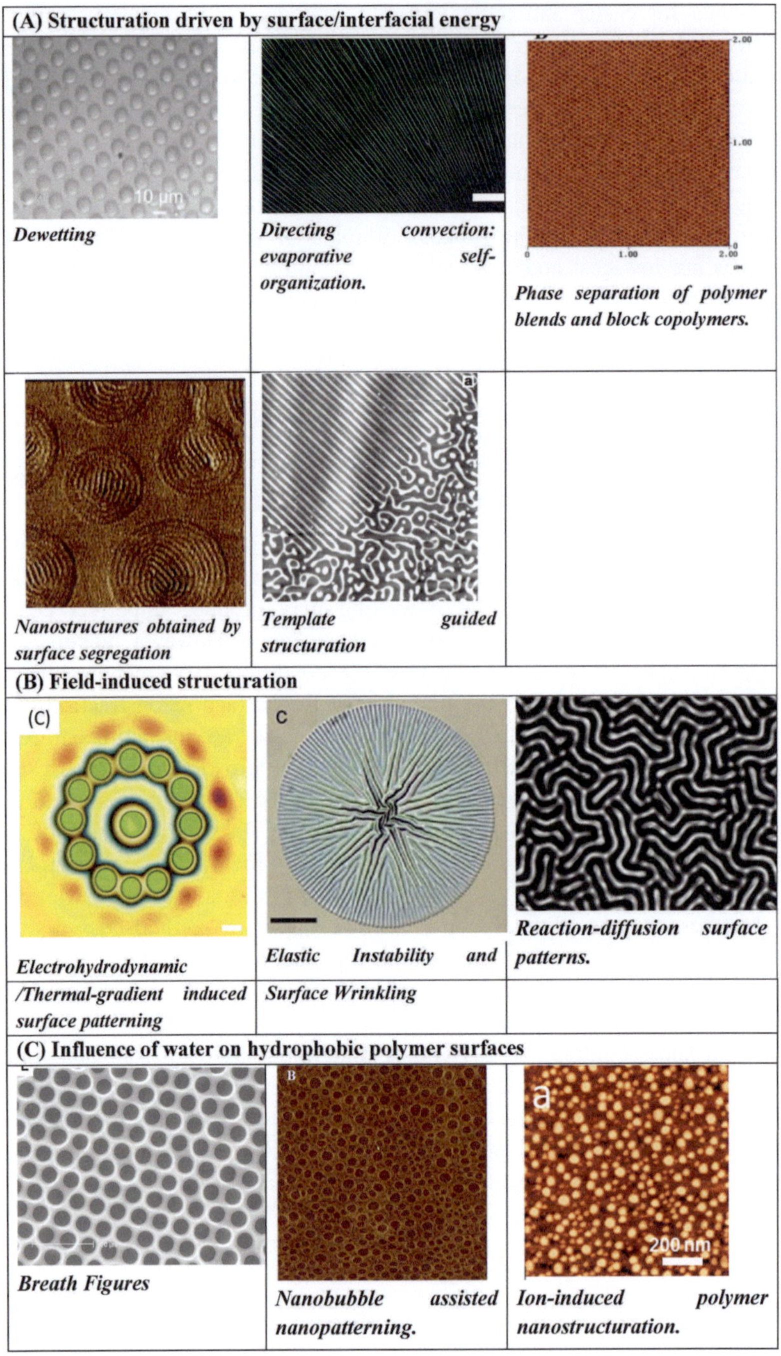

Fig. 1.1 Illustrative images of the surface patterns obtained by using different instability-based patterning approaches (**a**) Structuration driven by surface/interfacial energy [1–5] (**b**) Field-induced structuration [6–12] (**c**) Influence of water on hydrophobic polymer surfaces

1.3.1 Spontaneous Structuration Driven by Surface/Interfacial Energy

In this group can be included the structuration driven by dewetting, evaporative self-organization, block copolymer phase separation, segregation of additives to surfaces, or template guided surface structuration.

Dewetting may be induced by long-range attractive interaction forces intrinsic to the different interfaces present (e.g., polymer–air and polymer–substrate). Dewetting may occur for the case of thin films, but it is less important for thick films or bulk materials. After the seminal work of Reiter [13], a great deal of research has been done on this important subject. Often, the film rupture leads to the formation of disordered morphologies; [14–16] random distributions of holes, bicontinuous structures, polygon, and droplets among others have been observed [17, 18]. The control of dewetting can be used for patterning polymeric materials. A better degree of patterning control has been reached using heterogeneous substrates [19] fabricated using microcontact printing [20–22], vapor deposition [23], or photolithography [24]. Dewetting of thin films on periodic physically and chemically patterned surfaces can produce a large variety of surface patterns since the polymer film will wet and dewet specific areas. A description of this process is presented in Chap. 2.

Directing convection: evaporative self-organization. Ordered structures can be obtained by controlled evaporative and simultaneous self-assembly processes in a drying polymer solution. As described by Lin et al. [25] drying a drop of solution containing nonvolatile solutes on a solid surface is a simple, emergent technique to yield self-assembled, 1D or 2D structures with controlled dimensions (features of the order of few hundred nanometers), function, and topology. In this context, one of the most extended examples of surface structures formed by controlled evaporation processes is the case of "coffee ring" patterns. Coffee rings are observed when a spilled drop of a dispersed solute dries on a solid surface, leaving a dense, ring-like deposit along the perimeter. A large number of studies have evidenced the possibility of delicately manipulating the drying process in order to tune the assembly of a large variety of solutes that include inorganic nanoparticles, polymers, and biological molecules. A description of this process is presented in Chap. 3.

Phase separation of polymer blends and block copolymers. Confining polymer blends and block copolymers between surfaces may influence the phase separation process, as a consequence of the preferential affinity of one of the components for the interface. Since the pioneer works of Reich and Cohen [26] and later by Nesterov et al. [27], Ball and Essery [28], and Jones [29] amongst others much work has been done to understand the mechanisms of phase separation in polymer thin films. The presence of substrate–film and/or film–air interfaces introduces an additional complexity compared to bulk phase separation processes [30–35]. Complex structures can be produced by slight differences on parameters

[36] such as film thickness [26, 34, 37–41], affinity for the substrate employed [42–46], substrate patterning [5, 47], surface tension of the polymer [39, 48, 49], Flory–Huggins parameter [49–55], molecular weight of the polymer [56–58], component ratio [39, 52, 59–62], the relevant solvent parameter [38, 42, 63, 64], evaporation speed [65–69] and polymer concentration [67], chemical reaction in the blend [70], addition of additives [71], or even the temperature [72, 73]. Moreover, since the structures obtained are typically far from the equilibrium [34, 63] additional annealing steps will induce the formation of thermodynamically stable morphologies [36, 72]. Some examples of this method are discussed in Chap. 4.

Surface segregation is a thermodynamical phenomenon that directs the interfacial migration of certain components within a particular blend. In general, provided that there exists an appreciable difference in surface energy between the components of the blend, the lower energy component is preferred at the interface in order to minimize the surface free energy [38, 74–78]. In some cases the surface migration is accompanied by self-assembly processes (for instance using blends containing block copolymers) leading to surfaces with nanometer size features. This method is described in Chap. 5.

Template guided structuration include those approaches that involve an interface which direct the phase-separation process. Topographically and/or chemically patterned substrates [72] can modify and guide the phase separation process leading to a variety of morphologies. For instance, Steiner and coll. have shown that the phase separated domains created from a polymer blend in thin films can be guided into previously designed morphologies using chemically prepatterned substrates to create a surface with laterally varying surface energies [5]. The extensive literature in this important topic is discussed in Chap. 6.

1.3.2 Field-Induced and Dynamic Control of Surface Structuration

A polymer surface can also be perturbed by an externally applied force. Electric or magnetic fields, thermal gradients or mechanical stresses have been used to modify the morphology of polymer films. A reasonable understanding of some of these methods has been achieved. In this category we have included the following patterning methodologies:

Electrohydrodynamic patterning/Thermal-gradient induced surface patterning. A stable liquid polymer film becomes unstable when the surface tension is overcome by a destabilizing force (e.g., due to an externally applied electric field). Analogously, solid polymer films become unstable under sufficiently large strains. The characteristic size of the instability will depend on the film thickness, surface tension or bulk modulus, and the strength of the applied force. Unstable thin polymer films can reorganize in different architectures: e.g., holes, columns or channels. In the absence of plastic deformation, the developed instability will relax to the original unperturbed morphology after the destabilizing force is removed, in a time scale that will depend on the viscoelastic properties of the film. Several strategies have been developed to preserve the surface nanostructure, including chemical/UV cross-linking or rapid glassification (quenching) of a structured liquid polymer film.

In electrohydrodynamic patterning, a polymer interface is destabilized due to the effect of an applied electric field. If the generated electrical stress at the interface is sufficiently large, it can cause an electrohydrodynamic instability, as has been extensively discussed in the literature [6, 79–81]. Analogously, an externally imposed temperature gradient causes a surface undulation with a characteristic wavelength that finally leads to the film breakup into different morphologies. As depicted by Steiner and coworkers [82] the observed instability at the origin of the morphology change is linked to the diffusive heat transport through the polymer and air layers. Some of these methods are described in Chap. 7.

Elastic Instability and Surface Wrinkling. Surface wrinkles, creases, and even folding are common instability phenomena found in polymer films. [83–86] Wrinkles or folds appear when membranes or films are subjected to mechanical stress (by osmotic pressure, stretching or heating) above a certain limit. Even though wrinkles appear as a consequence of buckling instabilities, the outcome of the process can be controlled and used for surface patterning. It has been shown that the use of sandwiched structures with layers of different elastic properties allows for a better control and degree of complexity of the produced structures [85, 87–89]. This method is discussed in Chap. 8.

Reaction-diffusion surface patterns. We include in this group the processes that take advantage of the competition between reaction and diffusion of species to generate patterns, controlled by the reaction dynamics and the preferential diffusion of reactive species. Particularly interesting is the case of polymer gels where pattern formation has been obtained due to the volume phase transition. Katsuragi [90] described the preparation of patterned gels by reaction-diffusion where the wavelength of the patterns is related with the diffusion length and the final morphology can be modified by varying the experimental conditions. This method is discussed in detail in Chap. 9. Photoembossing can also be classified in this group. In this technique spatial polymer concentration is modulated based on the differentiated migration of reacting monomers. The reaction is typically initiated by patterned UV irradiation; polymerization is initiated in irradiated regions. Reaction-driven diffusion of the reactive monomers to the illuminated regions will then occur, generating differences of monomer concentration between exposed and unexposed areas. To finish the process and froze the generated pattern, polymerization of the whole film is triggered. The pattern obtained is determined by the competition between diffusion and polymerization, the surface tension of the polymer film and the operational parameters like film thickness, temperature, and exposure and polymerization times [91, 92].

1.3.3 Influence of Water on Hydrophobic Polymer Surfaces

A changing physicochemical environment can also be used to modify the arrangement of polymer materials. Exposure to different solvents, vapors, water, or electrolytes may trigger the reconstruction of polymer surfaces. Three examples are included in this volume.

The *Breath Figures* method is based on the formation of ordered water droplet arrays on surfaces [93–97] to finally produce porous films in which the size, functionality, and shape can be finely tuned. In general, breath figures are formed when a cold surface is brought in contact with moist air. If the surface in not wetted by the vapor, moisture condenses on the cold surface forming water droplets that grow during the evaporation giving rise to distinct water droplet arrangements on the surface. This process is driven both by a thermocapillary effect and Marangoni convection, in which two droplets of the same liquid do not coalesce if a sufficiently large temperature gradient exists [98].

One particular case of BF formation concerns the case of drying polymer solutions which, under appropriate conditions, can produce highly ordered porous materials, as first described by Francois et al. in 1994 [99]. This process regained popularity during the last decade [10, 98, 100–103]. Several studies reported the physical mechanisms [104, 105] underlying of the breath figures formation.

As described in Chap. 10, the preparation of porous films by using this approach is simple, fast, and does not require the removal of any template. In addition, this technique can be used with different materials [10, 98, 106] including mixtures of polymers [107–110] and can produce films with controlled pore dimensions [101, 111–114] and precise functional group distribution [107, 114–120]. Recent reviews have been published on this topic [9, 10, 103, 121, 122].

Ion-induced nanostructuration. When hydrophobic surfaces are in contact with water in ambient conditions a layer of reduced density is present at the interface. This effect, exclusive of hydrophobic surfaces [123], reduces the intimate contact between the two condensed phases. Nevertheless, the extent of this layer can be decreased by degassing the aqueous solution. As reported by Siretanu et al. the enhanced proximity between a hydrophobic polymer film and an aqueous solution, avoiding the presence of a gas layer, can induce a self-assembled nanostructure on the solid surface due to ionic adsorption. This instability and the self-assembled structure formed are controlled by the ion-induced hydrophobic surface charge [12]. A complete description of this technique is presented in Chap. 11.

Nanobubble assisted nanopatterning. If no efforts are taken to remove the gases dissolved, nanobubbles appear on hydrophobic surfaces after inmersion in aqueous media. The interaction of the nanobubbles with the hydrophobic surface can induce surface reconstruction [124]. Thus, the presence of nanobubbles can be employed to form different surface morphologies. For instance, short exposure times to aqueous solutions (several seconds) appear to be enough to create patterns on glassy polymers as polystyrene [11]. This process is described in Chap. 12.

Several strategies can be followed to further increase the degree of complexity of the patterns obtained from the processes mentioned above. First, chemically or topographically patterning a substrate (template guiding) can be used to direct the rupturing path followed by an unstable polymer film. Another level of complexity can be achieved if the starting polymer material is prone to specific self-assembly ordering, e.g., block copolymers. All these approaches become particularly appealing when they are combined with building blocks in the polymer material with interesting properties, allowing the design of functional materials. It has been shown that a simple chemical or morphological modification may enhance existing material properties or provide a seemingly simple material with exceptional capabilities. Many different properties have been targeted in the last decade: some examples are superhydrophobicity or superhydrophilicity, antireflectivity, anti-fouling, antisoiling, improved scratch, corrosion or abrasion resistance, bonding enhancement, self-cleaning, visual aspect, printability, iridescence, and light reflection and absorption. In this book we explore some of these strategies.

As a summary of this chapter, Tables 1.1, 1.2, and 1.3 compile the most significant aspects of the different patterning methods explored in this book, including the cause of the interface modification, the type and typical length scale of the patterns observed, and illustrative references with additional details of each approach.

Table 1.1 Structuration driven by surface/interfacial energy

Patterning method	Driving force	Pattern induced	Typical feature size	Observations	Selected references
Dewetting	Surface Forces	Isotropic, random structures (well-defined mean length scale)	From tens of micrometers down to ~100 nm	Occur in molten ultrathin films or films with low elastic modulus. Nucleation can destabilize thicker films	[125–131]
Directing Convection	Marangoni Convection/ Evaporation/Self-assembly	Coffee rings, polygonal network structures, fingering instabilities, cracks, chevron patterns, etc.	Nanometer to micrometer	Marangoni convection and stick–slip motion can determine the final patterns observed	[132–134]
Self-assembly and microphase separation	Phase separation of polymer blends and/or block copolymers	Random structures (polymer blends) and self-assembled nanostructures (block copolymers)	Nanometer to micrometer	Phase separation influenced by film thickness, temperature and substrate	[36, 130, 135–141]
Surface segregation	Surface/interfacial energy	Layered segregation	Nanometer	Block copolymer segregation from blends afford nano-structured interfaces	[4, 142, 143]
Template guided structuration	Surface-induced structure formation	Governed by template: hexagonal, stripes, squares	Nanometer to micrometer	Employed in combination with block copolymers and blends or dewetting	[144–148]

Table 1.2 Field-induced and dynamic control of surface structuration

Patterning method	Driving force	Pattern induced	Typical feature size	Observations	Selected references
Electrohydrodynamic	Electrical stress	Waves, holes, columns, or channels	Submicrometer to micrometer	Morphology controlled by film thickness, interelectrode distance, electrode patterning, applied field	[149–154]
Thermal-gradient	External thermal gradient	Similar to electrohydrodynamic patterning.	Submicrometer to micrometer	Morphology depends on film thickness, gap between the plates, thermal gradient and temperature of the plates	[82, 155–165]
Surface wrinkling, creasing, and folding	Mechanical stress	Controlled wrinkles	Down to micrometer	Wrinkle, creasing or folding orientation depends on direction and magnitude of the applied stress	[84, 85, 166–175]
Reaction-diffusion	Reaction-diffusion dynamics	Hexagonal, striped, and mixed patterns	Micrometer to millimeter	Patterns obtained by reaction-diffusion competition	[90, 176–179]

Table 1.3 Waterborne-methods

Patterning method	Driving force	Pattern induced	Typical feature size	Observations	Selected references
Breath figures	Condensation of water vapor	Porous films with variable surface distribution and pore sizes	Hundreds of nanometers up to ~20 μm	Water vapor condensation during solvent evaporation drives the formation of pores. Solvent, polymer concentration, temperature and relative humidity play a key role on the pore formation	[9, 10, 98, 103, 121]
Water-ions induced nanostructuration	Adsorption of ions at the interface	Spherical-cap-shaped bumps	Few tens of nanometers	Depend of the amount of gas dissolved in the water, ions in solution and temperature	[12, 180, 181]
Nanobubble assisted nanopatterning	Interfacial tension	Nanograins, net-like nanopatterns	Few tens of nanometers (below 50 nm in some cases)	Surface pattern obtained in seconds	[11, 182]

1.4 Conclusions

The use of nontraditional patterning approaches that resort to interfacial instabilities is a research area in which physicist and chemists have found increasing interest. This is due to the multiple types of patterns that can be achieved, in most of the cases employing easy-to-implement methodologies. In contrast to the existing, "traditional" patterning approaches that are closely sustained on technological advances, the methods described resort to the smart use of properties inherent to the materials to achieve a control on the surface characteristics (topography, chemistry, . . .).

This book covers those methodologies that, based on surface instabilities or control of dynamic processes, are able to guide to a certain extent the modification of polymeric surfaces. It is divided in different sections related to the groups of methodologies depicted in this chapter. Our aim is to provide a snapshot of a rapidly growing field; most of the methodologies described have been reported during the last two decades.

References

1. Lu, G., Li, W., Yao, J.M., Zhang, G., Yang, B., Shen, J.C.: Fabricating ordered two-dimensional arrays of polymer rings with submicrometer-sized features on patterned self-assembled monolayers by dewetting. Adv. Mater. **14**(15), 1049 (2002)
2. Byun, M., Han, W., Li, B., Xin, X., Lin, Z.: An unconventional route to hierarchically ordered block copolymers on a gradient patterned surface through controlled evaporative self-assembly. Angew. Chem. Int. Ed. **52**(4), 1122–1127 (2013)
3. Fukunaga, K., Elbs, H., Magerle, R., Krausch, G.: Large-scale alignment of ABC block copolymer microdomains via solvent vapor treatment. Macromolecules **33**(3), 947–953 (2000)
4. Ibarboure, E., Bousquet, A., Toquer, G., Papon, E., Rodriguez-Hernandez, J.: Tunable hierarchical assembly on polymer surfaces: Combining microphase and macrophase separation in copolymer/homopolymer blends. Langmuir **24**(13), 6391–6394 (2008)
5. Boltau, M., Walheim, S., Mlynek, J., Krausch, G., Steiner, U.: Surface-induced structure formation of polymer blends on patterned substrates. Nature **391**(6670), 877–879 (1998)
6. Schaffer, E., Thurn-Albrecht, T., Russell, T.P., Steiner, U.: Electrically induced structure formation and pattern transfer. Nature **403**(6772), 874–877 (2000)
7. Vandeparre, H., Gabriele, S., Brau, F., Gay, C., Parker, K.K., Damman, P.: Hierarchical wrinkling patterns. Soft Matter **6**(22), 5751–5756 (2010)
8. Guvendiren, M., Yang, S., Burdick, J.A.: Swelling-induced surface patterns in hydrogels with gradient crosslinking density. Adv. Funct. Mater. **19**(19), 3038–3045 (2009)
9. Muñoz-Bonilla, A., Fernández-García, M., Rodríguez-Hernández, J.: Towards hierarchically ordered functional porous polymeric surfaces prepared by the breath figures approach. Prog. Polym. Sci. **39**(3), 510–514 (2014)
10. Escale, P., Rubatat, L., Billon, L., Save, M.: Recent advances in honeycomb-structured porous polymer films prepared via breath figures. Eur. Polym. J. **48**(6), 1001–1025 (2012)
11. Tarabkova, H., Janda, P.: Nanobubble assisted nanopatterning utilized for ex situ identification of surface nanobubbles. J. Phys. Condensed Matter. **25**(18), 184001 (2013)

12. Siretanu, I., Chapel, J.P., Drummond, C.: Water-ions induced nanostructuration of hydrophobic polymer surfaces. ACS Nano **5**(4), 2939–2947 (2011)
13. Reiter, G.: Dewetting of thin polymer films. Phys. Rev. Lett. **68**(1), 75–78 (1992)
14. Xie, R., Karim, A., Douglas, J.F., Han, C.C., Weiss, R.A.: Spinodal dewetting of thin polymer films. Phys. Rev. Lett. **81**(6), 1251–1254 (1998)
15. Higgins, A.M., Jones, R.A.L.: Anisotropic spinodal dewetting as a route to self-assembly of patterned surfaces. Nature **404**(6777), 476–478 (2000)
16. Muller-Buschbaum, P.: Dewetting and pattern formation in thin polymer films as investigated in real and reciprocal space. J. Phys. Condensed Matter. **15**(36), R1549–R1582 (2003)
17. Reiter, G.: Dewetting of thin polymer-films. Phys. Rev. Lett. **68**(1), 75–78 (1992)
18. Wyart, F.B., Daillant, J.: Drying of solids wetted by thin liquid-films. Can. J. Phys. **68**(9), 1084–1088 (1990)
19. Kargupta, K., Sharma, A.: Dewetting of thin films on periodic physically and chemically patterned surfaces. Langmuir **18**(5), 1893–1903 (2002)
20. Zhang, Z.X., Wang, Z., Xing, R.B., Han, Y.C.: Patterning thin polymer films by surface-directed dewetting and pattern transfer. Polymer **44**(13), 3737–3743 (2003)
21. Xia, Y.N., Whitesides, G.M.: Extending microcontact printing as a microlithographic technique. Langmuir **13**(7), 2059–2067 (1997)
22. Evans, S.D., Flynn, T.M., Ulman, A.: Self-assembled multilayer formation on predefined templates. Langmuir **11**(10), 3811–3814 (1995)
23. Herminghaus, S., Jacobs, K., Mecke, K., Bischof, J., Fery, A., Ibn-Elhaj, M., Schlagowski, S.: Spinodal dewetting in liquid crystal and liquid metal films. Science **282**(5390), 916–919 (1998)
24. Calvert, J.M.: Lithographic patterning of self-assembled films. J. Vac. Sci. Technol. B **11**(6), 2155–2163 (1993)
25. Han, W., Lin, Z.: Learning from "coffee rings": ordered structures enabled by controlled evaporative self-assembly. Angew. Chemie Int. Ed. **51**(7), 1534–1546 (2012)
26. Reich, S., Cohen, Y.: Phase separation of polymer blends in thin films. J. Polymer Sci. A2 Polymer Phys. **19**(8), 1255–1267 (1981)
27. Nesterov, A.E., Lipatov, Y.S., Horichko, V.V., Gritsenko, O.T.: Filler effects on the compatibility and phase separation kinetics of poly(vinyl acetate)-poly(methyl methacrylate) mixtures. Polymer **33**(3), 619–622 (1992)
28. Ball, R.C., Essery, R.L.H.: Spinodal decomposition and pattern formation near surfaces. J. Phys. Condens. Matter **2**(51), 10303–10320 (1990)
29. Jones, R.A.L., Norton, L.J., Kramer, E.J., Bates, F.S., Wiltzius, P.: Surface-directed spinodal decomposition. Phys. Rev. Lett. **66**(10), 1326–1329 (1991)
30. Chung, H.J., Wang, H., Composto, R.J.: A morphology map based on phase evolution in polymer blend films. Macromolecules **39**(1), 153–161 (2006)
31. Wang, H., Composto, R.J.: Thin film polymer blends undergoing phase separation and wetting: Identification of early, intermediate, and late stages. J. Chem. Phys. **113**(22), 10386–10397 (2000)
32. Wang, H., Composto, R.J.: Understanding morphology evolution and roughening in phase-separating thin-film polymer blends. Europhys. Lett. **50**(5), 622–627 (2000)
33. Wang, H., Composto, R.J.: Kinetics of surface and interfacial fluctuations in phase separating polymer blend films. Macromolecules **35**(7), 2799–2809 (2002)
34. Wang, H., Composto, R.J.: Wetting and phase separation in polymer blend films: Identification of four thickness regimes with distinct morphological pathways. Interface Sci. **11**(2), 237–248 (2003)
35. Wang, H., Composto, R.J., Hobbie, E.K., Han, C.C.: Multiple lateral length scales in phase-separating thin-film polymer blends. Langmuir **17**(9), 2857–2860 (2001)
36. Xue, L., Zhang, J., Han, Y.: Phase separation induced ordered patterns in thin polymer blend films. Prog. Polymer Sci. (Oxf) **37**(4), 564–594 (2012)

37. Ogawa, H., Kanaya, T., Nishida, K., Matsuba, G.: Phase separation and dewetting in polystyrene/poly(vinyl methyl ether) blend thin films in a wide thickness range. Polymer **49**(1), 254–262 (2008)
38. Tanaka, K., Takahara, A., Kajiyama, T.: Film thickness dependence of the surface structure of immiscible polystyrene/poly(methyl methacrylate) blends. Macromolecules **29**(9), 3232–3239 (1996)
39. Ton-That, C., Shard, A.G., Teare, D.O.H., Bradley, R.H.: XPS and AFM surface studies of solvent-cast PS/PMMA blends. Polymer **42**(3), 1121–1129 (2001)
40. Li, X., Han, Y., An, L.: Annealing effects on the surface morphologies of thin PS/PMMA blend films with different film thickness. Appl. Surf. Sci. **230**(1–4), 115–124 (2004)
41. Tanaka, K., Yoon, J.S., Takahara, A., Kajiyama, T.: Ultrathinning-induced surface phase separation of polystyrene/poly(vinyl methyl ether) blend film. Macromolecules **28**(4), 934–938 (1995)
42. Walheim, S., Böltau, M., Mlynek, J., Krausch, G., Steiner, U.: Structure formation via polymer demixing in spin-cast films. Macromolecules **30**(17), 4995–5003 (1997)
43. Affrossman, S., O'Neill, S.A., Stamm, M.: Topography and surface composition of thin films of blends of polystyrene with brominated polystyrenes: Effects of varying the degree of bromination and annealing. Macromolecules **31**(18), 6280–6288 (1998)
44. Slep, D., Asselta, J., Rafailovich, M.H., Sokolov, J., Winesett, D.A., Smith, A.P., Ade, H., Strzhemechny, Y., Schwarz, S.A., Sauer, B.B.: Phase separation of polystyrene and bromo-polystyrene mixtures in equilibrium structures in thin films. Langmuir **14**(17), 4860–4864 (1998)
45. Raczkowska, J., Rysz, J., Budkowski, A., Lekki, J., Lekka, M., Bernasik, A., Kowalski, K., Czuba, P.: Surface patterns in solvent-cast polymer blend films analyzed with an integral-geometry approach. Macromolecules **36**(7), 2419–2427 (2003)
46. Cyganik, P., Bernasik, A., Budkowski, A., Bergues, B., Kowalski, K., Rysz, J., Lekki, J., Lekka, M.: Phase decomposition in polymer blend films cast on substrates patterned with self-assembled monolayers. Vacuum **63**(1–2), 307–313 (2001)
47. Kielhorn, L., Muthukumar, M.: Phase separation of polymer blend films near patterned surfaces. J. Chem. Phys. **111**(5), 2259–2269 (1999)
48. Jones, R.A.L., Kramer, E.J., Rafailovich, M.H., Sokolov, J., Schwarz, S.A.: Surface enrichment in an isotopic polymer blend. Phys. Rev. Lett. **62**(3), 280–283 (1989)
49. Walheim, S., Ramstein, M., Steiner, U.: Morphologies in ternary polymer blends after spin-coating. Langmuir **15**(14), 4828–4836 (1999)
50. Gutmann, J.S., Muller-Buschbaum, P., Stamm, M.: Complex pattern formation by phase separation of polymer blends in thin films. Faraday Discuss. **112**, 285–297 (1999)
51. Gutmann, J.S., Müller-Buschbaum, P., Schubert, D.W., Stribeck, N., Stamm, M.: Influence of the blend compatibility on the morphology of thin polymer blend films. J. Macromol. Sci. Phys. **38** (1999)
52. Affrossman, S., Henn, G., O'Neill, S.A., Pethrick, R.A., Stamm, M.: Surface topography and composition of deuterated polystyrene-poly(bromostyrene) blends. Macromolecules **29**(14), 5010–5016 (1996)
53. Raczkowska, J., Bernasik, A., Budkowski, A., Sajewicz, K., Penc, B., Lekki, J., Lekka, M., Rysz, J., Kowalski, K., Czuba, P.: Structures formed in spin-cast films of polystyrene blends with poly(butyl methacrylate) isomers. Macromolecules **37**(19), 7308–7315 (2004)
54. Genzer, J., Composto, R.J.: Effect of molecular weight on the interfacial excess, tension, and width in a homopolymer/binary polymer blend system. Macromolecules **31**(3), 870–878 (1998)
55. Cyganik, P., Budkowski, A., Raczkowska, J., Postawa, Z.: AFM/LFM surface studies of a ternary polymer blend cast on substrates covered by a self-assembled monolayer. Surf. Sci. **507–510**, 700–706 (2002)
56. Hariharan, A., Kumar, S.K., Russell, T.P.: Surface segregation in binary polymer mixtures: A lattice model. Macromolecules **24**(17), 4909–4917 (1991)

57. Zhao, X., Zhao, W., Sokolov, J., Rafailovich, M.H., Schwarz, S.A., Wilkens, B.J., Jones, R.A. L., Kramer, E.J.: Determination of the concentration profile at the surface of d-PS/h-PS blends using high-resolution ion scattering techniques. Macromolecules **24**(22), 5991–5996 (1991)
58. Li, X., Han, Y., An, L.: Surface morphology control of immiscible polymer-blend thin films. Polymer **44**(26), 8155–8165 (2003)
59. Müller-Buschbaum, P., Gutmann, J.S., Stamm, M.: Influence of blend composition on phase separation and dewetting of thin polymer blend films. Macromolecules **33**(13), 4886–4895 (2000)
60. Yao, L., Xuming, X., Qi, Z., Liming, T.: Formation of cylindrical phase structure in PMMA/HBP polymer blend films. Polymer **46**(25), 12004–12009 (2005)
61. Ton-That, C., Shard, A.G., Bradley, R.H.: Surface feature size of spin cast PS/PMMA blends. Polymer **43**(18), 4973–4977 (2002)
62. Ton-That, C., Shard, A.G., Daley, R., Bradley, R.H.: Effects of annealing on the5 surface composition and morphology of PS/PMMA blend. Macromolecules **33**(22), 8453–8459 (2000)
63. Hopkinson, I., Myatt, M.: Phase separation in ternary polymer solutions induced by solvent loss. Macromolecules **35**(13), 5153–5160 (2002)
64. Schmidt, J.J., Gardella Jr., J.A., Salvati Jr., L.: Surface studies of polymer blends. 2. An ESCA and IR study of poly(methyl methacrylate)/poly(vinyl chloride) homopolymer blends. Macromolecules **22**(12), 4489–4495 (1989)
65. Müller-Buschbaum, P., Gutmann, J.S., Wolkenhauer, M., Kraus, J., Stamm, M., Smilgies, D., Petry, W.: Solvent-induced surface morphology of thin polymer films. Macromolecules **34**(5), 1369–1375 (2001)
66. Strawhecker, K.E., Kumar, S.K., Douglas, J.F., Karim, A.: The critical role of solvent evaporation on the roughness of spin-cast polymer films [1]. Macromolecules **34**(14), 4669–4672 (2001)
67. Cui, L., Ding, Y., Li, X., Wang, Z., Han, Y.: Solvent and polymer concentration effects on the surface morphology evolution of immiscible polystyrene/poly(methyl methacrylate) blends. Thin Solid Films **515**(4), 2038–2048 (2006)
68. Kumacheva, E., Li, L., Winnik, M.A., Shinozaki, D.M., Cheng, P.C.: Direct imaging of surface and bulk structures in solvent cast polymer blend films. Langmuir **13**(9), 2483–2488 (1997)
69. Huraux, K., Narita, T., Frétigny, C., Lequeux, F.: Solution drying and phase separation morphology of polyacrylamide/polyethylene glycol/water system. Macromolecules **40**(23), 8336–8341 (2007)
70. Matsushita, Y., Furukawa, H., Okada, M.: Crystal-like pattern formation in polymerization-induced phase separation. Phys Rev E Stat Nonlin Soft Matter Phys **70**(4 Pt 1), 040501-1-040501-3 (2004)
71. Tanaka, H., Lovinger, A.J., Davis, D.D.: Pattern evolution caused by dynamic coupling between wetting and phase separation in binary liquid mixture containing glass particles. Phys. Rev. Lett. **72**(16), 2581–2584 (1994)
72. Krausch, G.: Surface induced self assembly in thin polymer films. Mat. Sci. Eng. R: Rep. **14** (1–2), 1–94 (1995)
73. Binder, K.: Phase transitions of polymer blends and block copolymer melts in thin films. Adv. Polym. Sci. **138**, 1–89 (1999)
74. Pan, D.H.K., Prest Jr., W.M.: Surfaces of polymer blends: X-ray photoelectron spectroscopy studies of polystyrene/poly(vinyl methyl ether) blends. J. Appl. Phys. **58**(8), 2861–2870 (1985)
75. Cowie, J.M.G., Devlin, B.G., McEwen, I.J.: A study of surface enrichment in polystyrene/poly(vinyl methyl ether) blends using attenuated total reflectance infra-red spectroscopy: 1. Polymer **34**(3), 501–504 (1993)

76. Cowie, J.M.G., Devlin, B.G., McEwen, I.J.: Surface enrichment in polystyrene/poly(vinyl methyl ether) blends. 3. An analysis of the near-surface composition profile. Macromolecules **26**(21), 5628–5632 (1993)
77. Cowie, J.M.G., Devlin, B.G., McEwen, I.J.: Surface enrichment in PS/PVME blends: 2. The effect of specific interactions in the bulk mixture. Polymer **34**(19), 4130–4135 (1993)
78. Forrey, C., Koberstein, J.T., Pan, D.H.: Surface segregation in miscible blends of polystyrene and poly(vinylmethyl ether): Comparison of theory and experiment. Interface Sci. **11**(2), 211–223 (2003)
79. Sarkar, J., Sharma, A.: A unified theory of instabilities in viscoelastic thin films: from wetting to confined films, from viscous to elastic films, and from short to long waves. Langmuir **26** (11), 8464–8473 (2010)
80. Seemann, R., Herminghaus, S., Jacobs, K.: Dewetting patterns and molecular forces: a reconciliation. Phys. Rev. Lett. **86**(24), 5534–5537 (2001)
81. Monch, W., Herminghaus, S.: Elastic instability of rubber films between solid bodies. Europhys. Lett. **53**(4), 525–531 (2001)
82. Schaffer, E., Harkema, S., Blossey, R., Steiner, U.: Temperature-gradient-induced instability in polymer films. Europhys. Lett. **60**(2), 255–261 (2002)
83. Chen, C.-M., Yang, S.: Wrinkling instabilities in polymer films and their applications. Polym. Int. **61**(7), 1041–1047 (2012)
84. Schweikart, A., Fery, A.: Controlled wrinkling as a novel method for the fabrication of patterned surfaces. Microchim. Acta **165**(3–4), 249–263 (2009)
85. Genzer, J., Groenewold, J.: Soft matter with hard skin: From skin wrinkles to templating and material characterization. Soft Matter **2**(4), 310–323 (2006)
86. McCormick, S.: Materials science - Exploiting wrinkle formation. Science **317**(5838), 605–606 (2007)
87. Bowden, N., Brittain, S., Evans, A.G., Hutchinson, J.W., Whitesides, G.M.: Spontaneous formation of ordered structures in thin films of metals supported on an elastomeric polymer. Nature **393**(6681), 146–149 (1998)
88. Chua, D.B.H., Ng, H.T., Li, S.F.Y.: Spontaneous formation of complex and ordered structures on oxygen-plasma-treated elastomeric polydimethylsiloxane. Appl. Phys. Lett. **76**(6), 721–723 (2000)
89. Huck, W.T.S., Bowden, N., Onck, P., Pardoen, T., Hutchinson, J.W., Whitesides, G.M.: Ordering of spontaneously formed buckles on planar surfaces. Langmuir **16**(7), 3497–3501 (2000)
90. Katsuragi, H.: Diffusion-induced spontaneous pattern formation on gelation surfaces. Europhys. Lett. **73**(5), 793–799 (2006)
91. Leewis, C.M., de Jong, A.M., van IJzendoorn, L.J., Broer, D.J.: Reaction–diffusion model for the preparation of polymer gratings by patterned ultraviolet illumination. J. Appl. Phys. **95** (8), 4125–4139 (2004)
92. Hermans, K., Tomatsu, I., Matecki, M., Sijbesma, R.P., Bastiaansen, C.W.M., Broer, D.J.: Highly efficient surface relief formation via photoembossing of a supramolecular polymer. Macromol. Chem. Phys. **209**(20), 2094–2099 (2008)
93. Aitkek, J.: Breath figures. Nature **86**, 516–517 (1911)
94. Aitkek, J.: Breath figures. Nature **90**, 619–621 (1912)
95. Giltay, J.W.: Breath figures. Nature **86**, 585–586 (1911)
96. Rayleigh, L.: Breath figures. Nature **86**, 416–417 (1911)
97. Rayleigh, L.: Breath figures. Nature **90**, 436–438 (1912)
98. Bunz, U.H.F.: Breath figures as a dynamic templating method for polymers and nanomaterials. Adv. Mater. **18**(8), 973–989 (2006)
99. Widawski, G., Rawiso, M., François, B.: Self-organized honeycomb morphology of star-polymer polystyrene films. Nature **369**(6479), 387–389 (1994)
100. Jenekhe, S.A., Chen, X.L.: Self-assembly of ordered microporous materials from rod-coil block copolymers. Science **283**(5400), 372–375 (1999)

101. Srinivasarao, M., Collings, D., Philips, A., Patel, S.: Three-dimensionally ordered array of air bubbles in a polymer film. Science **292**(5514), 79–83 (2001)
102. Sun, H., Wu, L.X.: Ordered honeycomb-patterned films via breath figures. Prog. Chem. **22** (9), 1784–1798 (2010)
103. Hernandez-Guerrero, M., Stenzel, M.H.: Honeycomb structured polymer films via breath figures. Polym. Chem. **3**(3), 563–577 (2012)
104. Barrow, M.S., Jones, R.L., Park, J.O., Srinivasarao, M., Williams, P.R., Wright, C.J.: Physical characterisation of microporous and nanoporous polymer films by atomic force microscopy, scanning electron microscopy and high speed video microphotography. Spectroscopy **18**, 577–585 (2004)
105. Kuo, C.T., Lin, Y.S., Liu, T.K., Liu, H.C., Hung, W.C., Jiang, I.M., Tsai, M.S., Hsu, C.C., Wu, C.Y.: Dynamics of single-layer polymer breath figures. Opt. Express **18**(17), 18464–18470 (2010)
106. Wu, D., Xu, F., Sun, B., Fu, R., He, H., Matyjaszewski, K.: Design and preparation of porous polymers. Chem. Rev. **112**(7), 3959–4015 (2012)
107. Stenzel-Rosenbaum, M.H., Davis, T.P., Fane, A.G., Chen, V.: Porous polymer films and honeycomb structures made by the self-organization of well-defined macromolecular structures created by living radical polymerization techniques. Angew. Chem. Int. Ed. **40**(18), 3428–3432 (2001)
108. Madej, W., Budkowski, A., Raczkowska, J., Rysz, J.: Breath figures in polymer and polymer blend films spin-coated in dry and humid ambience. Langmuir **24**(7), 3517–3524 (2008)
109. Kim, B.S., Basavaraja, C., Jo, E.A., Kim, D.G., Huh, D.S.: Effect of amphiphilic copolymer containing ruthenium tris(bipyridyl) photosensitizer on the formation of honeycomb-patterned film. Polymer **51**(15), 3365–3371 (2010)
110. Thinh, P., Basavaraja, C., Kim, D.G., Huh, D.: Characterization and electrochemical behaviors of honeycomb-patterned poly(N-vinylcarbazole)/polystyrene composite films. Polym. Bull. **69**(1), 81–94 (2012)
111. Maruyama, N., Karthaus, O., Ijiro, K., Shimomura, M., Koito, T., Nishimura, S., Sawadaishi, T., Nishi, N., Tokura, S.: Mesoscopic pattern formation of nanostructured polymer assemblies. Supramol. Sci. **5**(3–4), 331–336 (1998)
112. Zhang, R., Wang, J., Wang, M., He, Y.: Fabrication of honeycomb polyvinyl butyral film under humidity provided by super saturated salt solutions. J. Appl. Polym. Sci. **124**(1), 495–500 (2012)
113. Peng, J., Han, Y.C., Yang, Y.M., Li, B.Y.: The influencing factors on the macroporous formation in polymer films by water droplet templating. Polymer **45**(2), 447–452 (2004)
114. Cui, L., Wang, H., Ding, Y., Han, Y.: Tunable ordered droplets induced by convection in phase-separating P2VP/PS blend film. Polymer **45**(24), 8139–8146 (2004)
115. Bolognesi, A., DiGianvincenzo, P., Giovanella, U., Mendichi, R., Schieroni, A.G.: Polystyrene functionalized with EDOT oligomers. Eur. Polym. J. **44**(3), 793–800 (2008)
116. Yunus, S., Delcorte, A., Poleunis, C., Bertrand, P., Bolognesi, A., Botta, C.: A route to self-organized honeycomb microstructured polystyrene films and their chemical characterization by ToF-SIMS imaging. Adv. Funct. Mater. **17**(7), 1079–1084 (2007)
117. Stenzel, M.H., Barner-Kowollik, C., Davis, T.P.: Formation of honeycomb-structured, porous films via breath figures with different polymer architectures. J. Polym. Sci. A Polym. Chem. **44**(8), 2363–2375 (2006)
118. Ting, S.R.S., Min, E.H., Escalé, P., Save, M., Billon, L., Stenzel, M.H.: Lectin recognizable biomaterials synthesized via nitroxide-mediated polymerization of a methacryloyl galactose monomer. Macromolecules **42**(24), 9422–9434 (2009)
119. Escalé, P., Ting, S.R.S., Khoukh, A., Rubatat, L., Save, M., Stenzel, M.H., Billon, L.: Synthetic route effect on macromolecular architecture: from block to gradient copolymers based on acryloyl galactose monomer using RAFT polymerization. Macromolecules **44**(15), 5911–5919 (2011)

120. Muñoz-Bonilla, A., Ibarboure, E., Bordege, V., Fernandez-Garcia, M., Rodriguez-Hernandez, J.: Fabrication of honeycomb-structured porous surfaces decorated with glycopolymers. Langmuir **26**(11), 8552–8558 (2010)
121. Heng, L., Wang, B., Li, M., Zhang, Y., Jiang, L.: Advances in fabrication materials of honeycomb structure films by the breath-figure method. Materials **6**(2), 460–482 (2013)
122. Ma, H., Hao, J.: Ordered patterns and structures via interfacial self-assembly: superlattices, honeycomb structures and coffee rings. Chem. Soc. Rev. **40**(11), 5457–5471 (2011)
123. Agrawal, A., Park, J., Ryu, D.Y., Hammond, P.T., Russell, T.P., McKinley, G.H.: Controlling the location and spatial extent of nanobubbles using hydrophobically nanopatterned surfaces. Nano Lett. **5**(9), 1751–1756 (2005)
124. Janda, P., Frank, O., Bastl, Z., Klementova, M., Tarabkova, H., Kavan, L.: Nanobubble-assisted formation of carbon nanostructures on basal plane highly ordered pyrolytic graphite exposed to aqueous media. Nanotechnology **21**(9), 095707 (2010)
125. Govor, L.V., Reiter, G., Bauer, G.H., Parisi, J.: Treelike branched structures formed in dewetting thin films of a binary solution. Appl. Phys. Lett. **89**(13), 133126 (2006)
126. Sharma, A., Reiter, G.: Morphological phase transitions in spontaneous dewetting of thin films on homogeneous and heterogeneous surfaces. Phase Transit. **75**(4–5), 377–399 (2002)
127. Reiter, G.: Unstable thin polymer-films - rupture and dewetting processes. Langmuir **9**(5), 1344–1351 (1993)
128. Sharma, A., Reiter, G.: Instability of thin polymer films on coated substrates: rupture, dewetting, and drop formation. J. Colloid Interface Sci. **178**(2), 383–399 (1996)
129. Harkema, S., Schaffer, E., Morariu, M.D., Steiner, U.: Pattern replication by confined dewetting. Langmuir **19**(23), 9714–9718 (2003)
130. Muller-Buschbaum, P., Bauer, E., Wunnicke, O., Stamm, M.: The control of thin film morphology by the interplay of dewetting, phase separation and microphase separation. J. Phys. Condensed Matter **17**(9), S363–S386 (2005)
131. Mukherjee, R., Sharma, A., Steiner, U.: Surface instability and pattern formation in thin polymer films. In: del Campo, A., Arzt, E. (eds.) Generating Micro- and Nanopatterns on Polymeric Materials. Wiley-VCH, Weinheim (2011)
132. Deegan, R.D., Bakajin, O., Dupont, T.F., Huber, G., Nagel, S.R., Witten, T.A.: Capillary flow as the cause of ring stains from dried liquid drops. Nature **389**(6653), 827–829 (1997)
133. Nguyen, V., Stebe, K.: Patterning of small particles by a surfactant-enhanced Marangoni-Bénard instability. Phys. Rev. Lett. **88**(16), 164501 (2002)
134. Pauliac-Vaujour, E., Stannard, A., Martin, C., Blunt, M., Notingher, I., Moriarty, P., Vancea, I., Thiele, U.: Fingering instabilities in dewetting nanofluids. Phys. Rev. Lett. **100**(17), 176102 (2008)
135. Reiter, G., Castelein, G., Hoerner, P., Riess, G., Blumen, A., Sommer, J.U.: Nanometer-scale surface patterns with long-range order created by crystallization of diblock copolymers. Phys. Rev. Lett. **83**(19), 3844–3847 (1999)
136. Koehler, W., Krekhov, A., Zimmermann, W.: Thermal Diffusion in Polymer Blends: Criticality and Pattern Formation. In: *Complex Macromolecular Systems I*, Muller, A.H.E., Schmidt, H.W. (eds.) Vol. 227, 145–198 (2010)
137. Farrell, R.A., Fitzgerald, T.G., Borah, D., Holmes, J.D., Morris, M.A.: Chemical interactions and their role in the microphase separation of block copolymer thin films. Int. J. Mol. Sci. **10** (9), 3671–3712 (2009)
138. Kim, H.-C., Hinsberg, W.D.: Surface patterns from block copolymer self-assembly. J. Vac. Sci. Technol. A **26**(6), 1369–1382 (2008)
139. Sommer, J.-U., Reiter, G.: The formation of ordered polymer structures at interfaces: a few intriguing aspects. Adv. Polym. Sci. **200**, 1–36 (2006)
140. Abetz, V., Simon, P.F.W.: Phase behaviour and morphologies of block copolymers. In: Block Copolymers I, Abetz, V., (ed.) Vol. 189, pp 125–212, (2005)
141. Segalman, R.A.: Patterning with block copolymer thin films. Mat. Sci. Eng R Rep. **48**(6), 191–226 (2005)

142. Bousquet, A., Pannier, G., Ibarboure, E., Papon, E., Rodríguez-Hernández, J.: Control of the surface properties in polymer blends. J. Adhes. **83**(4), 335–349 (2007)
143. Bousquet, A., Ibarboure, E., Teran, F.J., Ruiz, L., Garay, M.T., Laza, J.M., Vilas, J.L., Papon, E., Rodríguez-Hernández, J.: PH responsive surfaces with nanoscale topography. J. Polym. Sci. A Polym. Chem. **48**(14), 2982–2990 (2010)
144. Karim, A., Douglas, J.F., Lee, B.P., Glotzer, S.C., Rogers, J.A., Jackman, R.J., Amis, E.J., Whitesides, G.M.: Phase separation of ultrathin polymer-blend films on patterned substrates. Phys. Rev. E Stat. Phys. Plasmas Fluids Relat. Interdiscip. Topics **57**(6), R6273–R6276 (1998)
145. Ermi, B.D., Nisato, G., Douglas, J.F., Rogers, J.A., Karim, A.: Coupling between phase separation and surface deformation modes in self-organizing polymer blend films. Phys. Rev. Lett. **81**(18), 3900–3903 (1998)
146. Nisato, G., Ermi, B.D., Douglas, J.F., Karim, A.: Excitation of surface deformation modes of a phase-separating polymer blend on a patterned substrate. Macromolecules **32**(7), 2356–2364 (1999)
147. Petera, D., Muthukumar, M.: Effect of patterned surface on diblock-copolymer melts and polymer blends near the critical point. J. Chem. Phys. **107**(22), 9640–9644 (1997)
148. Park, J.-W., Cho, Y.-H.: Surface-induced morphologies in thin films of a rod-coil diblock copolymer. Langmuir **22**(26), 10898–10903 (2006)
149. Deshpande, P., Pease, L.F., Chen, L., Chou, S.Y., Russel, W.B.: Cylindrically symmetric electrohydrodynamic patterning. Phys. Rev. E **70**(4 Pt 1), 041601 (2004)
150. Pease, L.F., Russel, W.B.: Electrostatically induced submicron patterning of thin perfect and leaky dielectric films: A generalized linear stability analysis. J. Chem. Phys. **118**(8), 3790–3803 (2003)
151. Pease, L.F., Russel, W.B.: Limitations on length scales for electrostatically induced submicrometer pillars and holes. Langmuir **20**(3), 795–804 (2004)
152. Pease III, L.F., Russel, W.B.: Charge driven, electrohydrodynamic patterning of thin films. J. Chem. Phys. **125**(18), 184716 (2006)
153. Wu, N., Pease III, L.F., Russel, W.B.: Toward large-scale alignment of electrohydrodynamic patterning of thin polymer films. Adv. Funct. Mater. **16**(15), 1992–1999 (2006)
154. Herminghaus, S.: Dynamical instability of thin liquid films between conducting media. Phys. Rev. Lett. **83**(12), 2359–2361 (1999)
155. Bormashenko, E., Balter, S., Pogreb, R., Bormashenko, Y., Gendelman, O., Aurbach, D.: On the mechanism of patterning in rapidly evaporated polymer solutions: Is temperature-gradient-driven Marangoni instability responsible for the large-scale patterning? J. Colloid Interface Sci. **343**(2), 602–607 (2010)
156. Dietzel, M., Troian, S.M.: Mechanism for spontaneous growth of nanopillar arrays in ultrathin films subject to a thermal gradient. J. Appl. Phys. **108**(7), 074308 (2010)
157. Janes, D.W., Katzenstein, J.M., Shanmuganathan, K., Ellison, C.J.: Directing convection to pattern thin polymer films. J. Polymer Sci. B Polymer Phy. **51**(7), 535–545 (2013)
158. Schaffer, E., Harkema, S., Roerdink, M., Blossey, R., Steiner, U.: Thermomechanical lithography: Pattern replication using a temperature gradient driven instability. Adv. Mater. **15**(6), 514–517 (2003)
159. Voit, A., Krekhov, A., Enge, W., Kramer, L., Kohler, W.: Thermal patterning of a critical polymer blend. Phys. Rev. Lett. **94**(21), 214501 (2005)
160. Weh, L.: Surface structures in thin polymer layers caused by coupling of diffusion-controlled Marangoni instability and local horizontal temperature gradient. Macromol. Mater. Eng. **290** (10), 976–986 (2005)
161. Peng, J., Wang, H.F., Li, B.Y., Han, Y.C.: Pattern formation in a confined polymer film induced by a temperature gradient. Polymer **45**(23), 8013–8017 (2004)
162. Nedelcu, M., Morariu, M.D., Harkema, S., Voicu, N.E., Steiner, U.: Pattern formation by temperature-gradient driven film instabilities in laterally confined geometries. Soft Matter **1** (1), 62–65 (2005)

163. Nmabu, T., Yamuachi, Y., Kushiro, T., Sakurai, S.: Micro-convection, dissipative structure and pattern formation in polymer blend solutions under temperature gradients. Faraday Discuss. **128**, 285–98 (2005)
164. Verma, R., Sharma, A., Banerjee, I., Kargupta, K.: Spinodal instability and pattern formation in thin liquid films confined between two plates. J. Colloid Interface Sci. **296**(1), 220–232 (2006)
165. Paronyan, T.M., Pigos, E.M., Chen, G., Harutyunyan, A.R.: Formation of ripples in graphene as a result of interfacial instabilities. ACS Nano **5**(12), 9619–9627 (2011)
166. Li, B., Cao, Y.-P., Feng, X.-Q., Gao, H.: Mechanics of morphological instabilities and surface wrinkling in soft materials: a review. Soft Matter **8**(21), 5728–5745 (2012)
167. Ebata, Y., Croll, A.B., Crosby, A.J.: Wrinkling and strain localizations in polymer thin films. Soft Matter **8**(35), 9086–9091 (2012)
168. Davis, C.S., Crosby, A.J.: Wrinkle morphologies with two distinct wavelengths. J Polymer Sci. B Polymer Phy. **50**(17), 1225–1232 (2012)
169. Chan, E.P., Smith, E.J., Hayward, R.C., Crosby, A.J.: Surface wrinkles for smart adhesion. Adv. Mater. **20**(4), 711 (2008)
170. Chan, E.P., Crosby, A.J.: Spontaneous formation of stable aligned wrinkling patterns. Soft Matter **2**(4), 324–328 (2006)
171. Kim, H.S., Crosby, A.J.: Solvent-responsive surface via wrinkling instability. Adv. Mater. **23**(36), 4188 (2011)
172. Harrison, C., Stafford, C.M., Zhang, W.H., Karim, A.: Sinusoidal phase grating created by a tunably buckled surface. Appl. Phys. Lett. **85**(18), 4016–4018 (2004)
173. Chiche, A., Stafford, C.M., Cabral, J.T.: Complex micropatterning of periodic structures on elastomeric surfaces. Soft Matter **4**(12), 2360–2364 (2008)
174. Lee, J.-H., Ro, H.W., Huang, R., Lemaillet, P., Germer, T.A., Soles, C.L., Stafford, C.M.: Anisotropic, hierarchical surface patterns via surface wrinkling of nanopatterned polymer films. Nano Lett. **12**(11), 5995–5999 (2012)
175. Chung, J.Y., Youngblood, J.P., Stafford, C.M.: Anisotropic wetting on tunable micro-wrinkled surfaces. Soft Matter **3**(9), 1163–1169 (2007)
176. Turing, A.M.: The chemical basis of morphogenesis. Philos. Trans. R. Soc. Lond. B Biol. Sci. **237**(641), 37–72 (1952)
177. Ouyang, Q., Swinney, H.L.: Transition from a uniform state to hexagonal and striped Turing patterns. Nature **352**(6336), 610–612 (1991)
178. Walgraef, D.: Reaction-diffusion approach to nanostructure formation and texture evolution in adsorbed monoatomic layers. Int. J. Quantum Chem. **98**(2), 248–260 (2004)
179. Bozzini, B., Lacitignola, D., Sgura, I.: A reaction-diffusion model of spatial pattern formation in electrodeposition. J. Phys. Conf. Ser. **96**, 012051 (2008)
180. Siretanu, I., Chapel, J.P., Drummond, C.: Substrate remote control of polymer film surface mobility. Macromolecules **45**(2), 1001–1005 (2012)
181. Burgo, T.A.L., Ducati, T.R.D., Francisco, K.R., Clinckspoor, K.J., Galembeck, F., Galembeck, S.E.: Triboelectricity: macroscopic charge patterns formed by self-arraying ions on polymer surfaces. Langmuir **28**(19), 7407–7416 (2012)
182. Darwich, S., Mougin, K., Vidal, L., Gnecco, E., Haidara, H.: Nanobubble and nanodroplet template growth of particle nanorings versus nanoholes in drying nanofluids and polymer films. Nanoscale **3**(3), 1211–1217 (2011)

Chapter 2
From Holes to Drops to Toroids: Conditions for the Transcription of Surface Patterns into Three-Dimensional Morphologies via Rim Instabilities in the Course of Dewetting

Samer Al Akhrass, Laurent Vonna, and Günter Reiter

2.1 Introduction

Intrinsic ordering processes on surfaces which allow to generate regular topographic patterns with a microscopic length-scale are of importance for many technological applications [1–3]. In this context, it is of interest to explore if dewetting of thin films, a simple and thus highly convenient process, is capable of creating regularly ordered topographical patterns. Dewetting causes changes in morphology which eventually transform a thin fluid film into a set of droplets [4–19]. This transition in morphology can be controlled by external forces or via patterned substrates and thus represents a highly promising approach for generating engineered topographical patterns on a meso- and nanoscale. This pattern formation process can be scaled up to large areas and thus may be of interest for various applications.

The physical laws governing the process of dewetting are fairly well understood and have been explored in various experiments. Ideally flat (smooth) and defect-free substrates have been used for testing these laws [4–19]. At a late stage of dewetting, i.e. after dewetting holes have been initiated and grown until their coalescence, filaments of fluids are formed. Their decay into an array of droplets via a variant of the Plateau–Rayleigh instability has been examined in detail. While

S. Al Akhrass (✉)
Ingénierie des Matériaux Polymères (IMP - UMR CNRS 5223), Université Claude Bernard Lyon 1, 15 Boulevard Latarjet, 69622 Villeurbanne Cedex, France
e-mail: samer.al-akhrass@univ-lyon1.fr

L. Vonna
Institut de Science des Matériaux de Mulhouse (IS2M - UMR CNRS 7361), 15, rue Jean Starcky, 68057 Mulhouse Cedex, France

G. Reiter
Institute of Physics, University of Freiburg, 79104 Freiburg, Germany

Freiburg Materials Research Centre, University of Freiburg, 79104 Freiburg, Germany

J. Rodríguez-Hernández, C. Drummond (eds.), *Polymer Surfaces in Motion*,
DOI 10.1007/978-3-319-17431-0_2

dewetting of thin films on flat and homogeneous substrates is able to generate droplets of fairly uniform size, their positioning on such substrates is not predictable and thus not regular. A random collection of isotropically distributed droplets is the outcome of such a non-guided dewetting process.

For all dewetting processes, the initial thickness (h) of the film plays a crucial role [4–19]. For example, h determines the strength of van der Waals forces acting across the thin film and h often indirectly is responsible for the mean distance between dewetting holes. In the case of slippage, the width of the dewetting rim and consequently the amount of energy dissipated by viscous friction within the rim, are related to h. In addition, besides h also interfacial properties determine both the capillary forces driving dewetting and the frictional losses which balance the driving force [4–19]. Average properties (e.g. the characteristic wavelength of surface fluctuations, the mean distance between nucleated holes or the average size of the final droplets) can be identified, but no long range order can be generated by non-guided dewetting processes. Thus, for the fabrication of devices for potential applications, the challenge is to introduce long-range order and alignment of droplets in dewetted structures [20–22].

Long range ordering of polymeric nano- or microstructures can emerge from periodic fluctuations occurring at the retracting triple-line of a dewetting fluid, the corresponding fluid flow within the dewetting rim and the consequently induced instabilities. Eventually, droplets may detach from such unstable rims in a way analogous to the Plateau–Rayleigh instability [23], as observed on slippery but homogeneous (non-patterned) substrates [24, 25]. A direct relation between droplet size and initial film thickness has been identified [26]. In dewetting polymer solutions, where concentration (Marangoni) effects [27] play an important role, even more complex morphologies induced in the retracting fluid can be produced via such fingering instabilities [28].

Potentially, dewetting therefore allows to transform a quasi-two-dimensional layer into self-organised, ordered arrays with three-dimensional topography [29–37]. However, without any guiding influence, dewetting only can generate randomly placed holes or droplets separated by a mean length-scale [4–19, 38–40]. Guiding the retracting contact line by a chemically or topographically patterned substrates has been investigated. However, such guidance was not always successful. Often guidance relies on the commensurability of the thickness of the polymer film with size and frequency of the imposed pattern, as discussed in theoretical [41] and experimental [42, 43] studies on wetting of structured surfaces by a fluid layer. The various pathways of dewetting and pattern formation in thin films on heterogeneous substrates have been investigated theoretically [38–40, 44–48], suggesting that ordered patterns are only possible when the characteristic length-scale of dewetting is close to commensurate with the imposed periodicity of the pattern [44–48]. On surfaces with chemical or topographical patterns, pinning and deformation of a moving contact-line can be guided and thus can lead to the formation of regular structures. Surface features can act either individually or collectively as pinning sites. Perturbations in the retraction velocity of the contact-line, with the ultimate case of local pinning, are influenced by the (characteristic) length-scale of surface features and the wettability contrast and topography (aspect ratio) of the surface pattern.

Structure formation via the control of the various stages of dewetting has been investigated both theoretically [44–50] and experimentally [32, 51–69]. For example, the structures can be aligned by dewetting on chemically [52–57] or topographically patterned substrates [58–62], and vertical confinement of the film under a patterned mould or stamp [63–69]. Often experiments and simulations of dewetting on topographically patterned surfaces have focused on one-dimensional patterns like parallel stripes [58–61] and on two-dimensional arrays of square pillars [62].

In a simplistic view, one therefore might expect that the resulting morphology is uniquely and unambiguously determined by the templating two-dimensional surface pattern. Here, we show that besides the surface pattern itself also other factors are controlling the resulting three-dimensional morphology. The goal is to understand the underlying physical mechanisms which govern the generation of various changes in morphologies for a given template. To that end, we report an experimental dewetting study on chemically patterned substrates exploring the conditions of dewetting of a thin polymer films which can lead to various morphologies. In particular, we focus on the formation of ordered arrays of cylindrical holes, microdroplets or toroids, guided by a predefined pattern of non-wettable patches of uniform size arranged with hexagonal symmetry.

2.2 Experimental Methods

2.2.1 *Preparation of the Silicon Substrate*

Highly polished silicon wafers were cleaned by immersion in a piranha solution (70 % volume fraction concentrated sulphuric acid and 30 % volume fraction hydrogen peroxide) at 80 °C for 30 min, followed by rinsing in deionized water and drying under nitrogen gas. This step served firstly to eliminate surface contaminations and, secondly, to allow for surface hydroxylation in order to reach a high areal density of silanol groups SiOH (~5 SiOH/nm^2) on the surface of the silicon wafer [70]. Subsequently, as an adhesive interlayer, a homogeneous self-assembled monolayer (SAM) was attached to the surface wafers. This SAM was achieved by immersing the wafer in a 1 mM toluene solution of (3-mercaptopropy) -triethoxysilane (MPS, Sigma-Aldrich Chemie, Germany) for approximately 24 h. Samples were rinsed several times with toluene and sonicated for about 30 s to remove molecules from the surface onto which they were only physisorbed. These so washed samples were dried under a stream of nitrogen gas. As the final step, by thermal evaporation at a rate of 0.1 nm/s at room temperature, gold layers with a thickness of ca. 40 nm were deposited onto the silicon wafers pre-coated with the MPS monolayer.

2.2.2 Pattering Using Micro-contact Printing

These so prepared substrates were subsequently chemically patterned through micro-contact printing. Elastomeric stamps, consisting of ca. 5 μm large pillars arranged on a hexagonal array with a mesh size of 8 μm, were fabricated by casting a silicone pre-polymer (polydimethylsilicone (PDMS) Sylgard 184, Dow Corning) into a "mould". The mould was prepared by e-beam lithography in silicon (CSEM, Neuchâtel/Switzerland) resulting in a pattern with the complementary relief structure. The pre-polymer was filled into the mould and cross-linked at 45 °C for 15 h. Finally, the solid stamp was peeled off the mould. The stamps were cleaned prior to use by ultrasonic treatment in alcohol for several minutes, followed by complete drying under nitrogen gas.

For micro-contact printing the stamp was soaked with reactive molecules by immersion in a 2 mM solution of hexadecanethiol (HDT, Sigma-Aldrich Chemie, Germany). After removal from the solution, the stamp was dried under a nitrogen gas stream. For transferring the pattern onto the silicon substrates, the stamp was gently brought into contact with the gold substrate, kept in contact for 30 s and then carefully lifted off. The imprinted pattern on the gold-coated substrate was characterised in the friction mode by atomic force microscopy (Digital Instruments Nanoscope IIIA/D3100). This allowed to determine the size of the imprinted domains and to verify the fidelity of the pattern replication process. Due to the methyl-termination of the HDT-molecules anchored on the gold substrate via the thiol end-groups, the resulting HDT patches were less wettable by polystyrene than the surrounding gold surface.

2.2.3 Polystyrene Film Formation and Dewetting

As the final step of sample preparation, polystyrene films were prepared by spin coating rather dilute toluene solutions of variable concentration onto gold substrates structured with the micro-pattern of a HDT monolayer. We have chosen a medium molecular weight polymer, PS120k, having a molecular weight of $M_w = 120{,}000$ g/mol and a polydispersity of $M_w/M_n = 1.05$, purchased from Polymer Standards Service, Mainz/Germany. The thickness h of the PS120k film was varied from ca. 15 nm to ca. 100 nm by changing the concentration of the solution.

Dewetting was initiated by heating the sample above the glass transition temperature of PS within a nitrogen-purged hot stage (Linkam TMS 91, Surrey, UK). In order to allow for dewetting velocities v of the order of some μm/min, annealing temperatures up to 220 °C have been used. The temperature was controlled within ± 0.5 °C. Real time dewetting was observed under an optical microscope (Leitz Metallux 3, Wetzlar, Germany). After dewetting, samples have been characterised by AFM (Digital Instruments Nanoscope IIIA/D3100) in the tapping

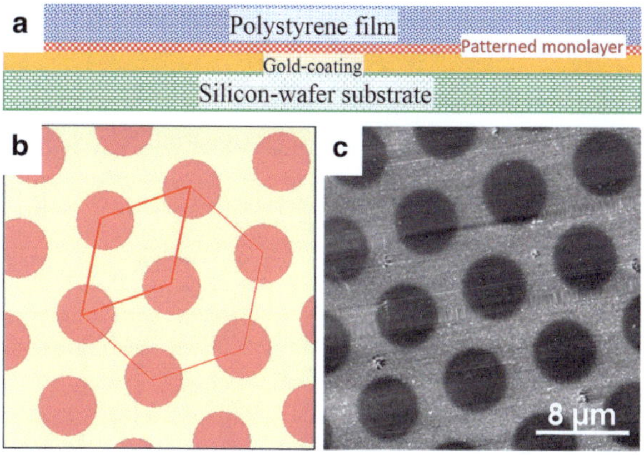

Fig. 2.1 (**a**) Schematic representation of the build-up of the samples used. (**b**) Scheme of the pattern which was imprinted on the gold-coated silicon wafer substrates in a hexagonal geometry (the unit cell is indicated by *red lines*). (**c**) AFM image in friction mode showing the imprinted hexadecanethiol (HDT) circular domains. The *dark discs* correspond to the HDT domains while the *light grey area* in between corresponds to the gold substrate

mode, yielding a higher lateral resolution than optical microscopy and providing in addition information of the topography of the pattern at a nanometer resolution.

In Fig. 2.1a a schematic representation of the build-up of the samples is shown. Figure 2.1b shows a two-dimensional schematic representation of the pattern imprinted onto the gold-coated silicon wafers. A typical AFM image in friction mode of the imprinted of hexadecanethiol (HDT) circular domains is shown in Fig. 2.1c. The size of the hexagonal pattern with a mesh size of 8 µm was chosen so that it was comparable to the size of the width w of the rim which developed during the dewetting process. As a function of the thickness of the spin-coated film, w was increasing from much less than the distance between patches up to values much larger than the mesh size.

2.3 Results and Discussion

2.3.1 *Thin PS120k Films (15 nm < h < 35 nm)*

The design of the pattern composed of HDT-patches surrounded by a gold matrix imposes a wettability contrast, i.e. the interfacial tension between polystyrene and the substrate is lower on the patches than on the surrounding matrix. Thus, the wettability of the HDT-patches is lower than on the surrounding gold. Accordingly, it is more likely that nucleation of holes and subsequent retraction, i.e. dewetting of polystyrene, will first occur on the HDT-patches. In Fig. 2.2a for a 15 nm thick PS120k film, it can be clearly seen that all dewetted cylindrical holes started to grow on the HDT-patches. Within an individual patch, no preferred location for the

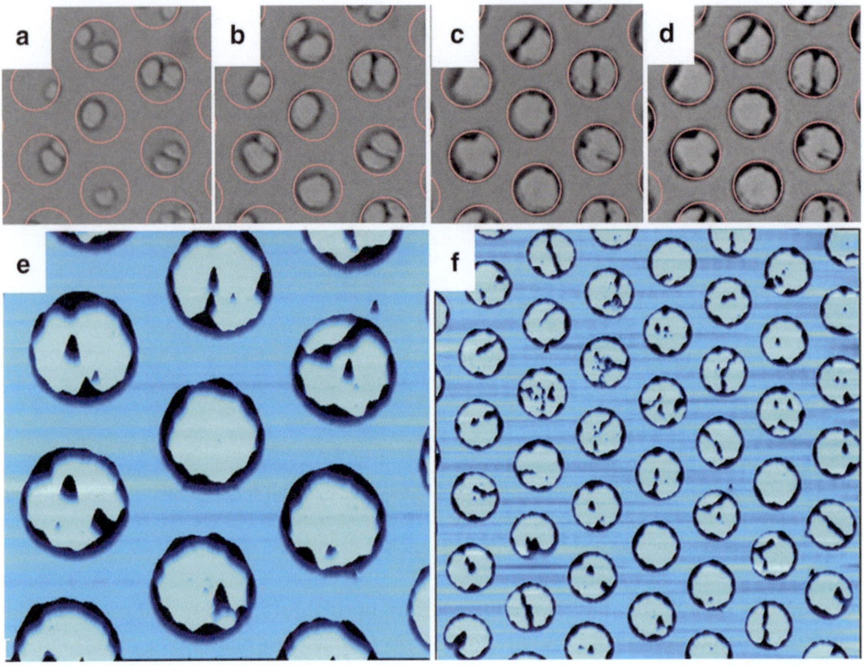

Fig. 2.2 (**a–d**): Optical micrographs ($23 \times 23\ \mu m^2$) showing the progression of dewetting in a ca. 15 nm thick PS120k film after (**a**): 60 s, (**b**): 90 s, (**c**): 195 s and (**d**): 240 s. Dewetting was performed at 130 °C. (**e**) + (**f**): 3D topographic AFM images of holes formed after ca. 100 min of dewetting at 130 °C in an ca. 15 nm thick PS film. Holes are circular with both a diameter and a distribution on the substrate identical to the pattern of the imprinted domains. The size of the images is (**e**): $20 \times 20\ \mu m^2$ and (**f**): $50 \times 50\ \mu m^2$, respectively

nucleation on the patch was detectable. Holes were randomly distributed within the patches. Moreover, for such thin films, at least one hole was nucleated on each patch. Sometimes, however, also two holes were nucleated on an individual patch. No holes were found outside these imprinted domains. This proves that the imprinted domains are significantly less wettable compared to the surrounding. Holes were nucleated on the imprinted domains which actually is expected since the polystyrene does not wet the methyl-terminated domain surface as much as it does the bare gold surface.

In the course of time (see Fig. 2.2a–d) these nucleated holes grew in size and eventually reached the periphery of the patch where dewetting stopped. The dewetted material form the patch was collected in a rim surrounding the dewetted patch. Thus, after dewetting reached the periphery of the patch and stopped, patches were enclosed by a circular rim, as can be clearly seen in Fig. 2.2e. The shape of the holes formed in such thin films was circular with both a diameter and a distribution on the substrate identical to the pattern of the imprinted domains. Some tiny droplets within the dewetted region resulted from the coalescence of rims of multiple holes nucleated within this area. Therefore, dewetting allowed to transform a quasi-two dimensional chemical pattern into a three-dimensional topography consisting of regularly arranged cylindrical containers (holes). It should be noted for the 15 nm thick film that even at the end of the dewetting process the width of the rims (see for example Fig. 2.2e) was small enough so that the rims of neighbouring holes did not touch each other and thus no coalescence of rims cold occur.

The amount of dewetted material increased by increasing the thickness h of the film. As can be seen from Fig. 2.3, such increase in h had several consequences. The most prominent observation is that dewetting proceeded faster for thinner films. This observation is not surprising. For dewetting on non-wettable surfaces like the HDT-patches one may expect slippage of the polymer [71]. Accordingly, the hole diameter D initially increased with time t with an exponent of roughly 2/3, i.e. $D \sim h^{-1/3} \cdot t^{2/3}$. However, when the periphery of the non-wettable patches was reached, dewetting slowed down drastically. Consequently, at the rather low

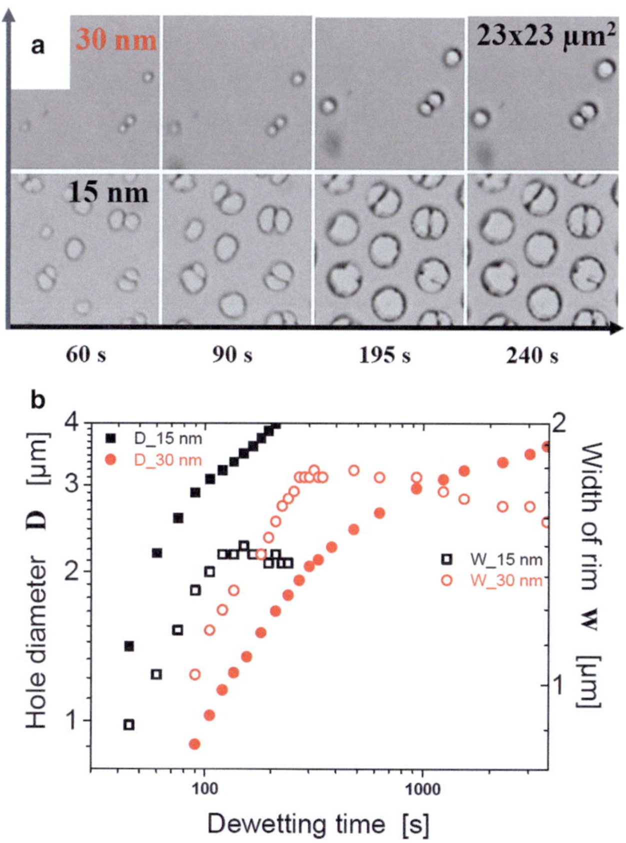

Fig. 2.3 (**a**) Optical micrographs showing the growth of dewetting holes in a 15 nm and a 30 nm thick PS120k film at 130 °C after times indicated in the figure. (**b**): Graphical representation of the temporal evolution of the hole diameter D and the width w of the rim surrounding these holes, corresponding to the images shown in (**a**)

dewetting temperature of 130 °C chosen in Fig. 2.3, only the area of the non-wettable patches was dewetted.

Moreover, the width w of the rim at this stage was mainly determined by the amount of polymer removed from the patches. As this amount is proportional to h, the width of the rim was increasing approximately with $h^{1/2}$, i.e. $w \sim h^{1/2}$. For slipping rims, the dewetting velocity v is inversely proportional to the width of the rim, i.e. $v = \mathrm{d}D/\mathrm{d}t \sim w^{-1} \sim h^{-1/2}$. Thus, in thicker films dewetting proceeds more slowly. It can also be seen that for a film which is twice as thick as the one of Fig. 2.2 not all patches were dewetted. This suggests that the probability of nucleating a hole on a non-wettable patch depends on h.

Alternatively to decreasing h for accelerating dewetting, one may also reduce the viscosity of the fluid by increasing temperature. As can be seen in Fig. 2.4,

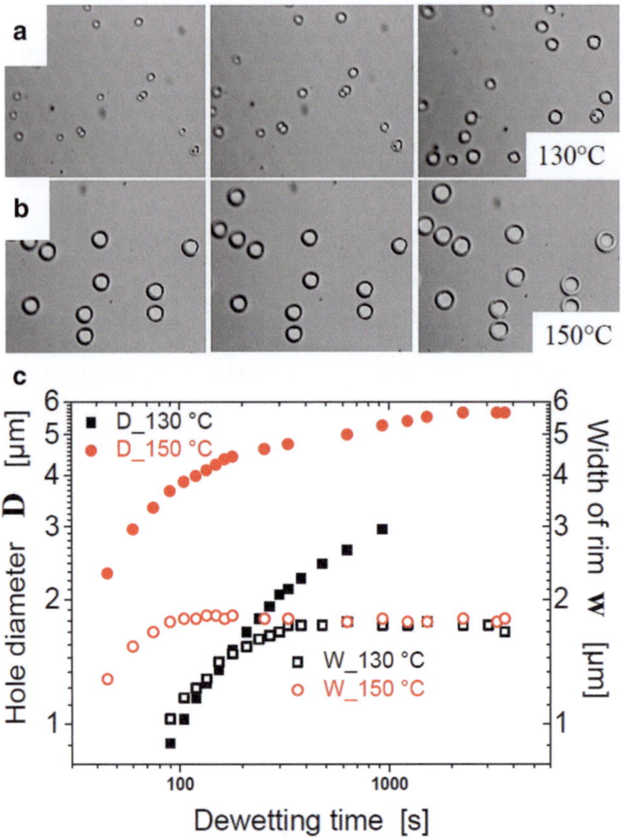

Fig. 2.4 Optical micrographs showing the growth of dewetting holes in a 30 nm PS120K film at (**a**) 130 °C and (**b**) 150 °C after (from left to right) 200 s, 400 s, and 3,600 s, respectively. The size of the images is 77x64 µm2. (**c**): Graphical representation of the temporal evolution of the hole diameter D and the width w of the rim surrounding these holes, corresponding to the images shown in (**a**) and (**b**), respectively

dewetting proceeded faster at the higher temperature. It is worthwhile to note that the width of the rim after the periphery of the non-wettable patches had been reached, is almost independent of dewetting temperature, as expected based on mass conservation. Of course, the periphery was reached faster when dewetting proceeded at a higher temperature.

Figure 2.3 suggested that the probability on nucleating a hole on a non-wettable patch depends on h. This hypothesis has been tested in Fig. 2.5. Increasing h led to a clear decrease in the number N_{Holes} of nucleated holes per unit area (here we used the area covered by 172 patches, i.e. 7,100 μm^2). Although the limited number of data points allow only for cautious and tentative conclusions, one is nonetheless able to deduce from the images in Fig. 2.5a that holes were generated rather randomly. No preferred direction and no correlations between nucleated holes were detectable. However, one may observe occasionally that there is a higher nucleation probability around some holes, suggesting that defects and inhomogeneities may allow for heterogeneous nucleation events.

A reduction in nucleation density of dewetted holes was also found for such polystyrene films on homogeneous substrates [5]. Thus, a threshold film thickness h_{T} can be determined for which the nucleation density of dewetted holes is identical to the number density of imprinted non-wettable domains. Obviously, $h < h_{\mathrm{T}}$ was required for observing a dense array of hexagonally arranged holes.

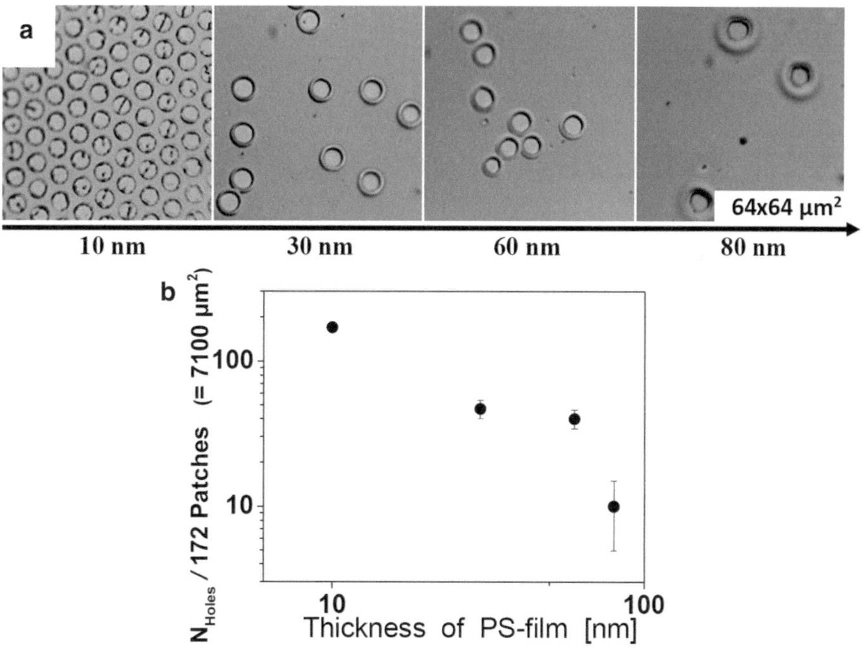

Fig. 2.5 (**a**) Optical micrographs showing holes after 1 min of dewetting at 180 °C for PS120k film of different thicknesses: 10, 30, 60 and 80 nm respectively. (**b**): Graphical representation of the decrease of the number of holes N_{Holes} for an area of 7,100 μm^2 with the increase of the thickness of the PS120k film

2.3.2 *PS120k Films of Intermediate Thickness (35 nm < h < 60 nm)*

A transition in morphology from holes to droplets, deposited at the non-wettable patches, was observed when the film thickness was increased to values above $h_{\min}$ of ca. 35 nm. At such thicknesses, rather few holes were nucleated on the non-wettable patches. Thus, the average distance between holes was several times the spacing between patches, as can be seen in Fig. 2.6. When keeping the annealing temperature at values below ca. 160 °C, even after many hours of annealing, these few nucleated holes did not increase in diameter significantly beyond the size of the patch. However, when the annealing temperature was increased above ca. 180 °C, the three-phase contact line at the border of the dewetting holes slowly advanced, i.e. dewetting also proceeded on the gold surface. This implies that at such high temperature the adsorption energy for polystyrene onto gold was not high enough to assure wetting of gold by polystyrene.

Initially, the holes kept a circular contact line. When the hole diameter and, in particular, the width of the rim became large enough the contact line touched the periphery of neighbouring non-wettable patches. Then, the contact line got practically pinned at the periphery of these non-wettable patches while it continuously advanced on the gold surface. Concomitantly with this onset of the deformation of the contact line, initiating the formation of finger-like protrusions, both rim width and rim height also started to change. These changes varied locally and thus caused changes in shape of the rim induced by the corresponding mass flow within the rim.

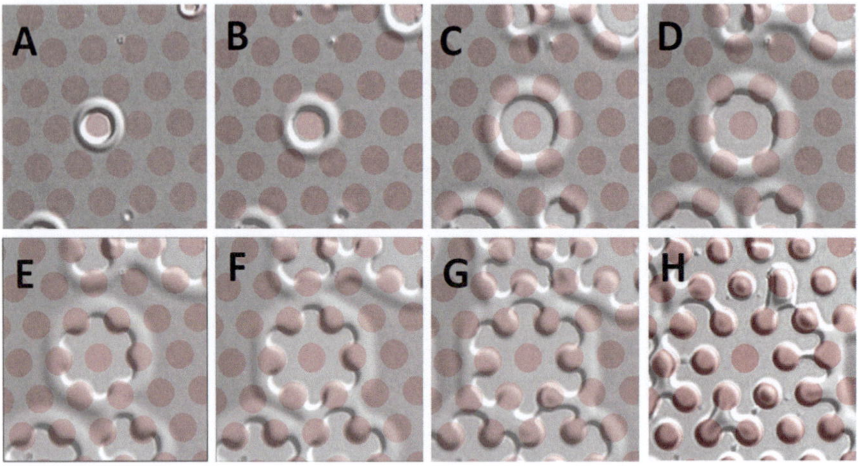

Fig. 2.6 Optical micrographs showing the temporal evolution of a hole in a ca. 40 nm thick PS120k film for (**a**) 180 s, (**b**) 255 s, (**c**) 600 s, (**d**) 615 s, (**e**) 630 s, (**f**) 645 s, (**g**) 700 s and (**h**) 715 s of dewetting at 220 °C. The *red discs* superimposed on the snapshots help to visualise the underlying imprinted domains. The size of the images is 40 × 40 μm^2

In fact, polymer melt was drawn onto the patches, leading to an increase in rim height. This change in shape of the rim occurred on a rather fast time scale, τ_{RC}. As a first approximation, we assume that during τ_{RC} the contact line did not advance significantly. Because of the varying radii of curvature along the rim, the change in shape is accompanied by the generation of gradients in the Laplace pressure which induced a flow of polymers in the orthoradial direction. This so induced flow caused the formation of thicker and thinner parts of the rim, with the thicker parts being located on the non-wettable patches. As a consequence, the rim was getting narrower at locations in between non-wettable patches but wider across the non-wettable patches. One has to note, provided that the contact angle stays constant, i.e. that there is no contact line pinning, that width and height are directly related, i.e. when a rim gets narrower it also reduces its height. This redistribution of material along the rim, together with the advancement of the three-phase contact line, led to the formation of fingers (see Fig. 2.6g) and eventually to the deposition of droplets (see Fig. 2.6h).

In Fig. 2.7, we demonstrate the various stages in the formation and deposition of droplets on the non-wettable patches, both using experimental results and a schematic representation. In particular, the detachment process can be clearly seen in Fig. 2.7d. As indicated in the scheme of Fig. 2.7b, two length-scales are relevant: the characteristic distance (λ) between non-wettable patches and the local width (w) of the rim. In the course of the dewetting process, the advancement of the contact line got slowed down when a non-wettable patch was reached while at locations in

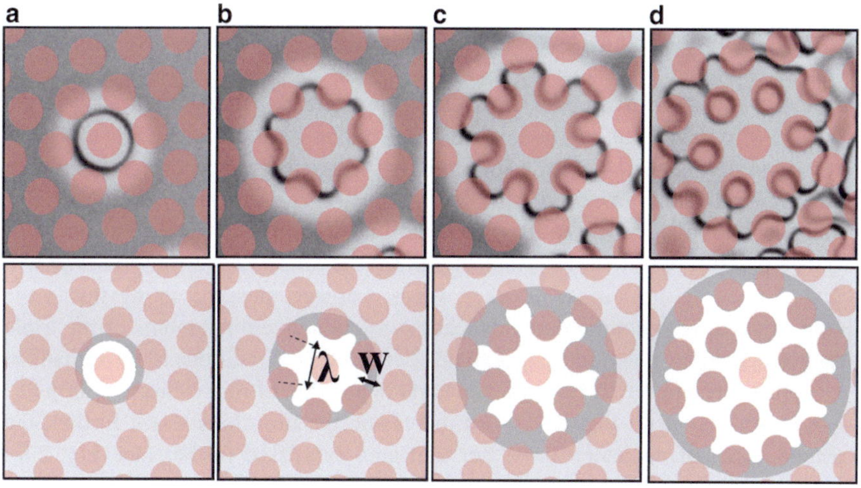

Fig. 2.7 (**a–d**) Optical micrographs showing the growth kinetics of a dewetting hole in a 40 nm thick PS120k film 2 annealed at 220 °C. Time interval between snapshots is 660 s. The *red discs* superimposed on the snapshots of the upper row help to visualise the underlying imprinted HDT domains. In the lower row, schematic representations of the snapshots of the upper row are presented. The *light red discs* represent again the imprinted HDT domains, while the *dark red discs* represent such domains covered by a polystyrene droplet. The light (*whitish*) regions in between represent the dewetted areas, limited by the dewetting rim represented in *grey*

between these patches, it advanced steadily. For slipping polymer films the dewetting velocity (v) depends inversely on w. Accordingly, a reduction of w for locations in between patches led to a local increase in $\mathbf{v}$. Eventually, at such locations the rim became extremely thin while its width increased on the non-wettable patches so that the gradients in Laplace pressure caused a strong orthoradial flow along the rim until a droplet got detached from the moving rim (Fig. 2.7d). The shape and morphology of the rim is governed by the locally acting Laplace pressure (Γ). If Γ is constant along the rim, an almost symmetric rim profile results, which can be schematically represented by a semi-cylinder. Variations of Γ along the cylinder, generated by variations in the contact angle Θ, result in flow along the cylinder and thus will cause corresponding variations in the width and height of the rim.

In addition, the length-scale of the non-wettable pattern imprinted on the substrate causes periodic fluctuation of the triple-line which, in turn, can trigger the Plateau–Rayleigh type instability [23]. This instability amplifies most rapidly a wavelength proportional to the width of the rim. All other wavelengths, even if they were initiated by external stimuli, will thus loose the competitive growth and eventually disappear. Based on this simple argument, we expect that a Plateau–Rayleigh type instability will be most pronounced when the width of the rim is comparable to the distance between imprinted HDT domains. It was shown both experimentally [26] and theoretically [72] that the rim instability leading to the formation of the droplets is characterised by a wavelength which is about 2.1 to 2.7 times the width of the rim. In the example shown in Fig. 2.6 the width of the rim is around 4 μm. Accordingly, the Plateau–Rayleigh should yield a characteristic wavelength of 8–11 μm. This length is compatible with the mesh size of the imprinted non-wettable pattern.

Thicker films form broader rims. Thus, by adjusting the film thickness the wavelength of the Plateau–Rayleigh instability can be increased even to values larger than the distance between imprinted HDT domains. Accordingly, rims can span over more than one imprinted HDT domain. Consequently, these domains cannot act individually in stimulating a triple-line deformation like for narrow rims. Laplace pressure will tend to establish a mean-curvature of the rim and thus will dampen these deformations. In other words, Laplace pressure acting inside the liquid rim is able to couple domains leading to collective movements of the triple line averaged over neighbouring domains. Under such conditions, the mechanism for droplet formation will be eliminated and deposition of micro-droplets stops.

In accordance with Fig. 2.5, when increasing the film thickness h further, the number N_{Holes} of nucleated holes per unit area decreased further (see Fig. 2.8). Thus, the average distance between holes increased and the droplet deposition process could be followed over longer distances, i.e. the number of deposited droplets within a dewetted area increased. Interestingly, the width of the rim was found to be almost constant in time because the fluid added in the course of progressing dewetting was compensated by the continuous detachement of micro-droplets.

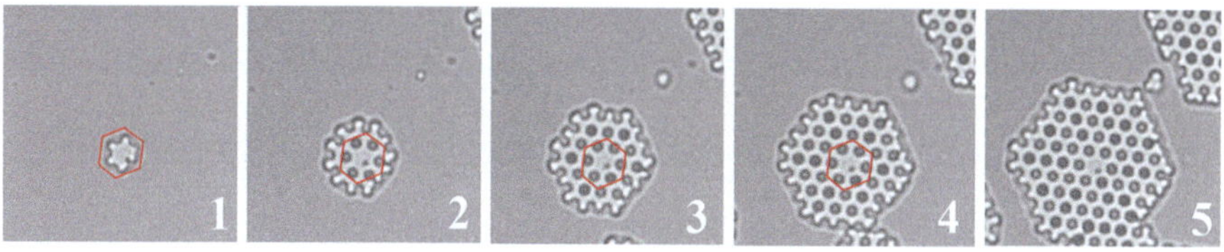

Fig. 2.8 Optical micrographs of a PS120k film (ca. 50 nm thick) showing the top view morphology of a dewetted region for annealing at 220 °C at increasing times. Time interval between each snapshots is 260 s. The envelope of the dewetted region shows hexagonal symmetry

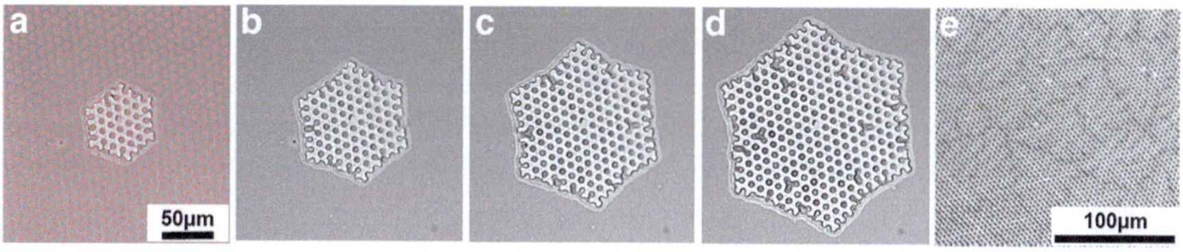

Fig. 2.9 (**a–d**): Top view schematic of the increase of the dewetted area in a ca. 70 nm thick PS120k film annealed at 220 °C. Droplets were deposited regularly on the imprinted regions on the substrate, schematically represented by *red discs* in (**a**). The time interval between the images is 300 s. The envelope of the dewetted region shows a hexagonal symmetry. (**e**) In the final state, reached after ca. 1 h of annealing at 220 °C, all HDT domains were covered with a single PS120k droplet

As can be seen in Fig. 2.8, all the many deposited droplets were replicating the symmetry and the periodicity of the imprinted pattern of non-wettable patches. Interestingly, the boundaries of the dewetted area containing all the deposited droplets became straight, eventually forming a hexagonally shaped envelope around the deposited droplets. This hexagonal shape became even more pronounced when h was increased further as can be seen in Fig. 2.9. As can be seen from Fig. 2.9d, eventually, the whole film was transformed into a regular array of droplets all deposited onto non-wettable patches. Only faint shadows of lines, where the dewetted areas coalesced, were visible.

While optical microcopy is certainly able to follow the droplet deposition process clearly over large areas, atomic force microscopy (AFM) provides, besides a higher lateral resolution, also detailed information on thickness variations generated by the dewetting process. In Fig. 2.10, the detachment process was visualised. In particular, the flow of polymer molecules onto the patches can be identified through the increase in height of the forming droplets. In the sequence of Fig. 2.10c–f, one can clearly see that the maximum height of the two droplets, which eventually got detached, increased significantly while the bridges, which connected them to the advancing rim, progressively got thinner. It also can be observed from Fig. 2.10 that the rim at the boundaries of the dewetted area was always significantly lower in height than the final droplets. Thus, the droplets effectively "suck-up" most of the polymers from the advancing rim. As a

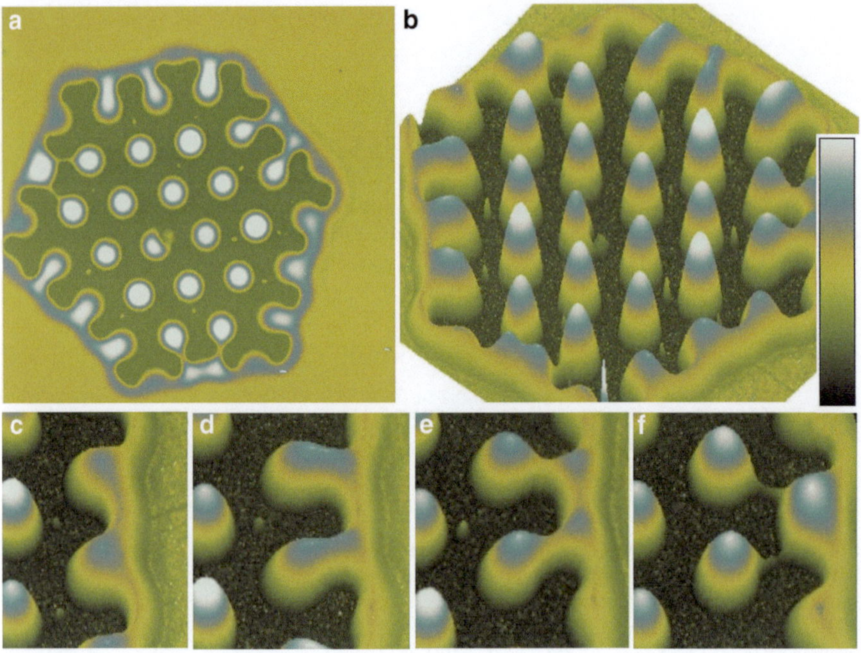

Fig. 2.10 (**a**) 2D and (**b**) 3D topographic AFM image ($60 \times 60\ \mu m^2$, height range is 250 nm) of a hexagonal dewetted area in a 50 nm thick PS120K film annealed at 220 °C. (**c–f**): Topographic 3D AFM images of successive stages of dewetting rim morphologies showing the growth dynamics at 220 °C of a dewetted area in a 50 nm thick PS120K film, focussing in particular on the detachment process of droplets. The size along the *y*-axis of the images is 15 μm

consequence the rim did not grow in width during dewetting because the material from the dewetted area was transferred and finally deposited as droplets onto the non-wettable patches.

2.3.3 *Thick PS120k Films* (h > *60 nm*)

Assuming that there exists an equilibrium contact angle $\Theta < \pi/2$ for polystyrene on the non-wettable patches, then there exists a maximum amount of polystyrene which can be deposited on these patches. A deposited drop has the shape of a spherical cap with a circular base of diameter 2a and a height H. The volume of such a cap is then $V = \pi \cdot H \cdot (3a^2 + H^2)/6$. For small values of Θ (we have measured values of $\Theta \approx 0.1$), we can approximate the height H of the cap by $H \approx a \cdot \Theta$. This leads to a volume of approximately (in first order of Θ) $V \approx \pi \cdot a^3 \cdot \Theta/2$. If all the material from the area of one patch and the surrounding wettable gold surface (together that approximates to an area of ca. $\pi(2a)^2 \rightarrow V \approx \pi \cdot a^3 \cdot \Theta/2 = 4\pi a^2 \cdot h$) has to be deposited on a single patch, this yield a maximum value for the thickness of the film: $h_{max} = a \cdot \Theta/8$. Plugging in the values used in this study ($a = 5$ μm and $\Theta = 0.1$), this yields a value of $h_{max} = 63$ nm. We note that this value is only an approximation taking into account all the assumption made. It is interesting to note that for $h \approx h_{max}$ the average value of w (7 ± 1 μm) was very similar to the mesh size of the domain pattern (8 μm).

For films thicker than h_{max}, not all the dewetted material could be deposited on the non-wettable patches. Accordingly, in the course of dewetting the width of the rim had to grow continuously as not all material could be left behind. As the rim was getting wider, the variations in width induced by the non-wettable patches were getting smaller and the orthoradial flow within the rim was reduced. Accordingly, although the contact line was deformed with a period determined by the mesh size, these undulations were never amplified and thus did not lead to droplet deposition.

Thus, it is not surprising that for films thicker than ca. 60 nm we observed a transition to a third type of pattern generated by dewetting, where almost no polymer was left behind in the dewetted area. In Fig. 2.11 we give a typical example. It can be clearly seen that the width of the rim was significantly larger than the period of the patches. Interestingly, as can be clearly seen in the AFM

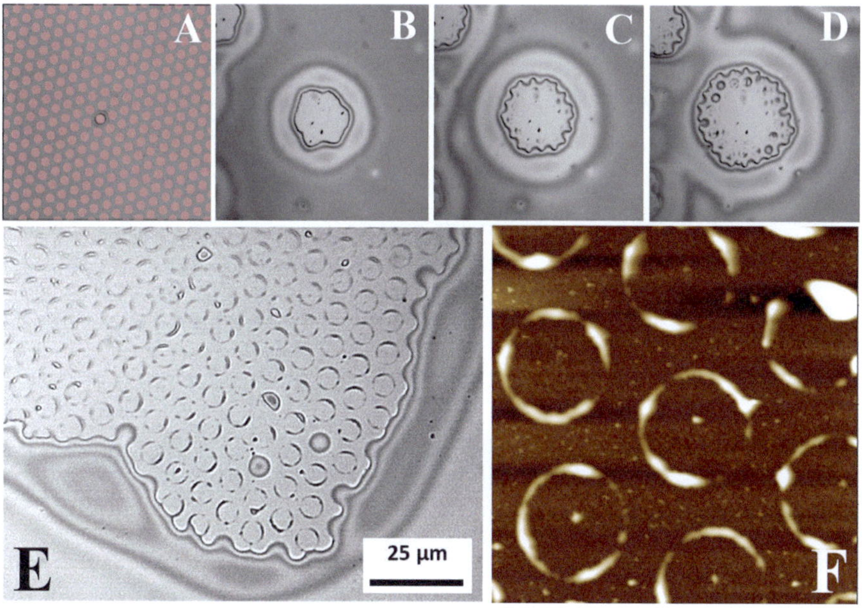

Fig. 2.11 (**a–d**): Optical micrographs (120 × 120 μm^2) of a PS120k film (ca. 90 nm thick) showing the top view morphology of dewetted area for annealing 180 °C (5 min time interval between subsequent images). The *rose discs* in (**a**) represent schematically the imprinted HDT-pattern. (**e**) Part of the dewetted area after ca. 60 min of annealing at 180 °C. (**f**). AFM topography image (20 × 20 μm^2) of a section inside the dewetted area. No micro-droplets but circular toroids have been deposited at the periphery of the imprinted domains. The toroids at the periphery of the imprinted domains are attributed to polymers remaining after dewetting of the PS120k film

image shown in Fig. 2.11f, a little amount of polymer was left behind at the periphery of the non-wettable patches, forming a regular array of toroidal structures deposited on the patches. This observation suggests that polystyrene was somehow anchored at the interface between the non-wettable domains of circular geometry and the bare gold surface. As a possible explanation for such anchoring effect one may identify the vertical component of the capillary force acting at the contact line which locally may compete with the Laplace pressure, characterised by the local slope of the polymer rim and the lateral driving force, i.e. the uncompensated Young force characterised by the deviation of the dynamic contact angle from its equilibrium value.

Occasionally also a fraction of a droplet or a whole droplet were deposited on this patches. As one may expect, the probability for depositing full droplets increased when decreasing *h*. In the example shown in Fig. 2.12, both deposition modes can be observed, denoted by the yellow and red arrows. At locations between patches where the width of the advancing rim became thinner than the diameter of the patch, detachment of a droplet occurred. For the opposite case, i.e. when the width of the rim was always wider than the diameter of the patches, the pinning of the contact line of the retracting rim was not strong enough to keep all material on the patch. As can be seen in particular in Fig. 2.12g, h, the contact line cut loose from the patch and rapidly retracted to catch up with the more advanced parts of the rim. Within less than ca. 100 s an almost straight contact line was achieved. This has to be compared with the ca. four times slower mean retraction velocity of the rim.

In Fig. 2.13, we demonstrate that for an increase in film thickness the transition from a regular array of deposited droplets to regularly deposited toroids is rather

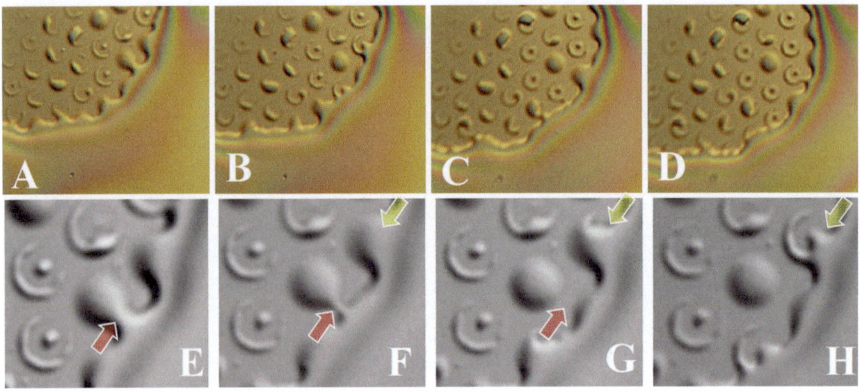

Fig. 2.12 (**a–d**): Optical micrographs of a PS120k film (ca. 85 nm thick) annealed at 220 °C for increasing times (time interval 200 s) showing the top view morphology of dewetted area for the borderline case with only few droplets deposited on the substrate. The size of the images is 52 × 60 μm^2. (**e–h**): Zoom-in into above images (time interval 50 s). The size of the images is 20 × 23 μm^2. The *red arrows* indicate the process of necking in the rim leading to the deposition of a droplet. The *yellow arrows* point out the process of toroid formation where the rim drag most of the PS120k molecules off the patch

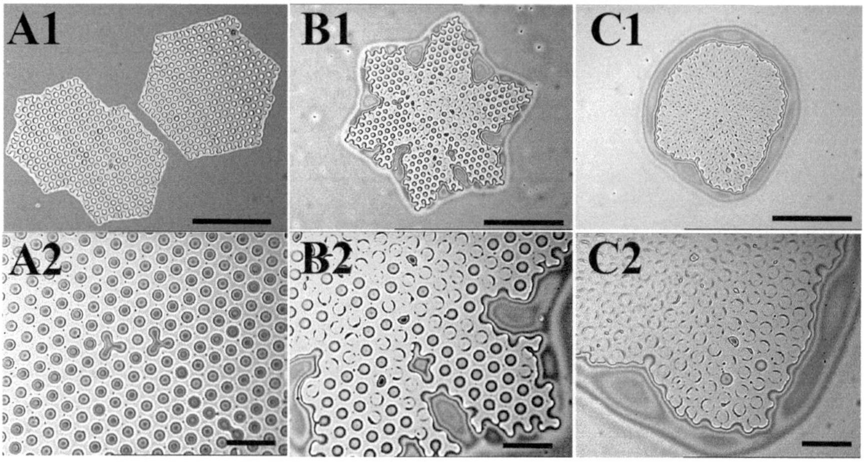

Fig. 2.13 Characteristic optical micrographs for dewetted areas in a PS120k film having a thickness of about (**a**) 50 nm, (**b**) 75 nm and (**c**) 90 nm. The second row shows a zoom-in into the dewetted areas shown in the *top row*. The scale bars represent 100 μm (*top row*) and 25 μm (*bottom row*), respectively

continuous. While for films of ca. 50 nm only droplets were deposited (see Fig. 2.13a), almost exclusively toroids remained for a ca. 90 nm thick film (see Fig. 2.13c). For films with a thickness in between these two cases, a coexistence of droplets and toroids was observed (see Fig. 2.13b). It also can be seen that the probability for droplet deposition was the highest at locations where the width of the rim was the smallest, which typically were the corners of the hexagonally shaped dewetted area. Interestingly, for the thickest films studied the width of the rim around the dewetted area became rather uniform and approached a circular shape (see Fig. 2.13c).

2.4 Conclusions

The experimental results presented here impressively demonstrate that a simple dewetting process of a thin polymer film on a chemically patterned substrate allows to transform a smooth layer of material into various structures. For increasing thickness of the layer the chemical pattern produces the following sequence (see Fig. 2.14): Cylindrical holes—droplets with the shape of a spherical cap—toroids.

The central element responsible for this sequence is the dewetted fluid collected the rim, characterised mainly by its width w. Thus, w may be considered as the crucial parameter determining the final 3D morphology. The volume V of this rim is exclusively determined by polymer from the dewetted area, i.e. by the dewetted distance and the thickness h of the film. w has to be related to the separation distance of the non-wettable patches and the local contact angle Θ, taking into account that Θ

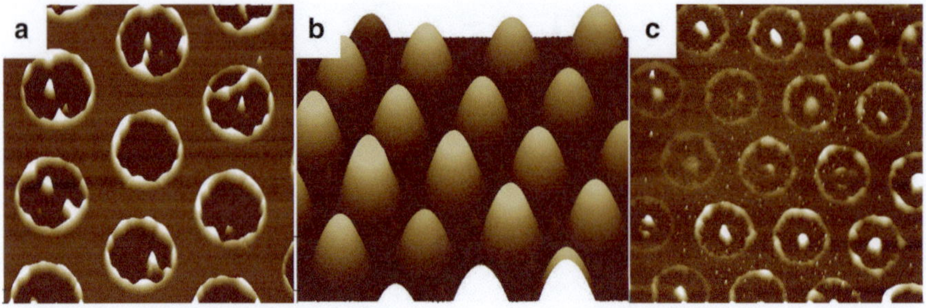

Fig. 2.14 (**a**) Topographic AFM image ($25 \times 25\ \mu m^2$) of holes formed in a ca. 15 nm thick PS120k film. The height range is 100 nm. Holes are circular with both a diameter and a surface distribution identical to that of the imprinted domains pattern. (**b**) Topographic AFM image ($30 \times 30\ \mu m^2$) of droplets formed in a ca. 50 nm thick PS120k film. The height range is 250 nm. The circular droplets with a diameter identical to that of the imprinted domains are distributed regularly on the substrate. (**c**) Topographic AFM image ($30 \times 30\ \mu m^2$) of patterns formed in a ca. 90 nm thick PS120k film. The height range is 40 nm. Circular toroids with a diameter identical to that of the imprinted domains are distributed regularly on the substrate. These toroids at the periphery of the imprinted domains were generated by polymers remaining after dewetting at the boundaries of the domains

is varying along the chemical pattern of the substrate. The area of a cross-sectional area A of the rim is proportional to Θ and w, i.e. $A = w^2 \cdot \Theta$. Accordingly, A is larger on the non-wettable patches. During dewetting the contact line passes from a weakly non-wettable surface to a less-wettable surface, requiring an increase in Θ and thus an increase in A. When the rim is spanning regions of varying Θ, fluid will be redistributed from the regions of lower values of Θ to the ones having a higher value of Θ. Such induced flow will lead to a decrease of material on the more wettable parts of the substrate and to an increase of material on the non-wettable patches. For a given finite volume V, i.e. for a certain h, the thinning of the rim on the more wettable regions of the substrate may even cause rupture of the rim at these locations and thus cause the detachment of droplets. This is the case for films with $h_{\min} < h < h_{\max}$. For films thinner than $h_{\min}$, w is never getting so large that the rim can span the region in between non-wettable patches. For films thicker than $h_{\max}$, the induced weak flow of material between different regions is not able to create regions thin enough to allow for the detachment of droplets. Then, the shape of the dewetted area, which is enclosed by a wide rim, becomes circular.

In summary, a dewetting process is capable to transform a featureless and smooth film into a three-dimensional topography when a thin coating of finite thickness is deposited onto a chemical two-dimensional wettability pattern. Of course, the characteristic length-scale of the pattern and its periodicity are reproduced in the topographical structure.

Acknowledgments The authors acknowledge the Centre Suisse d'Electronique et de Microtechnique SA (CSEM, Neuchâtel, Switzerland) for the fabrication of micro-structured moulds. Help of Hanan ASSAF at an initial stage of this work is highly appreciated.

References

1. Xia, Y.N., Kim, E., Zhao, X.M., Rogers, J.A., Prentiss, M., Whitesides, G.M.: Science **273**, 347 (1996)
2. Sirringhaus, H., Tessler, N., Friend, R.H.: Science **280**, 1741 (1998)
3. Jager, E.W.H., Smela, E., Inganas, O.: Science **290**, 1540 (2000)
4. Brochard-Wyart, F., Daillant, J.: Can. J. Phys. **68**, 1084 (1990)
5. Reiter, G.: Phys. Rev. Lett. **68**, 75 (1992)
6. Brochard-Wyart, F., di Meglio, J.-M., Quéré, D., De Gennes, P.-G.: Langmuir **7**, 335 (1991)
7. Yerushalmi-Rozen, R., Kerle, T., Klein, J.: Science **285**, 1254 (1999)
8. Steiner, U., Klein, J., Eiser, E., Budkowski, A., Fetters, L.J.: Science **258**, 1126 (1992)
9. Steiner, U., Klein, J., Fetters, L.J.: Phys. Rev. Lett. **72**, 1498 (1994)
10. Morariu, M.D., Schaffer, E., Steiner, U.: Phys. Rev. Lett. **92**, 156102 (2004)
11. Seeman, R., Herminghaus, S., Jacobs, K.: Phys. Rev. Lett. **86**, 5534 (2001)
12. Becker, J., Grun, G., Seemann, R., Mantz, H., Jacobs, K., Mecke, K.R., Blossey, R.: Nat. Mater. **2**, 59 (2003)
13. Sharma, A., Reiter, G.: J. Colloid Interface Sci **178**, 383 (1996)
14. Sharma, A., Khanna, R.: Phys. Rev. Lett. **81**, 3463 (1998)
15. Ghatak, A., Khanna, R., Sharma, A.: J. Colloid Interface Sci. **212**, 483 (1999)
16. Reiter, G., Khanna, R., Sharma, A.: Phys. Rev. Lett. **85**, 1432 (2000)
17. Sharma, A.: Eur. Phys. J. E. **12**, 397 (2003)
18. Verma, R., Sharma, A.: Ind. Eng. Chem. Res. **46**, 3108 (2007)
19. Reiter, G., Hamieh, M., Damman, P., Sclavons, S., Gabriele, S., Vilmin, T., Raphael, E.: Nat. Mater. **4**, 754 (2005)
20. Cavallini, M., Facchini, M., Massi, M., Biscarini, F.: Synth. Met. **146**, 283 (2004)
21. Luan, S., Cheng, Z., Xing, R., Wang, Z., Yu, X., Han, Y.: J. Appl. Phys. **97**, 086102 (2005)
22. Wang, J.Z., Zheng, Z.H., Li, H.W., Huck, W.T.S., Sirringhaus, H.: Nat. Mater. **3**, 171 (2004)
23. Brochard-Wyart, F., Redon, C.: Langmuir **8**(9), 2324 (1992)
24. Gabriele, S., Sclavons, S., Reiter, G., Damman, P.: Phys. Rev. Lett. **96**, 4 (2006)
25. Reiter G., Sharma A.: Phys. Rev. Lett., 2001, 8716, (16), art. no.-166103.
26. Reiter, G.: J. Adhes. **81**, 381 (2005)
27. Karthaus, O., Grasjo, L., Maruyama, N., Shimomura, M.: Chaos **9**, 308 (1999)
28. Leizerson, I., Lipson, S.G., Lyushnin, A.V.: Langmuir **20**, 291 (2004)
29. Meyer, E., Braun, H.G.: Macromol. Mater. Eng. **276**, 44 (2000)
30. Lenz, P., Bechinger, C., Schafle, C., Leiderer, P., Lipowsky, R.: Langmuir **17**, 7814 (2001)
31. Xie, R., Karim, A., Douglas, J.F., Han, C.C., Weiss, R.A.: Phys. Rev. Lett. **81**, 1251 (1998)
32. Higgins, A.M., Jones, R.A.L.: Nature **404**, 476 (2000)
33. Rehse, N., Wang, C., Hund, M., Geoghegan, M., Magerle, R., Krausch, G.: Eur. Phys. J. E. **4**, 69 (2001)
34. Rockford, L., Liu, Y., Mansky, P., Russell, T.P., Yoon, M., Mochrie, S.G.J.: Phys. Rev. Lett. **82**, 2602 (1999)
35. Krausch, G., Kramer, E.J., Rafailovich, M.H., Sokolov, J.: Appl. Phys. Lett. **64**, 2655 (1994)
36. Boltau, M., Walheim, S., Mlynek, J., Krausch, G., Steiner, U.: Nature **391**, 877 (1998)
37. Zhang, Z.X., Wang, Z., Xing, R.B., Han, Y.C.: Surf. Sci. **539**, 129 (2003)
38. Brusch, L., Kühne, H., Thiele, U., Bär, M.: Phys. Rev. E **66**, 011602 (2002)
39. Thiele, U., Brusch, L., Bestehorn, M., Bär, M.: Eur. Phys. J. E. **11**, 255 (2003)
40. Kao, J.C.T., Golovin, A.A., Davis, S.H.: J. Colloid Interface Sci. **303**, 532 (2006)
41. Lenz, P., Lipowsky, R.: Phys. Rev. Lett. **80**, 1920 (1998)
42. Gau, H., Herminghaus, S., Lenz, P., Lipowsky, R.: Science **283**, 46 (1999)
43. Kumar, A., Whitesides, G.M.: Science **263**, 60 (1994)
44. Konnur, R., Kargupta, K., Sharma, A.: Phys. Rev. Lett. **84**, 931 (2000)
45. Kargupta, K., Konnur, R., Sharma, A.: Langmuir **16**, 10243 (2000)
46. Zope, M., Kargupta, K., Sharma, A.: J. Chem. Phys. **114**, 7211 (2001)

47. Kargupta, K., Sharma, A.: Phys. Rev. Lett. **86**, 4536 (2001)
48. Kargupta, K., Sharma, A.: Langmuir **19**, 5153 (2002)
49. Sharma, A., Konnur, R., Kargupta, K.: Physica A **318**, 262 (2003)
50. Simmons, D., Chauhan, A.: J. Colloid Interface Sci. **295**, 472 (2006)
51. Zhang, X., Xie, F., Tsui, O.K.C.: Polymer **46**, 8416 (2005)
52. Braun, H.G., Mayer, E.: Macromol. Mater. Eng. **276/277**, 44 (2000)
53. Sehgal, A., Ferreiro, V., Douglas, J.F., Amis, E.J., Karim, A.: Langmuir **18**, 7041 (2002)
54. Julthongpiput, D., Zhang, W., Douglas, J.F., Karim, A., Fasolka, M.J.: Soft Matter **3**, 613 (2007)
55. Zhang, Z., Wang, Z., Xing, R., Han, Y.: Polymer **44**, 3737 (2003)
56. Wang, X., Ostblom, M., Johansson, T., Inganas, O.: Thin Solid Films **449**, 125 (2004)
57. Wang, X., Tvingstedt, K., Inganas, O.: Nanotechnology **16**, 437 (2005)
58. Mukherjee, R., Bandyopadhyay, D., Sharma, A.: Soft Matter **4**, 2086 (2008)
59. Geoghegan, M., Wang, C., Rhese, N., Magerle, R., Krausch, G.: J. Phys. Condens. Matter **17**, S389 (2005)
60. Luo, C., Xing, R., Zhang, Z., Fu, J., Han, Y.: J. Colloid Interface Sci. **269**, 158 (2004)
61. Mukherjee, R., Gonuguntla, M., Sharma, A.: J. Nanosci. Nanotechnol. **7**, 2069 (2007)
62. Xing, R., Luo, C., Wang, Z., Han, Y.: Polymer **48**, 3574 (2007)
63. Suh, K.Y., Lee, H.H.: J. Chem. Phys. **115**, 8204 (2001)
64. Suh, K.Y., Park, J., Lee, H.H.: J. Chem. Phys. **116**, 7714 (2002)
65. Suh, K.Y., Lee, H.H.: Adv. Mater. **14**, 346 (2002)
66. Harkema, S., Schaffer, E., Morariu, M.D., Steiner, U.: Langmuir **19**, 9714 (2003)
67. Luo, C., Xing, R., Han, Y.: Surf. Sci. **552**, 139 (2004)
68. Kim, Y.S., Lee, H.H.: Adv. Mater. **15**, 332 (2003)
69. Kim, D.H., Kim, M.J., Park, J.-Y., Lee, H.H.: Adv. Funct. Mater. **15**, 1445 (2005)
70. Fleith, S., Ponche, A., Bareille, R., Amédée, J., Nardin, M.: Colloids Surf. B: Biointerfaces **44**, 15 (2005)
71. Reiter, G.: R Khanna. Langmuir **16**, 6351 (2000)
72. Münch, A., Wagner, B.: Phys D Nonlinear Phenom **209**, 178 (2005)

Chapter 3
Directing Convection to Pattern Thin Polymer Films: Coffee Rings

Bo Li, James Iocozzia, and Zhiqun Lin

3.1 Introduction

In daily life, coffee rings are commonly seen after a droplet of coffee has dried on a table or dish. The mechanism of this phenomenon was first proposed by Deegan in 1997 [1]. It was found that the "coffee ring" of solute (e.g., colloidal particles) was formed because the suspended solutes were carried by radial flow towards the edge due to larger evaporative flux at the contact line with the surface. For applications such as surface coatings, it is generally desirable to have a uniform deposition of material across the entire surface. Consequently, the coffee ring effect is a problematic issue for many applications such as inkjet printing. Thus, numerous studies on evaporation-induced patterning of a droplet composed of volatile solvent and nonvolatile solutes have been performed in an attempt to eliminate the coffee ring effect entirely and thus enhance inkjet printing efficiency and quality. Though it seems such a phenomenon would always be considered a nuisance, research into the coffee ring effect has since redeemed it as both an interesting and useful property.

Drying droplets containing nonvolatile solutes (polymers, nanospheres, nanoparticles, DNA, etc.) on a solid substrate have been utilized to yield self-assembled structures (e.g., "coffee rings" [1], fingering instabilities [2], cellular structures [3]), possessing dimensions from a few hundred nanometers and larger. One simple route to such interesting structures, based on the coffee ring effect, is the evaporative self-assembly of nonvolatile solutes from a sessile droplet on a solid substrate [4]. However, these dissipative structures created by evaporation are often irregular and not in equilibrium (e.g., randomly organized convection patterns within the drop, and stochastically distributed multi-rings). Yet for applications in microelectronics, data storage devices and biotechnology [5–8], it is prerequisite

B. Li • J. Iocozzia • Z. Lin (✉)
School of Materials Science and Engineering, Georgia Institute of Technology, Atlanta, GA 30332, USA
e-mail: zhiqun.lin@mse.gatech.edu

J. Rodríguez-Hernández, C. Drummond (eds.), *Polymer Surfaces in Motion*,
DOI 10.1007/978-3-319-17431-0_3

to achieve surface patterns with a highly ordered spatial arrangement at large scale (square centimeters or larger). Therefore, fully controlling evaporative self-assembly as a simple route to creating regular structures for various applications strongly depends on precise control over several factors, such as evaporative flux, radial and Marangoni flow, and interactions between substrate and solutes.

To date, several studies have focusing on establishing a means of controlling the drying process of an evaporating droplet to produce highly regular structures. For example, controlled evaporative self-assembly (CESA) in a restricted geometry is a simple, rational preparation route for the creation of microscopic structures having high fidelity and regularity [9, 10]. In addition, flow-enabled evaporative-induced self-assembly (FESA) stands out as an extremely simple route to creating intriguing one- or two-dimensional structures. In this strategy, the evaporation flux and the interfacial interaction between the solute and substrate are precisely controlled by programmable motion of the substrate on a motorized linear translation stage, thereby producing well-ordered structures with high fidelity and regularity.

In this chapter, recent progress on the theoretical study of a drying droplet both with and without solutes is first reviewed. Then, the advances in highly ordered structures crafted by controlled evaporative self-assembly are discussed. In particular, the focus is on controlled evaporative self-assembly in a "curve-on-flat" geometry and flow-enabled evaporative-induced self-assembly (FESA).

3.2 Understanding "Coffee Rings": Mechanism and Theoretical Models

A ring-like deposition can be commonly noticed after the evaporation of a solution composed of volatile solvent and nonvolatile solutes. In general, the formation of coffee-ring structures of nonvolatile solutes at the contact line requires the completion of four steps:

(a) A capillary flow is generated towards the contact line by an evaporation rate gradient at the air–solvent interface close to the contact line. It is noteworthy that the first step is often referred to as the "coffee-ring" effect proposed by Deegan in 1997 [1]. However, the entire four-step process is more accurately described as such with this merely being the first part in an essentially iterative process throughout the duration of evaporation.
(b) The solute is effectively transported by the radial flows to the contact line [4, 11].
(c) Due to the attraction between substrate and solute, the solute deposits and forms a continuous line along the contact line. Simultaneously, the contact line of the solvent is pinned at the deposition of the solute. This is known as the "pinning" process.
(d) Due to the continuous evaporation of the droplet, the force induced by surface tension of the extracted liquid surface overcomes the pinning force. As a result, the contact line depins from the ring of deposited solute and jumps

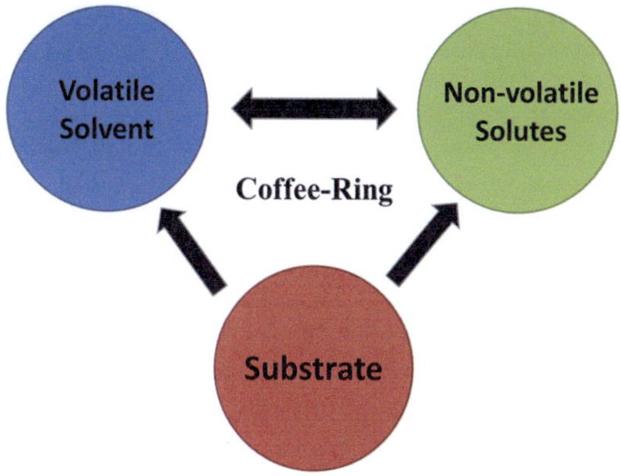

Scheme 3.1 The three essential components for the formation of "coffee-ring" structures: volatile solvent, nonvolatile solutes, and substrate

inward. This is known as the "depinning" process. As this process proceeds, a concentric ring-like deposition pattern is left behind with each retreating depinned contact line.

The formation of ring-like deposition pattern requires fulfillment of all the four steps in sequence. Thus the "coffee ring" cannot form if any of the four steps is not satisfied, thereby resulting in suppression of coffee-ring formation. The following sections review the theoretical principles underlying each step as well as the related experimental studies employing them.

In addition to a requisite series of steps, there are also three essential components for the successful formation of "coffee-ring" structures. These include the volatile solvent, nonvolatile solutes, and the substrate (Scheme 3.1). The influence of both the nonvolatile solute and substrate on the flow inside a volatile solvent drop is related to steps *a* and *b*, while the influence of volatile solvents and substrates to the deposition of nonvolatile solutes is related to steps *c* and *d*.

3.2.1 The Evaporation Flux of a Droplet Evaporating on a Substrate (i.e., Step a, the Evaporation Rate Difference Along the Meniscus)

The evaporation of droplets has been studied for a long time [12], particularly, the evaporation of an aerosol droplet in air [13]. Intuitively, one expects that the evaporation rate of a spherical droplet, which is proportional to the surface area and thus the radius of the droplet, decreases linearly with time. However, this is only the case for evaporation of a spherical droplet in a vacuum [14]. From experiments, it is found that the squared radius of a macroscopic droplet decrease linearly with time [15]. The reason is that the evaporation is determined by the diffusion of the solvent molecules into its quiescent gas phase, rather than the vacuum [13, 16]. Notably, the evaporation flux of an aerosol droplet in air is uniform all over its surface, which is given by the Hertz–Knudsen relation [16, 17]:

$$j = \alpha \sqrt{\frac{k_B T_s}{2\pi M}} \frac{\rho_s - \rho_v}{\rho_l} \tag{3.1}$$

where k_B is the Boltzmann constant, T_s is the surface temperature, M is the mass of a liquid molecule, ρ_s is the vapor density at the surface, ρ_v is the homogeneous vapor density, ρ_l is the vapor density of the liquid, and α is the accommodation coefficient describing the possibility of phase change, which normally has a value between 0.1 and 1 [18–20]. If the diffusion process is quasi-stationary and the temperature T is homogeneous, then the density ρ_V of the vapor obeys $\Delta\rho_V = 0$ with a fixed boundary conditions $\rho_V = \rho_\infty$ far from the drop and $\rho_V(R) = \rho_s$ just above the interface [17]. To this end, the evaporative flux can be given by the simplified equation [17]:

$$j = \frac{j_0}{R} \tag{3.2}$$

where R is the radius of the droplet and j_0 is the evaporation parameter defined as

$$j_0 = D_m \frac{\rho_s - \rho_\infty}{\rho_L} \tag{3.3}$$

where D_m is the diffusion coefficient of the vapor molecules in the atmosphere. From Eq. (3.2), it can be seen that the evaporation flux for a spherical droplet is related to the droplet size [21]. Furthermore, the evaporative flux of a hemispherical droplet on a substrate can also be assumed to be $j = \frac{j_0}{R}$. When the contact angle θ is smaller than 90°, the evaporation flux was calculated by an analogous electrostatic problem of a charged lens at fixed potential by Deegan [1]

$$j(r) \propto (R - r)^{-\lambda} \tag{3.4}$$

where $\lambda = \dfrac{\pi - \theta}{2\pi - \theta}$, R is the radius of the pinned droplet, and r is the distance to the center of a droplet. From Eq. (3.4), it can be seen that the evaporative flux increase nonlinearly from the center of the droplet to the edge (i.e., the contact line). Notably, when the contact angle is sufficiently small, λ will be ~0.5. Hence, the evaporation of a thin droplet (i.e., with small contact angle) is analogous to the evaporation of a thin film. This point is discussed later. Hu and Larson achieved better fit for Eq. (3.4), given by [22]

$$j = j_0 \frac{J_0(\theta)}{R} \left(1 - \left(\frac{r}{R}\right)^2\right)^{-\lambda} \tag{3.5}$$

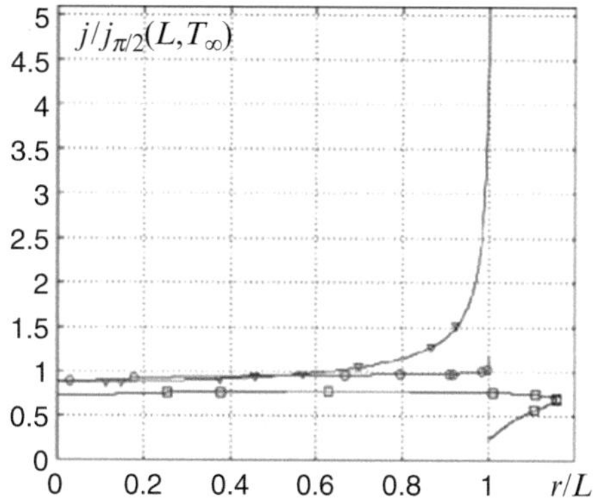

Fig. 3.1 Calculated distributions of local normal evaporative flux over the droplet surface, $L = 1$ mm. circles: $\theta = \pi/2$; squares: $\theta = 2\pi/3$; triangles: $\theta = 2\pi/9$ [23]. Adapted with permission from ref. [23], Copyright© 2012 Elsevier Ltd

where

$$J_0(\theta) = \left(0.27\theta^2 + 1.30\right)\left(0.6381 - 0.2239\left(\theta - \frac{\pi}{4}\right)^2\right) \tag{3.6}$$

Importantly, for a droplet with contact angle larger than 90° (i.e., evaporation of a droplet on a hydrophobic surface), the evaporative flux will be much different (as shown in Fig. 3.1) [23]. When the contact angle is smaller than 90°, the evaporative flux increases towards the contact line; when the contact angle is equal to 90°, the evaporative flux is constant all over the surface; when the contact angle is larger than 90°, the evaporative flux decreases when close to the contact line. A more detailed study with similar results was also reported elsewhere [24].

It is noteworthy that the radius and volume change of a droplet [21, 25–28], and the dynamics of the moving contact line [29–31] are of particular physical interests; however, they are not discussed in this review. The discussion focuses only on the evaporative flux along the surface of the droplet, which results in the radial flow inside the droplet.

It has also been reported that thermal conductivity of the substrate can strongly influence the evaporative flux throughout (Fig. 3.2) [32].

A distinct approach for describing the evaporation of volatile thin films can also be employed for calculating the evaporation flux of a thin droplet (or a thin disc) [33]. The difference between thin film and thin droplet is that the height (i.e., thickness) of a droplet changes with the distance to the center, while the thickness of thin film is uniform. In this approach, the influence of the gas phase on the evaporative flux is neglected. Thus, the liquid phase and vapor phase are decoupled in the calculation. Such a model is referred to as a nonequilibrium one-sided (NEOS) model while the abovementioned model developed by Deegan is referred to as the lens model. Based on the Clausius–Clapeyron law [34], which is used to relate the temperature and the pressure, the boundary condition at the liquid–gas interface is given as [19]

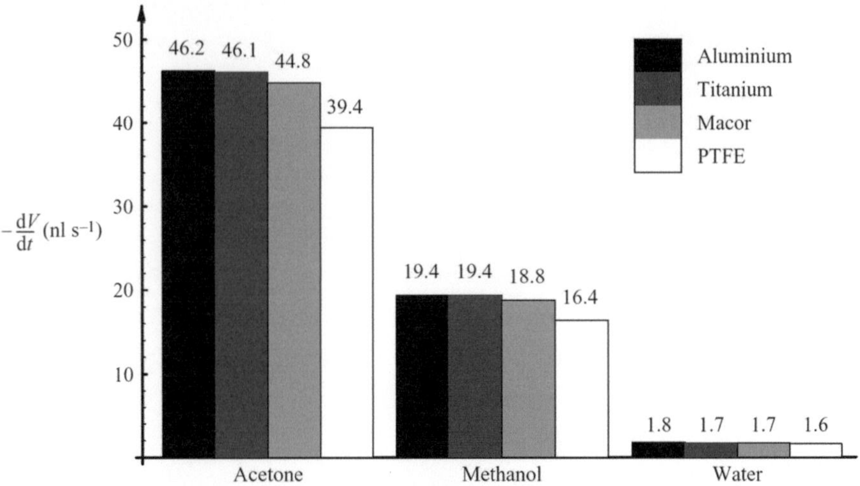

Fig. 3.2 Summary of theoretically predicted values for the total evaporation rate of a droplet of acetone, methanol and water of radius $R = 1.35$ mm on substrates of Al, Ti, Macor, and PTFE [32]. Adapted with permission from ref. [32], Copyright© 2009 Cambridge University Press

$$j(h) = \frac{1}{h + K + W} \tag{3.7}$$

where K is the nonequilibrium parameter [35], W is the thermal effect given by $W = \frac{kd_s}{k_s d_0}$, where k is the liquid thermal conductivity, k_s is the thermal conductivity of substrate, d_0 is the liquid thickness and d_s is the thickness of substrate. By assuming that the shape of a droplet is still a spherical cap, the thickness h can therefore be determined by r, the distance to the droplet center. Furthermore, the evaporative flux is also dependent on the distance to the droplet center. According to the lens model (i.e., Eqs. 3.4 and 3.5), the evaporative flux diverges at the contact line. However, based on the NEOS model, for $h = 0$ at the contact line, the evaporative flux is still a finite number. The difference of evaporative flux predicted by the two models is shown in Fig. 3.3 [35]. It can be seen that, in the lens model the evaporative flux is significantly larger close to the contact line, compared to NEOS model. In addition, for the evaporation of a water droplet on Si wafer, it was found that the volume change with time predicted by NEOS model fitted the experimental results well, while the lens model overestimated the volumes because of the singularity of evaporation flux at the contact line [35].

Clearly, even though the lens model allows the prediction of evaporative flux with contact angle ranging from 0° to 90°, the divergence of the evaporative flux at the contact is not physically true and needs to be corrected [36]. One way to solve this problem is to assume the formation of a thin film at the contact line region. Therefore, the evaporation flux at the contact line can be calculated using the NEOS model (i.e., Eq. 3.7) [37, 38]. It is noteworthy that the assumption of the presence of

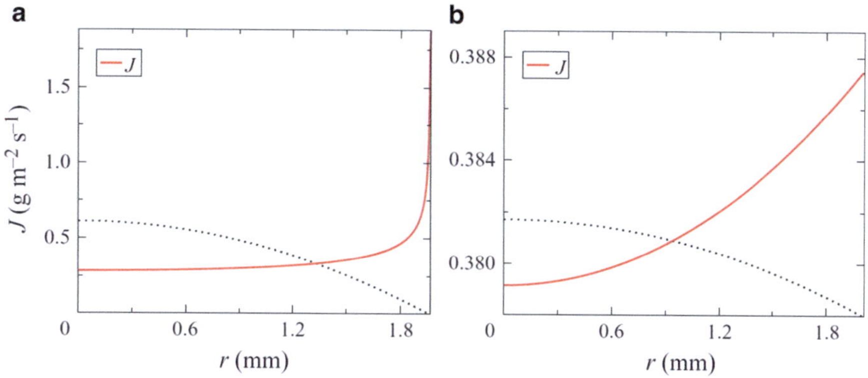

Fig. 3.3 Evaporative flux j of a water droplet as a function of the radial coordinate r predicted by (**a**) the lens model and (**b**) the NEOS model. The *solid red lines* represent j, and the *dotted lines* represent the corresponding drop profile [35]. Adapted with permission from ref. [35], Copyright© 2011 Cambridge University Press

a thin liquid film formed upon receding of contact line has been indirectly shown to exist by using a grafted substrate [39].

The theoretical models of the evaporation of a droplet on a substrate have been intensively studied during the past years [40–50]. However, the evaporative flux along the droplet surface has been rarely measured experimentally. Typically, the global evaporation rate over the entire surface of a droplet was provided in experimental data. To this end, the singularity of evaporative flux at the contact line remains a question. Thanks to the development of digital holographic interferometry, the local evaporative flux of an evaporating droplet can now be determined (Fig. 3.4) [51]. It is clear shown that the evaporative flux did not significantly increase close to the contact line and the evaporative flux at the contact line region predicted by a modified lens model failed to show good agreement with experimental results [51].

3.2.2 *The Flow Inside of the Droplet (i.e., Step* **b***)*

It is straightforward that the evaporative flux at the contact line will result in radial flow inside of a droplet to compensate the liquid loss at the contact line. However, the flow inside of an evaporating droplet turns out to be more complicated. It was found that radial flow, which facilitates coffee-ring formation, and Marangoni flow, which reverses coffee-ring formation, could occur during evaporation and ultimately lead to entirely different deposition patterns.

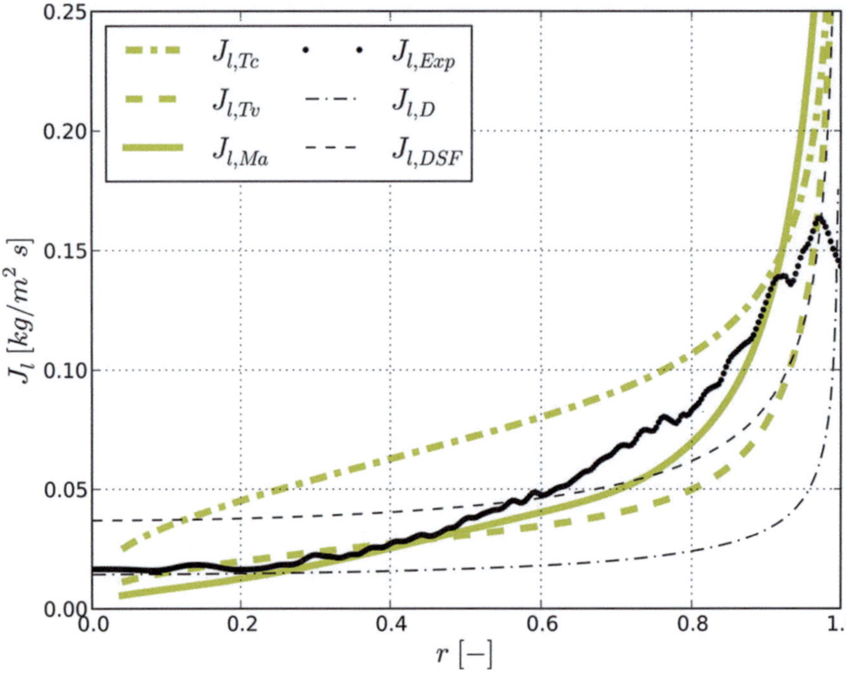

Fig. 3.4 Local evaporation rates versus r, the dimensionless radial distance to the center of a droplet. The *black dotted line* shows experimental results while the others are calculated results from models [51]. Adapted with permission from ref. [51], Copyright© 2014 American Chemical Society

3.2.2.1 Radial Flow

During the drying of a sessile droplet, the mass loss of liquid evaporating from the contact line is replenished by the migration of liquid from the interior, thereby causing a radial flow which carries solutes towards the contact line in the process. This is the reason for coffee-ring formation first noticed by Deegan in 1997 [1]. In this work, the mass of solute accumulated at the contact line, $M(R, t)$ was found to follow the expression $M(R,t)^{\sim} t^{1.37}$ [1], where t is the time of evaporation. Furthermore, the expression of evaporative flux allows for the prediction of the solute transport to the contact line $v(r, t)$ (i.e., the velocity of the radial flow at time t), which was given by the following equation [11]

$$v(r,t) = \frac{-1}{\rho r h}\int_{0}^{r} r\left(j(r)\sqrt{1+\left(\frac{\partial h}{\partial r}\right)^{2}} + \rho\frac{\partial h}{\partial t}\right) \mathrm{d}r \qquad (3.8)$$

where ρ is the density of the liquid and h is the position of the air–liquid interface. However, as the velocity of radial flow is obtained by the expression of evaporative

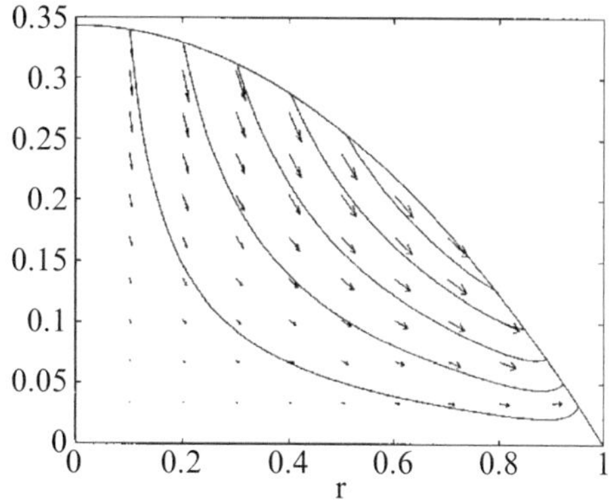

Fig. 3.5 Representative streamlines and velocity vectors for a pinned droplet in which the evaporative mass flux is given by Eq. (3.9). Fluid flows from the center of the droplet toward the contact line [52]. Adapted with permission from ref. [52], Copyright© 2001 American Chemical Society

flux, the singularity at the contact line remains problematic. To this end, Fisher employed the NEOS approximation to determine the radial flow at the contact line [52]. Notably, the evaporative flux was assumed to decrease exponentially near the contact line due to the effect of solute deposition at the contact line, and the evaporative flux expression was thus modified to [52]:

$$j(h) = \frac{1}{h+K}\left[1 - \mathrm{e}^{-A(r-1)^2}\right] \tag{3.9}$$

By assuming the contact line was always pinned during evaporation, the velocity of radial flow was given by

$$v = -\frac{1}{\mathrm{Ca}}\frac{\partial}{\partial r}\left[\frac{1}{r}\frac{\partial}{\partial r}\left(r\frac{\partial h}{\partial r}\right)\right]\left(\frac{1}{2}z^2 - hz\right) \tag{3.10}$$

where Ca is the capillary number and z is the vertical distance to the substrate. Thus, the velocity vectors can be obtained (Fig. 3.5) [52]. In general, as the evaporative flux was higher at the contact line, the capillary force created a radial flow towards the contact line [1]. Moreover, Popov also proposed an analytical solution to the evaporation of volatile solvent with low solute concentrations by assuming that solute occupied finite volume [53]. In addition, both the height and width of the deposition ring as a function of time were evaluated. By taking into account the convection, diffusion, and adsorption of the solute, an extended model for predicting the radial flow was developed with improved accuracy [36]. Notably, the deposition pattern left by the evaporation of a droplet with an unpinned contact line was also provided.

Recently the radial flow was measured experimentally using particle image velocimetry (PIV) analysis [54]. Interestingly, the velocity of a radial flow increased from zero at the center to its maximum at around 70 % of the radius of the droplet, which remained pinned during evaporation (Fig. 3.6) [54]. Interestingly, the intermediate radial position of maximum velocity within the droplet is not

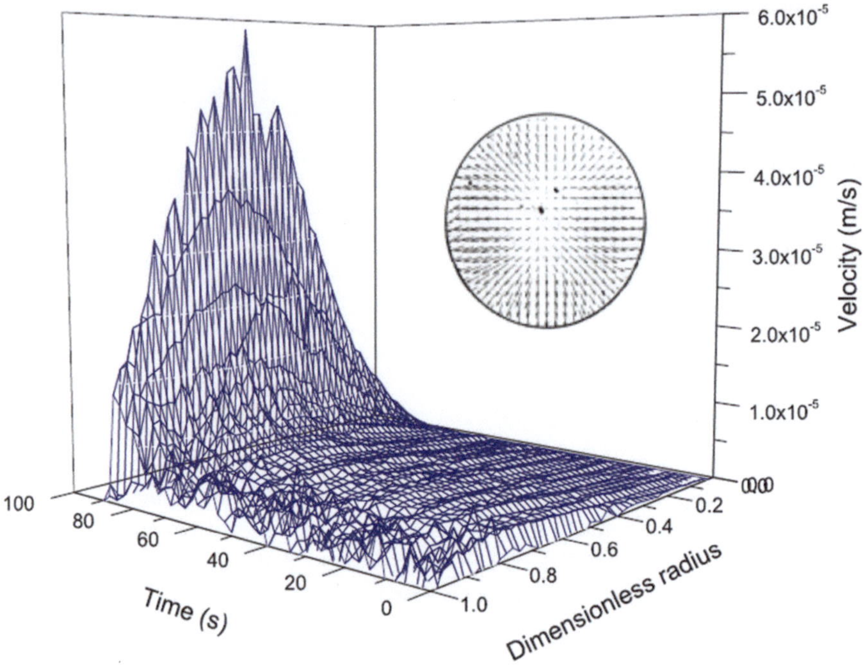

Fig. 3.6 Three-dimensional Spatiotemporal evolution of the flow field, measured just above substrate, in an evaporating water droplet. *Inset* shows that the velocity vectors are those of a radially outward flow [54]. Adapted with permission from ref. [54], Copyright© 2011 American Physical Society

considered in all the previously mentioned models. All previous models predicted a maximum velocity at the contact line. Moreover, the maximum velocity increased by an order of magnitude in the final moments of the pinned evaporating droplet. According to experimental results, at the early stage of the evaporation of a droplet the velocity can be well predicted by models that assumed evaporation occurs only at the contact line. In contrast, in the final stage of droplet evaporation (at approximately 70 % of the entire evaporation time) models that assumed evaporation occurred uniformly over the entire droplet agreed well with experimental data [54]. To this end, it can be concluded that the theoretical models proposed by Deegan [1, 4, 11], Fischer [52], and Popov [53] still provide good predictions of the velocity of the radial flow at the early stage of evaporation. Furthermore, a similar study also observed that the velocity of radial flow reached its maximum at an inner radial position (at around 80 % of the radius of the droplet) from the contact line, instead of at the contact line [55]. The concentration of solute was found to strongly influence the velocity of the radial flow as it affects the viscosity of solution [55, 56].

3.2.2.2 Marangoni Flow

Marangoni flow was first described in the early 1900s [57]. The effect has been observed for centuries and is the cause of the "tears of wine" phenomenon. The effect is induced by a surface tension gradient resulting from a concentration or temperature gradient along the liquid surface [58]. Different from the radial flow induced by evaporative flux at the contact line, Maragoni flow is circular in nature and carries solutes near the liquid surface inward toward the top of the droplet and then plunges them downward to the bottom-center of the droplet where they are carried along the substrate to the edge where they are recirculated back to the top (Fig. 3.7) [59]. This surface tension-induced circular flow has been intensively studied in interesting "self-organized" flat liquid structures called Bernard cells [3, 60–62].

In an evaporating droplet, the nonuniform evaporation flux will produce a temperature gradient, thereby generating a surface tension gradient along the surface and inducing a Marangoni flow. To date, Marangoni flow has been observed in various liquid droplets (e.g., ethanol, methanol, acetone) [63–65]. The theoretical study of Marangoni flow was proposed by Pearson and Nield [66, 67]. The Marangoni number (i.e., Ma) was introduced to characterize the importance of surface tension forces caused by the temperature gradient. If the Marangoni number exceeds a critical value, a Marangoni flow will be induced [68]. The Marangoni number is defined as [69, 70].

$$Ma = \frac{\Delta\gamma L}{\eta D} \tag{3.11}$$

where $\Delta\gamma$ is the surface tension gradient, L is the size of the system where $\Delta\gamma$ (i.e., surface tension gradient) exists, η is the viscosity of solution, and D is the diffusion

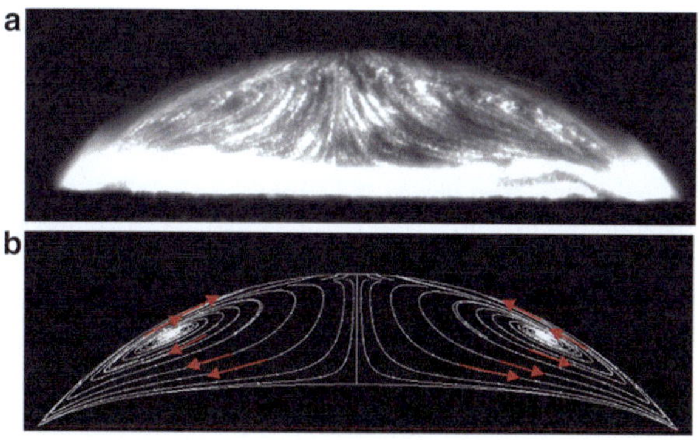

Fig. 3.7 Flow field in a drying octane droplet: (**a**) imaged experimentally, and (**b**) predicted. To observe a clear Marangoni vortex, the illumination plane was moved forward about 0.66 mm from the symmetrical axis of the droplet [59]. Adapted with permission from ref. [59], Copyright© 2006 American Chemical Society

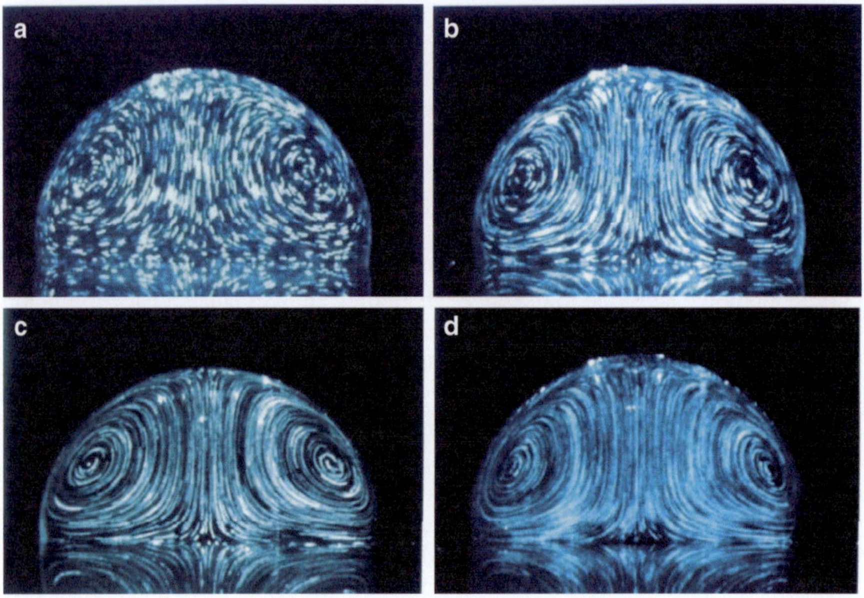

Fig. 3.8 Flow inside evaporating droplets with different NaCl concentrations: (**a**) 0.01 wt% (evaporation time (ET) ~ 39 min); (**b**) 0.1 wt% (ET ~ 45.7 min); (**c**) 1 wt% (ET ~ 50.5 min); and (**d**) 10 wt% (ET ~ 63 min). Exposure time was 20 s for (**a**), (**b**), and (**c**) and 2 s for (**d**). The substrate was placed on a glass substrate coated with the amorphous fluoropolymer Teflon [76]. Adapted with permission from ref. [76], Copyright© 2013 American Institute of Physics

coefficient of solvent. It is important to note that Marangoni flow can be strongly influenced by the contact angle of a droplet and that this is not considered in Eq. (3.11) [71].

However, it remains questionable as to whether the Pearson–Nield model can be accurately applied to volatile liquid systems [72]. This is because experimental results show that energy is transported from the vapor phase to the liquid–vapor interface which is opposite to the assumption made in the Pearson–Nield model. Moreover, it was predicted that thermally driven Marangoni flow would exist in an evaporating water droplet, which was not observed in experiments [58, 73]. The suppression of Marangoni flow in a water droplet may be attributed to the surfactant contaminants [74]. In addition, whether the temperature at the contact line is cooler than the top of the droplet is also unclear [11, 75].

Almost all the circular flow observed in evaporating droplets can be attributed to Marangoni flow. However, it was found recently that Rayleigh convection can also induce circular flow inside of a water droplet on a hydrophobic substrate (Fig. 3.8) [76]. Rayleigh convection is generated by the concentration gradient of solutes (e.g., NaCl) induced by the evaporation of water droplet [76].

3.2.3 *The Pinning and Depinning of the Contact Line (i.e., Steps* c *and* d)

Due to radial flow carrying solutes towards the contact line, the solutes deposit and form a coffee ring-like pattern resulting in the pinning of the contact line and preventing its recession inwards during evaporation. In some circumstance, even though solutes are successfully transported to the contact line by radial flow and negligible Marangoni flow, the solutes may still not be able to deposit onto the substrate. This in turn fails to pin the contact. As discussed in Scheme 3.1, either the interaction between the solutes or the interaction between the solutes and substrate will affect the deposition of solutes on the substrate, thereby determining the pinning of the contact line. For example, if the spherical colloidal particles are replaced by elongated ones, a uniform deposition will be left after evaporation of the droplet, because of the jamming of a monolayer of elongated colloids at the surface of the droplet suppressing the "coffee-ring" formation [77]. Moreover, by changing the colloidal particle surface charges or van de Waals interactions between particles, the coffee-ring formation can also be suppressed [78]. If there is a repulsion between substrate and solutes (i.e., negative Hamaker constant), the solute transported to the edge will move with the receding contact line and no deposition is observed. In contrast, rather than coffee-ring pattern, strong attractions between solutes and substrate (i.e., DLVO interactions) yielded uniform deposition of solutes on the substrate [79, 80].

Depinning is the process whereby the liquid phase detaches from the deposited solute ring. However, detailed theoretical studies of the depinning process have not yet been reported [53]. The mechanism of the depinning process is straightforward. The depinning force is determined by the surface tension of the liquid meniscus at the contact line. During the evaporation process, the depinning force gradually increases while the contact angle continues to decrease until a critical contact angle is reached. Upon further decreasing of the contact angle, the pinning force at the contact line cannot compensate the depinning force, and as a result, the droplet edge depins from the contact line and moves back [81]. Recently, by programmable crafting solute (i.e., diblock copolymer micelle) stripes with a series of different widths, it was found that the pinning force is proportional to the width of the deposited ring [82]. The increase of the width of the deposition ring (i.e., stripe) leads to the decrease of the critical contact angle and an increase in the depinning force.

3.3 Controlling "Coffee Rings": Highly Ordered Structures by Controlled Evaporation

Pinned drying droplets containing nonvolatile solutes (e.g., polymers, viruses, DNAs, microspheres, nanoparticles, carbon nanotubes) yield intriguing one or two-dimensional patterns after complete evaporation of solvent. To date, there have been several impressive studies investigating the use of evaporation to form well-ordered structures rapidly and cheaply over large areas by deliberately controlling the evaporation process [9, 81, 83–95]. Notably, several approaches have been successfully used to generate patterns by confining the evaporating solution in different geometries, such as cylindrical tubes [96, 97], crossed cylindrical mica [98], "curve-on-flat" geometry and nearly parallel-plate geometry. Here we only focus on the last two approaches and review the recent progress.

3.3.1 Controlled Evaporative Self-Assembly in a "Curve-on-Flat" Geometry

Controlled evaporative self-assembly in a "curve-on-flat" geometry has been shown as a simple, rational preparation route for the creation of microscopic structures having high fidelity and regularity [9, 10, 99, 100]. If a droplet of solution is allowed to evaporate in a "curve-on-flat" geometry composed of a curved upper surface situated on a flat substrate (i.e., forming a capillary-held solution), self-assembled structures of high regularity can be generated [101]. Similar to the free evaporation of a droplet on substrate, the evaporation flux also reaches its extremity at the region close to the contact line, leading to highly ordered deposition due to the geometrical restriction [9, 10, 81, 93, 95]. In addition to concentric rings of polymers, nanoparticles [81, 102], carbon nanotubes [90], graphenes [103], and other ordered yet complicated structures can also be produced in the "sphere-on-flat" geometry, including spokes [81], fingers [92, 93], "snake-skin" [104], and serpentines [91] by carefully tuning the experimental parameters such as the concentration and the interaction between solutes and substrates.

Due to the various potential applications of poly[2-methoxy-5-(2-ethylhexyloxy)-1,4-phenylenevinylene] (MEH-PPV) in light emitting diodes (LED), photovoltaic cells (PVCs), and thin-film transistors (TFTs) [105], the synthesis of highly regular concentric rings of MEH-PPV has been investigated by controlled evaporation in a sphere-on-flat geometry (Fig. 3.9) [9]. Since the slowly drying front is arrested (pinned) in the geometrical restriction, the evaporation flux reached the extremity at the region close to contact line. During the evaporation of the solution, the contact line was pinned by MEH-PPV deposition (i.e., "stick"), thereby yielding the formation of a MEH-PPV "coffee ring." The contact angle gradually decreased to a critical contact angle, at which point the depinning force overcomes the pinning force leading to the contact line jumping inward to a new position where the initial contact angle is restored (i.e., "slip"). As shown in Fig. 3.9b, the "slip" distance (i.e., $\lambda_{c\text{-}c}$) slowly decreased with increasing proximity to the sphere/Si contact center. Notably, theoretical predication of the height of the concentric rings based on the Navier–Stokes equation with lubrication approximation provided a good agreement with the experimental data [9]. In addition, the solution concentration and solvent vapor pressure were also found to have strong influence on $\lambda_{c\text{-}c}$ and the heights of concentric rings [9].

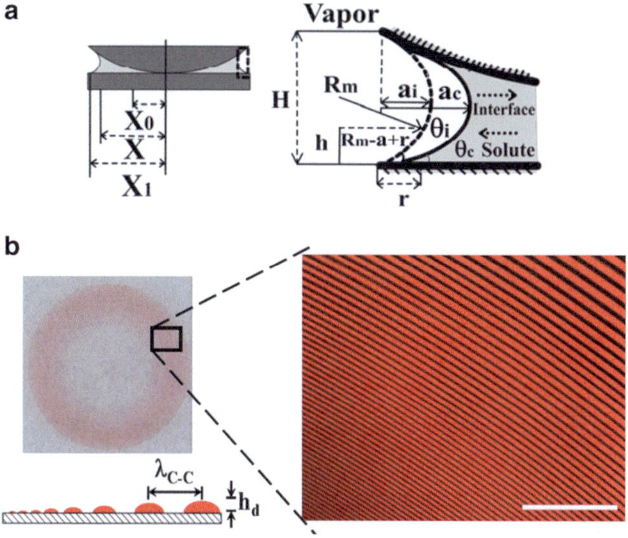

Fig. 3.9 (**a**) *Upper left*: Schematic cross section of a capillary-held solution containing a nonvolatile solute placed in a "sphere-on-flat" geometry. X_1, X, and X_0 are the radii of outermost, intermediate, and innermost rings from the sphere/flat contact center, respectively. *Upper right*: close-up of the capillary edge marked in the *left panel*. (**b**) *Bottom left*: Digital image of entire gradient concentric rings formed by the deposition of MEH-PPV in the geometry shown in (**a**). *Bottom right*: A small zone of the fluorescent image of MEH-PPV rings in *red* is shown. Scale bar = 200 mm. As the solution front moves inward, the rings become smaller and the height decreases as illustrated in lower left schematic [9]. Adapted with permission from ref. [9], Copyright© 2006 American Physical Society

It is also noteworthy that, due to the applications of organometallic polymer in magnetic data storage [106–108], photonic devices [109, 110], and redox-active materials [111, 112], highly ordered concentric rings of organometallic polymers (e.g., poly(ferrocenyldimethylsilane) (PFDMS)) was also obtained by confining PFDMS/toluene solutions in sphere-on-flat geometry [89, 95]. The subsequent pyrolysis of concentric PFDMS rings yielded ferromagnetic ceramics containing α-Fe nanoparticles [113]. Interestingly, once the low concentration PFDMS toluene solution was used for pattern, the PFDMS rings would break into dots due to the surface tension-driven Rayleigh instability at the contact line [89, 95].

Quantum dots (QDs) are highly emissive nanoparticles [114–116] due to their quantum-confinement property, offering promising opportunities in light emitting diodes (LED) [117–120], photovoltaic cells (PVCs) [121–123], biosensors [124–126], and bio-imaging [127, 128]. By passivation of the vacancies and trap sites on CdSe surface with ZnS, the prepared CdSe/ZnS core/shell QDs have bio-applications [129, 130] due to their strong photoluminescence [131–133]. To this end, QDs with two sizes (i.e., 4.4 and 5.5 nm in diameter) were prepared by passivation with a monolayer of tri-*n*-octylphosphine oxide (TOPO) to enhance solubility in toluene while retaining the spectroscopic properties and preventing

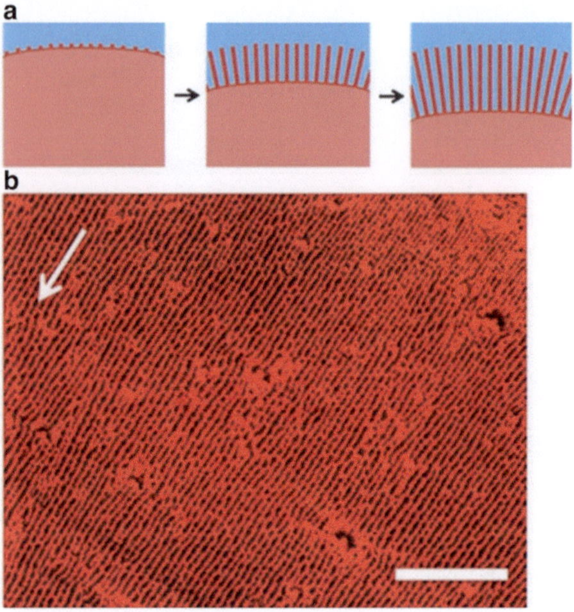

Fig. 3.10 (**a**) Formation of spoke patterns upon evaporation from the capillary bridge in the sphere-on-flat geometry. (**b**) Optical micrograph showing the spokes formed by drying 4.4-nm CdSe/ZnS toluene solution ($c = 0.25$ mg/mL). The scale bar is 100 mm. The *arrow* on the *upper left* indicates the direction of the movement of the solution [81]. Adapted with permission from ref. [81], Copyright© 2007 WILEY-VCH

aggregation [81]. Notably, the evaporation-induced self-assembly of larger CdSe/ZnS core shells QDs (i.e., 5.5 nm in diameter) yielded concentric rings through the use of sphere-on-flat geometry confinement [81]. It was also found that the presence of excess surface capping ligand TOPO lead to no concentric ring pattern formation. Whereas concentric rings formed using larger QDs under optimized conditions, spokes were generated exclusively during the drying of the solution when smaller QDs were used (i.e., 4.4 nm in diameter) (Fig. 3.10). The formation of spokes was caused by fingering instabilities at the liquid front [84, 134–136]. The reason for the distinct deposition patterns formed by large and small QDs was attributed to the moving speed of the liquid front during evaporation [81]. The slow moving speed induced fingering instability at the liquid front [135]. However, a theoretical model for the stripe-spoke transformation has not been proposed.

Moreover, gradient concentric rings (i.e., stripes) of asymmetric comb block copolymer (CBCP) were yielded by constraining CBCP toluene solution in a wedge-on-flat geometry during evaporation (Fig. 3.11) [10]. The height of the wedge strongly affected the deposition pattern of CBCP. In general, large wedge height lead to straight stripes of CBCP while small wedge heights lead to jagged stripes [10]. Intriguingly, hierarchically ordered patterns of CBCP were produced

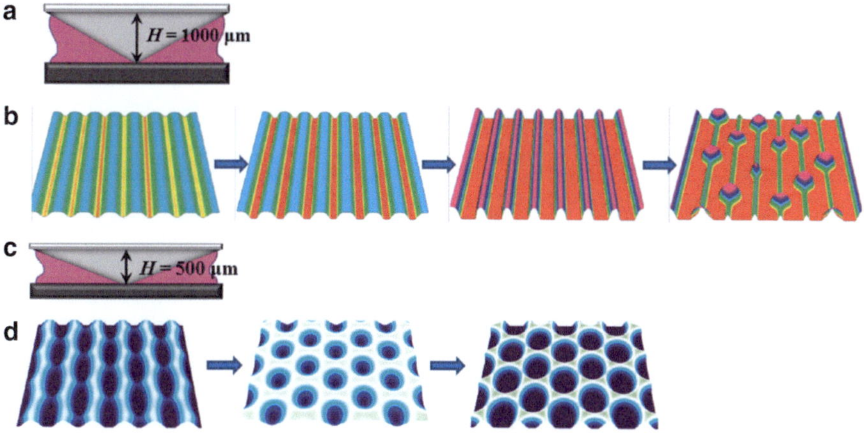

Fig. 3.11 (**a**) Schematic illustration of the wedge-on-flat geometry on a silicon surface. The height of the wedge, H was 1000 μm. (**b**) Stepwise representation of the morphological evolution of CBCP stripes as a function of solvent vapor annealing: *first panel*, as-prepared; *second panel*, annealed for 5 h; *third panel*, annealed for 10 h; *last panel*, annealed for 15 h. (**c**) Schematic illustration of the wedge-on-Si geometry with H of 500 μm. (**d**) Stepwise representation of the morphological evolution of CBCP stripes as a function of solvent vapor annealing time: *left panel*, as-prepared; *central panel*, annealed for 12 h; *right panel*, annealed for 16 h [10]. Adapted with permission from ref. [10], Copyright© 2013 American Chemical Society

by subsequent solvent vapor annealing, because that solvent vapor induced unfavorable interfacial interaction between CBCP rings and the Si substrate lead to the destabilization of CBCP substrate at the microscopic scale [10]. Moreover, within the microscopic stripes, CBCP phase separation was observed at the nanoscale (i.e., CBCP nanocylinders). The nanocylinders oriented either vertically or horizontally to the substrate, depending on the vapor annealing time.

On the other hand, chemically patterned surfaces consisting of gradient stripes of poly(methyl methacrylate) (PMMA) segments created by a wedge-on-flat geometry can be used for directing hierarchical ordered block copolymer structures [137]. First, PMMA stripes were generated by controlled evaporation in the wedge-on-flat geometry; then PMMA segments (i.e., ultrathin PMMA stripes) chemically adsorbed on the Si substrate were obtained after extensive washing with solvent. Finally, asymmetric diblock copolymer, polystyrene-*block*-poly(ethylene oxide) (PS-*b*-PEO) thin film was spin-coated on the prepared chemically patterned surfaces. After mixed-solvent vapor annealing, PS-*b*-PEO stripes would segregate on the ultrathin PMMA stripes [137]. Such chemically patterned surfaces may serve as templates for incorporating other functional nanomaterials or studying cell adhesion [137].

Due to promising applications in photonics, biosensors, electronics, and solar cells, conjugated polymers (e.g., poly(3-hexylthiophene) (P3HT) and poly(3-butylthiophene) (P3BT)) have attracted great attention. To this end, stripes of two conjugated homopolymers (P3HT and P3BT) and one all-conjugated diblock copolymer (P3BHT) were crafted by controlled evaporation in a cylinder-on-flat

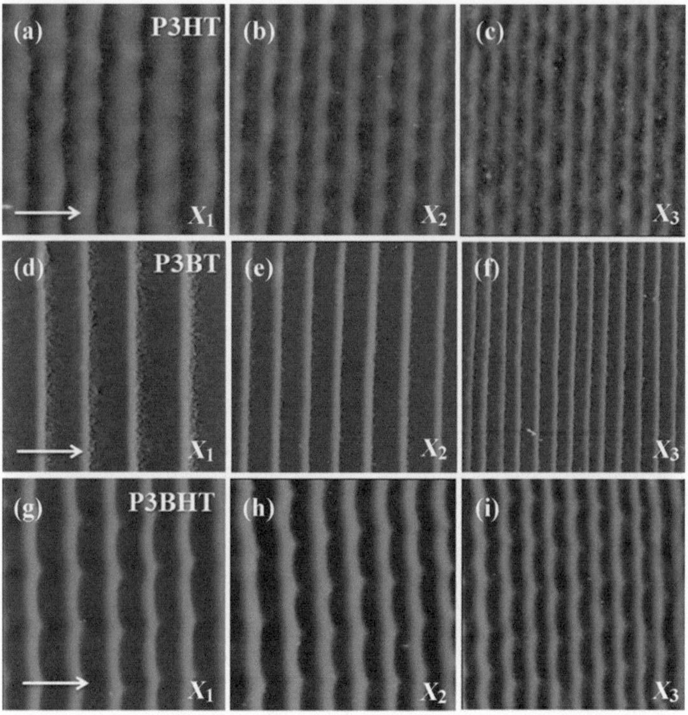

Fig. 3.12 Representative AFM height images of conjugated polymer stripes taken in the three different regions ($X_1 - X_3$). (**a–c**) P3HT; (**d–f**) P3BT; (**g–i**) P3BHT. X_1 is the outermost region, X_2 is an intermediate region, and X_3 is the innermost region, where X is the distance away from the cylinder/flat substrate contact. Image size = 80 × 80 μm [2]. Z range = 400 nm for (**a**, **d**, **g**); 300 nm for (**b**, **e**, **h**); and 200 nm for (**c**, **f**, **i**) [138]. Adapted with permission from ref. [138], Copyright© 2013 WILEY-VCH

geometry (Fig. 3.12). Due to the different interfacial interactions between conjugated polymers and substrate, microscopic stripes (straight or wavy) consisting of nanoscale fiber-like or nodule-like domains were formed. Furthermore, the improved crystallinity of all-conjugated diblock copolymer stripes by vapor annealing could lead to a fourfold increase in electrical conductivity [138]. In addition, hierarchically assembled structures can be realized such as amphiphilic diblock copolymer (poly(styrene)-*block*-poly(4-vinylpyridine) (PS-*b*-P4VP)) stripes produced via controlled evaporation in a cylinder-on-flat geometry. Such stripe structures processed morphology on the microscopic scale as well as self-assembly of PS-*b*-P4VP micelles arrays on the nanometer scale [139].

In addition to synthetic polymer structures, highly aligned DNA nanowires (i.e., spokes) were successfully created by confining aqueous DNA solutions to a curve-on-flat geometry (i.e., sphere-on-flat or cylinder-on-flat) during evaporation (Fig. 3.13) [140]. Unlike coffee-ring structures (or stripes), DNA nanowires formed perpendicular to the contact line. The extremities of DNA were able to anchor to the

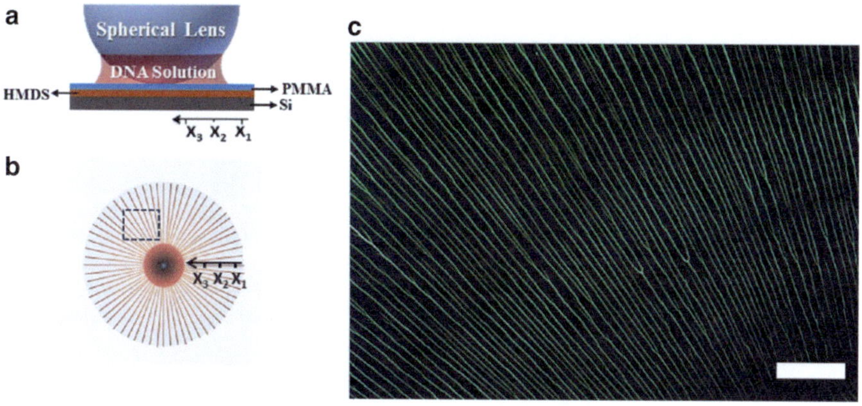

Fig. 3.13 (**a**) Schematic illustration of sphere-on-flat geometry (*side view*), where a drop of DNA solution is constrained, bridging the gap between the spherical lens and the PMMA-coated Si substrate (i.e., a thin PMMA film was spin-coated on HMDS-coated Si). (**b**) Schematic illustration of formation of DNA spokes (*top view*). (**c**) A small zone of DNA spokes (*dashed box* in **b**) obtained at pH = 6.2, $T = 65$ °C, and DNA solution concentration = 8 μg/mL, emitting green fluorescence. Scale bar = 300 μm [140]. Adapted with permission from ref. [140], Copyright© 2013 American Chemical Society

hydrophobic substrate under certain range of pH. Thus, analogous to aligning seaweeds on the beach during the ebbing tide, DNA molecules were preferentially accumulated at the fingers of the liquid front and stretched straight by the capillary force exerted on DNA molecules during the recession of contact line [140]. In addition, it was also found that the pH and temperature of aqueous DNA solutions strongly influenced the formation of DNA nanowires. For example, increasing the pH of the solution lead to a transition from spoke-like patterns to coffee-ring-like patterns [140].

In general, controlled evaporative self-assembly in a "curve-on-flat" geometry is a simple, rational preparation route for the creation of ordered self-assembled structures having high fidelity and regularity. The confinement of evaporating solutions turns out to be an effective means of achieving regular patterns. However, a restricted geometry always results in regular pattern with gradient parameters (e.g., progressive decrease in center-to-center distance between the deposits ($\lambda_{c\text{-}c}$) as the solution front moves to the center of the restricted geometry). Furthermore, the variables that dictate the formation of regular patterns are limited by the choice of concentration, temperature, and solvent at the outset of an experiment.

3.3.2 *Flow-Enabled Self-Assembly in Two-Plate Geometry*

The concept of flow coating came from "slot coating," which has been widely used in the coating industry since 1960s [141–145]. The thickness of the coated liquid layer is set by the prescribed flow rate fed into the coating die and is

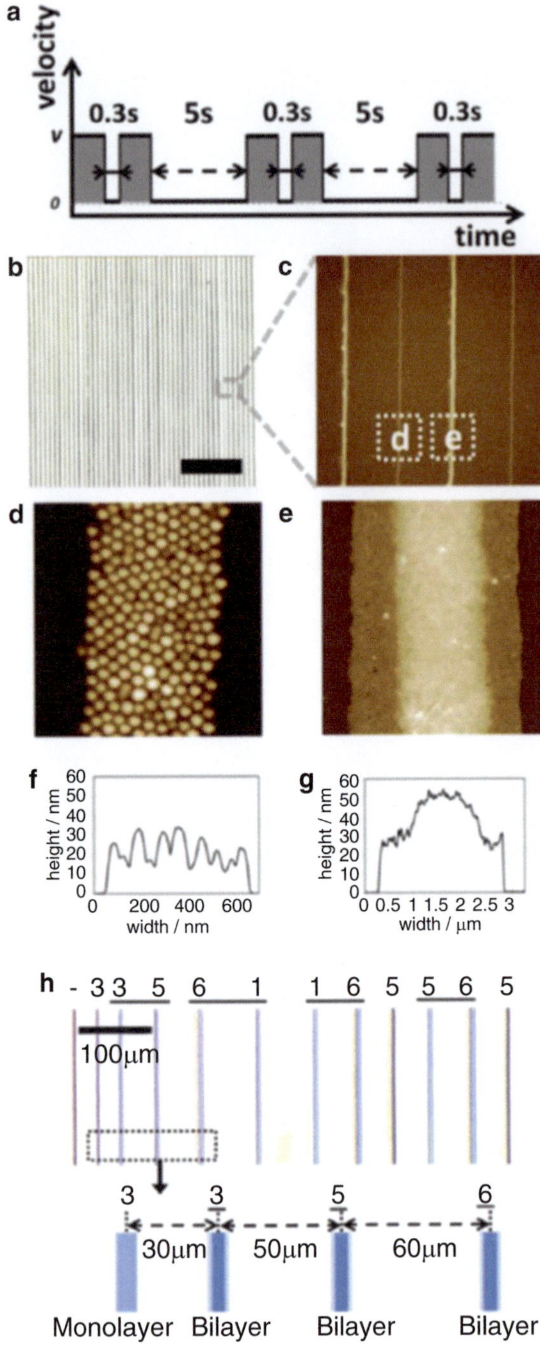

Fig. 3.14 (**a**) Programed movement of the Si substrate mounted on the translational stage in which the velocity is plotted as function of time. The *shaded areas* represent each moving distance of 20 μm after the alternative stop of the Si substrate for 0.3 and 5 s. (**b**) Representative optical

independent of other process variables. This makes this method ideal for high-precision coating. Starting in the late 1990s, the National Institute of Standards and Technology (NIST) developed the flow coating technique for producing gradients. Termed "flow coating," it is a modified blade-casting technique [146–149]. In a typical flow coating process a highly concentrated polymer solution (1–5 % mass fraction) is injected into the gap between a doctor blade positioned over a flat substrate (e.g., silicon wafer) mounted on a computer-controlled translation stage [147].

If the concentration is low, rather than thin films dissipative structures (e.g., convection patterns, fingering instabilities) are produced [87]. Such work was subsequently used to topographically control neurite extension on stripe-patterned polymer films [150], fabrication of periodic micro-structured honeycomb films having multiple periodicities, and polymer nanoparticles [151]. Large-scale ordering was observed. This was defined by the periodic thickness modulation of a block-copolymer film due to the self-organization of the receding contact line [152].

For example, hierarchically assembled amphiphilic diblock copolymer (e.g., PS-*b*-P4VP) micelles were successfully produced by subjecting PS-*b*-P4VP micelles solutions to evaporate in between two nearly parallel plates [82]. Unlike the curve-on-flat geometry, the lower substrate was mounted on a programmable motorized linear translational stage. Highly regular and parallel threads of PS-*b*-P4VP micelles were crafted by a "stop-and-move" procedure [82]. First, the lower substrate remains still for a period of time (i.e., "stop") to allow a thread of micelles to form at the contact line. Then the lower substrate is moved (i.e., "move") at high speed to the next position set by computer, allowing for the generation of the next thread. Thus, periodic parallel threads can be created by these repeated "stop-and-move" cycles. In general, longer stopping times lead to larger widths of threads. Larger move distance leads to larger spacing between threads. Importantly, both the stopping time and the move distance can be controlled, thereby yielding programmable deposition patterns on the substrate (Fig. 3.14). Notably, by varying the stop time, monolayer and monolayer/bilayer thickness threads can be crafted [82]. Interestingly, by controlling the formation of threads using the substrate, the jumping distance of a depinned contact line can be measured [82]. In addition, the precisely positioned PS-*b*-P4VP micelles can be converted into

Fig. 3.14 (continued) micrographs of periodic threads of PS-*b*-P4VP micelles formed on the Si substrate by flow-enabled self-assembly (FESA) based on the program designed in (**a**). The scale bar = 500 μm. Representative AFM images of (**c**) four threads (i.e., two monolayers and two coexisting monolayers/bilayers), (**d**) single thread containing a monolayer of PS-*b*-P4VP micelles, and (**e**) single thread containing a coexisting monolayer/bilayer of PS-*b*-P4VP micelles (i.e., monolayers at both edges with and bilayer in the center). Corresponding height profiles of (**f**) the monolayer-thick thread in (**d**) and of (**g**) the monolayer/bilayer-thick thread in (**e**). (**h**) Encoded thread pattern on the substrate representing the first sentence of the traditional Chinese song Jasmine. The image sizes are 80 μm × 80 μm in (**c**), 800 nm × 800 nm in (**d**), and 3 μm × 3 μm in (**e**) [82]. Adapted with permission from ref. [82], Copyright© 2014 American Chemical Society

Au nanoparticle arrays by exposing PS-*b*-P4VP micelles to Au precursors under oxygen plasma.

An important distinction to be made is that instead of being induced by spontaneous evaporation of solvent, the movement of the contact line will be controlled by the motion of the lower substrate, resulting in a programmable "stick–slip" motion in FESA. Therefore, ordered patterns (e.g., thin films, stripes and spokes) can be readily produced in a controllable manner. Notably, the transition between the different patterns may be observed by gradually tuning each variable individually (e.g., velocity of lower substrate). The influence of each variable on the evaporative self-assembly process can also be scrutinized separately.

3.4 Conclusions and Outlook

In this chapter, an overview of the mechanism of "coffee-ring" formation; including theoretical studies on evaporative flux of a volatile droplet, radial and Marangoni flow inside a droplet, and the pinning and depinning process during evaporation, were provided. In addition, highly ordered structures created by confining evaporation in restricted geometry were reviewed. The ability to craft well-defined, dissipative structures from a variety of materials at low cost opens up avenues to novel advanced functional systems and devices in the near future.

Acknowledgment We gratefully acknowledge funding support from NSF (CBET-1332780).

References

1. Deegan, R.D., Bakajin, O., Dupont, T.F., Huber, G., Nagel, S.R., Witten, T.A.: Capillary flow as the cause of ring stains from dried liquid drops. Nature **389**, 827–829 (1997)
2. Pauliac-Vaujour, E., Stannard, A., Martin, C.P., Blunt, M.O., Notingher, I., Moriarty, P.J., Vancea, I., Thiele, U.: Fingering instabilities in dewetting nanofluids. Phys. Rev. Lett. **100**, 176102 (2008)
3. Nguyen, V.X., Stebe, K.J.: Patterning of small particles by a surfactant-enhanced Marangoni-Benard instability. Phys. Rev. Lett. **88**, 164501 (2002)
4. Deegan, R.D.: Pattern formation in drying drops. Phys. Rev. E **61**, 475 (2000)
5. Lin, Y., Böker, A., He, J., Sill, K., Xiang, H., Abetz, C., Li, X., Wang, J., Emrick, T., Long, S.: Self-directed self-assembly of nanoparticle/copolymer mixtures. Nature **434**, 55–59 (2005)
6. Kim, Y., Han, H., Kim, Y., Lee, W., Alexe, M., Baik, S., Kim, J.K.: Ultrahigh density array of epitaxial ferroelectric nanoislands on conducting substrates. Nano Lett. **10**, 2141–2146 (2010)
7. Park, S., Lim, J.H., Chung, S.W., Mirkin, C.A.: Self-assembly of mesoscopic metal-polymer amphiphiles. Science **303**, 348–351 (2004)
8. Whitesides, G.M., Grzybowski, B.: Self-assembly at all scales. Science **295**, 2418–2421 (2002)

9. Xu, J., Xia, J.F., Hong, S.W., Lin, Z.Q., Qiu, F., Yang, Y.L.: Self-assembly of gradient concentric rings via solvent evaporation from a capillary bridge. Phys. Rev. Lett. **96**, 066104 (2006)
10. Byun, M., Bowden, N.B., Lin, Z.Q.: Hierarchically organized structures engineered from controlled evaporative self-assembly. Nano Lett. **10**, 3111–3117 (2010)
11. Deegan, R.D., Bakajin, O., Dupont, T.F., Huber, G., Nagel, S.R., Witten, T.A.: Contact line deposits in an evaporating drop. Phys. Rev. E **62**, 756–765 (2000)
12. Langmuir, I.: The evaporation of small spheres. Phys. Rev. **12**, 368–370 (1918)
13. Poulard, C., Guéna, G., Cazabat, A.M.: Diffusion-driven evaporation of sessile drops. J. Phys. Condens. Matter **17**, S4213 (2005)
14. Fuchs, N.A.: Evaporation and Droplet Growth in Gaseous Media. Elsevier, New York (2013)
15. Houghton, H.G.: A study of the evaporation of small water drops. J. Appl. Phys. **4**, 419–424 (1933)
16. Knacke, O., Stranski, I.N.: The mechanism of evaporation. Prog. Metal Phys. **6**, 181–235 (1956)
17. Cazabat, A.-M., Guena, G.: Evaporation of macroscopic sessile droplets. Soft Matter **6**, 2591–2612 (2010)
18. Prosperetti, A., Plesset, M.S.: The stability of an evaporating liquid surface. Phys. Fluids **27**, 1590–1602 (1984)
19. Burelbach, J.P., Bankoff, S.G., Davis, S.H.: Nonlinear stability of evaporating/condensing liquid films. J. Fluid Mech. **195**, 463–494 (1988)
20. Marek, R., Straub, J.: Analysis of the evaporation coefficient and the condensation coefficient of water. Int. J. Heat Mass Transfer. **44**, 39–53 (2001)
21. Frohn, A., Roth, N.: Dynamics of Droplets. Springer, New York (2000)
22. Hu, H., Larson, R.G.: Evaporation of a sessile droplet on a substrate. J. Phys. Chem. B **106**, 1334–1344 (2002)
23. Semenov, S., Starov, V.M., Rubio, R.G., Velarde, M.G.: Instantaneous distribution of fluxes in the course of evaporation of sessile liquid droplets: computer simulations. Colloid Surf. A **372**, 127–134 (2010)
24. Nguyen, T.A.H., Nguyen, A.V., Hampton, M.A., Xu, Z.P., Huang, L., Rudolph, V.: Theoretical and experimental analysis of droplet evaporation on solid surfaces. Chem. Eng. Sci. **69**, 522–529 (2012)
25. Anderson, D.M., Davis, S.H.: The spreading of volatile liquid droplets on heated surfaces. Phys. Fluids **7**, 248–265 (1995)
26. Shahidzadeh-Bonn, N., Rafaï, S., Azouni, A., Bonn, D.: Evaporating droplets. J. Fluid Mech. **549**, 307–313 (2006)
27. Birdi, K.S., Vu, D.T., Winter, A.: A study of the evaporation rates of small water drops placed on a solid surface. J. Phys. Chem. **93**, 3702–3703 (1989)
28. Dettre, R.H., Johnson, R.E.: Contact angle hysteresis. IV. Contact angle measurements on heterogeneous surfaces 1. J. Phys. Chem. **69**, 1507–1515 (1965)
29. Snoeijer, J.H., Andreotti, B.: Moving contact lines: scales, regimes, and dynamical transitions. Annu. Rev. Fluid Mech. **45**, 269–292 (2013)
30. Rowan, S.M., Newton, M.I., McHale, G.: Evaporation of microdroplets and the wetting of solid surfaces. J. Phys. Chem. **99**, 13268–13271 (1995)
31. Picknett, R.G., Bexon, R.: The evaporation of sessile or pendant drops in still air. J. Colloid Interface Sci **61**, 336–350 (1977)
32. Dunn, G.J., Wilson, S.K., Duffy, B.R., David, S., Sefiane, K.: The strong influence of substrate conductivity on droplet evaporation. J. Fluid Mech. **623**, 329–351 (2009)
33. Oron, A., Davis, S.H., Bankoff, S.G.: Long-scale evolution of thin liquid films. Rev. Mod. Phys. **69**, 931–980 (1997)
34. Clausius, R.: Ueber die bewegende Kraft der Wärme und die Gesetze, welche sich daraus für die Wärmelehre selbst ableiten lassen. Ann. Phys. Berlin **155**, 368–397 (1850)

35. Murisic, N., Kondic, L.: On evaporation of sessile drops with moving contact lines. J. Fluid Mech. **679**, 219–246 (2011)
36. Siregar, D.P., Kuerten, J.G.M., van der Geld, C.W.M.: Numerical simulation of the drying of inkjet-printed droplets. J. Colloid Interface Sci **392**, 388–395 (2013)
37. Berteloot, G., Pham, C.T., Daerr, A., Lequeux, F., Limat, L.: Evaporation-induced flow near a contact line: consequences on coating and contact angle. Europhys. Lett. **83**, 14003 (2008)
38. Cachile, M., Bénichou, O., Cazabat, A.M.: Evaporating droplets of completely wetting liquids. Langmuir **18**, 7985–7990 (2002)
39. Guéna, G., Poulard, C., Voué, M., De Coninck, J., Cazabat, A.M.: Evaporation of sessile liquid droplets. Colloid Surf. A **291**, 191–196 (2006)
40. Joshi, A., Sun, Y.: Wetting dynamics and particle deposition for an evaporating colloidal drop: a lattice Boltzmann study. Phys. Rev. E **82**, 041401 (2010)
41. Janeček, V., Nikolayev, V.S.: Contact line singularity at partial wetting during evaporation driven by substrate heating. Europhys. Lett. **100**, 14003 (2012)
42. Doumenc, F., Guerrier, B.: Drying of a solution in a meniscus: a model coupling the liquid and the gas phases. Langmuir **26**, 13959–13967 (2010)
43. Eggers, J., Pismen, L.M.: Nonlocal description of evaporating drops. Phys. Fluids **22**, 112101 (2010)
44. Hołyst, R., Litniewski, M., Jakubczyk, D., Kolwas, K., Kolwas, M., Kowalski, K., Migacz, S., Palesa, S., Zientara, M.: Evaporation of freely suspended single droplets: experimental, theoretical and computational simulations. Rep. Prog. Phys. **76**, 034601 (2013)
45. Crafton, E.F., Black, W.Z.: Heat transfer and evaporation rates of small liquid droplets on heated horizontal surfaces. Int. J. Heat Mass Transfer **47**, 1187–1200 (2004)
46. Cioulachtjian, S., Launay, S., Boddaert, S., Lallemand, M.: Experimental investigation of water drop evaporation under moist air or saturated vapour conditions. Int. J. Therm. Sci. **49**, 859–866 (2010)
47. Chini, S.F., Amirfazli, A.: Understanding the evaporation of spherical drops in quiescent environment. Colloid Surf. A **432**, 82–88 (2013)
48. Briones, A.M., Ervin, J.S., Putnam, S.A., Byrd, L.W., Jones, J.G.: A novel kinetically-controlled de-pinning model for evaporating water microdroplets. Int. Commun. Heat Mass. **39**, 1311–1319 (2012)
49. Gelderblom, H., Bloemen, O., Snoeijer, J.H.: Stokes flow near the contact line of an evaporating drop. J. Fluid Mech. **709**, 69–84 (2012)
50. Weon, B.M., Je, J.H., Poulard, C.: Convection-enhanced water evaporation. AIP Advances **012102** (2011)
51. Dehaeck, S., Rednikov, A., Colinet, P.: Vapor-based interferometric measurement of local evaporation rate and interfacial temperature of evaporating droplets. Langmuir **30**, 2002–2008 (2014)
52. Fischer, B.J.: Particle convection in an evaporating colloidal droplet. Langmuir **18**, 60–67 (2001)
53. Popov, Y.: Evaporative deposition patterns: spatial dimensions of the deposit. Phys. Rev. E **71**, 036313 (2005)
54. Hamamoto, Y., Christy, J., Sefiane, K.: Order-of-magnitude increase in flow velocity driven by mass conservation during the evaporation of sessile drops. Phys. Rev. E **83**, 051602 (2011)
55. Bodiguel, H., Leng, J.: Imaging the drying of a colloidal suspension: velocity field. Chem. Eng. Process. **68**, 60–63 (2013)
56. Maki, K.L., Kumar, S.: Fast evaporation of spreading droplets of colloidal suspensions. Langmuir **27**, 11347–11363 (2011)
57. Scriven, L., Sternling, C.: The Marangoni effects. Nature **187**, 186–188 (1960)
58. Hu, H., Larson, R.G.: Analysis of the effects of Marangoni stresses on the microflow in an evaporating sessile droplet. Langmuir **21**, 3972–3980 (2005)
59. Hu, H., Larson, R.G.: Marangoni effect reverses coffee-ring depositions. J. Phys. Chem. B **110**, 7090–7094 (2006)

60. Assenheimer, M., Steinberg, V.: Observation of coexisting upflow and downflow hexagons in Boussinesq Rayleigh-Bénard convection. Phys. Rev. Lett. **76**, 756–759 (1996)
61. Morris, S.W., Bodenschatz, E., Cannell, D.S., Ahlers, G.: Spiral defect chaos in large aspect ratio Rayleigh-Bénard convection. Phys. Rev. Lett. **71**, 2026–2029 (1993)
62. Block, M.J.: Surface tension as the cause of Benard cells and surface deformation in a liquid film. Nature **178**, 650–651 (1956)
63. Buffone, C., Sefiane, K.: Investigation of thermocapillary convective patterns and their role in the enhancement of evaporation from pores. Int. J. Multiphase Flow **30**, 1071–1091 (2004)
64. Buffone, C., Sefiane, K., Christy, J.R.E.: Experimental investigation of self-induced thermocapillary convection for an evaporating meniscus in capillary tubes using micro-particle image velocimetry. Phys. Fluids **17**, 052104 (2005)
65. Buffone, C., Sefiane, K.: IR measurements of interfacial temperature during phase change in a confined environment. Exp. Therm. Fluid. Sci. **29**, 65–74 (2004)
66. Pearson, J.R.A.: On convection cells induced by surface tension. J. Fluid Mech. **4**, 489–500 (1958)
67. Nield, D.A.: Surface tension and buoyancy effects in cellular convection. J. Fluid Mech. **19**, 341–352 (1964)
68. Barmi, M.R., Meinhart, C.D.: Convective flows in evaporating sessile droplets. J. Phys. Chem. B **118**, 2414–2421 (2014)
69. Hu, Y.-C., Zhou, Q., Ye, H.-M., Wang, Y.-F., Cui, L.-S.: Peculiar surface profile of poly (ethylene oxide) film with ring-like nucleation distribution induced by Marangoni flow effect. Colloid Surf. A **428**, 39–46 (2013)
70. Barash, L., Bigioni, T., Vinokur, V., Shchur, L.: Evaporation and fluid dynamics of a sessile drop of capillary size. Phys. Rev. E **79**, 046301 (2009)
71. Xu, X., Luo, J., Guo, D.: Criterion for reversal of thermal Marangoni flow in drying drops. Langmuir **26**, 1918–1922 (2009)
72. Thompson, I., Duan, F., Ward, C.: Absence of Marangoni convection at Marangoni numbers above 27,000 during water evaporation. Phys. Rev. E **80**, 056308 (2009)
73. Hu, H., Larson, R.G.: Analysis of the microfluid flow in an evaporating sessile droplet. Langmuir **21**, 3963–3971 (2005)
74. Savino, R., Paterna, D., Favaloro, N.: Buoyancy and Marangoni effects in an evaporating drop. J. Thermophys. Heat Transfer. **16**, 562–574 (2002)
75. Steinchen, A., Sefiane, K.: Self-organised Marangoni motion at evaporating drops or in capillary menisci–thermohydrodynamical model. J. Non Equil. Thermody. **30**, 39–51 (2005)
76. Kang, K.H., Lim, H.C., Lee, H.W., Lee, S.J.: Evaporation-induced saline Rayleigh convection inside a colloidal droplet. Phys. Fluids **25**, 042001 (2013)
77. Yunker, P.J., Still, T., Lohr, M.A., Yodh, A.G.: Suppression of the coffee-ring effect by shape-dependent capillary interactions. Nature **476**, 308–311 (2011)
78. Larson, R.G.: Re-shaping the coffee ring. Angew. Chem. Int. Ed. **51**, 2546–2548 (2012)
79. Bhardwaj, R., Fang, X., Somasundaran, P., Attinger, D.: Self-assembly of colloidal particles from evaporating droplets: role of DLVO interactions and proposition of a phase diagram. Langmuir **26**, 7833–7842 (2010)
80. Anyfantakis, M., Baigl, D.: Dynamic photocontrol of the coffee-ring effect with optically tunable particle stickiness. Angew. Chem. Int. Ed. **126**, 14301–14305 (2014)
81. Xu, J., Xia, J., Lin, Z.: Evaporation-induced self-assembly of nanoparticles from a sphere-on-flat geometry. Angew. Chem. Int. Ed. **119**, 1892–1895 (2007)
82. Li, B., Han, W., Jiang, B., Lin, Z.: Crafting threads of diblock copolymer micelles via flow-enabled self-assembly. ACS Nano **8**, 2936–2942 (2014)
83. Harris, D.J., Hu, H., Conrad, J.C., Lewis, J.A.: Patterning colloidal films via evaporative lithography. Phys. Rev. Lett. **98**, 148301 (2007)
84. Gleiche, M., Chi, L.F., Fuchs, H.: Nanoscopic channel lattices with controlled anisotropic wetting. Nature **403**, 173–175 (2000)

85. Chi, L.F., Rakers, S., Hartig, M., Gleiche, M., Fuchs, H., Schmid, G.: Monolayers of nanosized Au-55-clusters: preparation and characterization. Colloid Surf. A **171**, 241–248 (2000)
86. Prevo, B.G., Velev, O.D.: Controlled, rapid deposition of structured coatings from micro- and nanoparticle suspensions. Langmuir **20**, 2099–2107 (2004)
87. Yabu, H., Shimomura, M.: Preparation of self-organized mesoscale polymer patterns on a solid substrate: continuous pattern formation from a receding meniscus. Adv. Funct. Mater. **15**, 575–581 (2005)
88. Hong, S.W., Byun, M., Lin, Z.Q.: Robust self-assembly of highly ordered complex structures by controlled evaporation of confined microfluids. Angew. Chem. Int. Ed. **48**, 512–516 (2009)
89. Hong, S.W., Giri, S., Lin, V.S.Y., Lin, Z.Q.: Simple route to gradient concentric metal and metal oxide rings. Chem. Mater. **18**, 5164–5166 (2006)
90. Hong, S.W., Jeong, W., Ko, H., Kessler, M.R., Tsukruk, V.V., Lin, Z.Q.: Directed self-assembly of gradient concentric carbon nanotube rings. Adv. Funct. Mater. **18**, 2114–2122 (2008)
91. Hong, S.W., Wang, J., Lin, Z.Q.: Evolution of ordered block copolymer serpentines into a macroscopic, hierarchically ordered web. Angew. Chem. Int. Ed. **48**, 8356–8360 (2009)
92. Hong, S.W., Xia, J.F., Byun, M., Zou, Q.Z., Lin, Z.Q.: Mesoscale patterns formed by evaporation of a polymer solution in the proximity of a sphere on a smooth substrate: molecular weight and curvature effects. Macromolecules **40**, 2831–2836 (2007)
93. Hong, S.W., Xia, J., Lin, Z.: Spontaneous formation of mesoscale polymer patterns in an evaporating bound solution. Adv. Mater. **19**, 1413–1417 (2007)
94. Hong, S.W., Xu, J., Lin, Z.Q.: Template-assisted formation of gradient concentric gold rings. Nano Lett. **6**, 2949–2954 (2006)
95. Hong, S.W., Xu, J., Xia, J.F., Lin, Z.Q., Qiu, F., Yang, Y.L.: Drying mediated pattern formation in a capillary-held organometallic polymer solution. Chem. Mater. **17**, 6223–6226 (2005)
96. Abkarian, M., Nunes, J., Stone, H.A.: Colloidal crystallization and banding in a cylindrical geometry. J. Am. Chem. Soc. **126**, 5978–5979 (2004)
97. Wang, H., Li, X., Nakamura, H., Miyazaki, M., Maeda, H.: Continuous particle self-arrangement in a long microcapillary. Adv. Mater. **14**, 1662–1666 (2002)
98. Lin, Z., Granick, S.: Patterns formed by droplet evaporation from a restricted geometry. J. Am. Chem. Soc. **127**, 2816–2817 (2005)
99. Han, W., Li, B., Lin, Z.: Drying-mediated assembly of colloidal nanoparticles into large-scale microchannels. ACS Nano **7**, 6079–6085 (2013)
100. Byun, M., Han, W., Li, B., Hong, S.W., Cho, J.W., Zou, Q., Lin, Z.: Guided organization of λ-DNA into microring arrays from liquid capillary bridges. Small **7**, 1641–1646 (2011)
101. Han, W., Lin, Z.: Learning from "Coffee Rings": ordered structures enabled by controlled evaporative self-assembly. Angew. Chem. Int. Ed. **51**, 1534–1546 (2012)
102. Myunghwan, B., Jun, W., Zhiqun, L.: Massively ordered microstructures composed of magnetic nanoparticles. J. Phys. Condens. Matter **21**, 264014 (2009)
103. Kim, T.Y., Kwon, S.W., Park, S.J., Yoon, D.H., Suh, K.S., Yang, W.S.: Self-organized graphene patterns. Adv. Mater. **23**, 2734–2738 (2011)
104. Byun, M., Laskowski, R.L., He, M., Qiu, F., Jeffries-El, M., Lin, Z.: Controlled evaporative self-assembly of hierarchically structured regioregular conjugated polymers. Soft Matter **5**, 1583–1586 (2009)
105. Schwartz, B.J.: Conjugated polymers as molecular materials: how chain conformation and film morphology influence energy transfer and interchain interactions. Annu. Rev. Phys. Chem. **54**, 141–172 (2003)
106. Berenbaum, A., Ginzburg-Margau, M., Coombs, N., Lough, A.J., Safa-Sefat, A., Greedan, J.E., Ozin, G.A., Manners, I.: Ceramics containing magnetic Co–Fe alloy nanoparticles from the pyrolysis of a highly metallized organometallic polymer precursor. Adv. Mater. **15**, 51–55 (2003)

107. Clendenning, S.B., Aouba, S., Rayat, M.S., Grozea, D., Sorge, J.B., Brodersen, P.M., Sodhi, R.N.S., Lu, Z.H., Yip, C.M., Freeman, M.R., Ruda, H.E., Manners, I.: Direct writing of patterned ceramics using electron-beam lithography and metallopolymer resists. Adv. Mater. **16**, 215–219 (2004)
108. Clendenning, S.B., Fournier-Bidoz, S., Pietrangelo, A., Yang, G., Han, S., Brodersen, P.M., Yip, C.M., Lu, Z.-H., Ozin, G.A., Manners, I.: Ordered 2D arrays of ferromagnetic Fe/Co nanoparticle rings from a highly metallized metallopolymer precursor. J. Mater. Chem. **14**, 1686–1690 (2004)
109. MacLachlan, M.J., Ginzburg, M., Coombs, N., Coyle, T.W., Raju, N.P., Greedan, J.E., Ozin, G.A., Manners, I.: Shaped ceramics with tunable magnetic properties from metal-containing polymers. Science **287**, 1460–1463 (2000)
110. Paquet, C., Cyr, P.W., Kumacheva, E., Manners, I.: Polyferrocenes: metallopolymers with tunable and high refractive indices. Chem. Commun. **234–235** (2004)
111. Whittell, G.R., Manners, I.: Metallopolymers: new multifunctional materials. Adv. Mater. **19**, 3439–3468 (2007)
112. Manners, I.: Poly(ferrocenylsilanes): novel organometallic plastics. Chem. Commun. **857–865** (1999)
113. Ginzburg, M., MacLachlan, M.J., Yang, S.M., Coombs, N., Coyle, T.W., Raju, N.P., Greedan, J.E., Herber, R.H., Ozin, G.A., Manners, I.: Genesis of nanostructured, magnetically tunable ceramics from the pyrolysis of cross-linked polyferrocenylsilane networks and formation of shaped macroscopic objects and micron scale patterns by micromolding inside silicon wafers. J. Am. Chem. Soc. **124**, 2625–2639 (2002)
114. Leutwyler, W.K., Bürgi, S.L., Burgl, H.: Semiconductor clusters, nanocrystals, and quantum dots. Science **271**, 933–937 (1996)
115. Peng, X., Schlamp, M.C., Kadavanich, A.V., Alivisatos, A.P.: Epitaxial growth of highly luminescent CdSe/CdS Core/Shell nanocrystals with photostability and electronic accessibility. J. Am. Chem. Soc. **119**, 7019–7029 (1997)
116. Murray, C.B., Norris, D.J., Bawendi, M.G.: Synthesis and characterization of nearly monodisperse CdE (E = sulfur, selenium, tellurium) semiconductor nanocrystallites. J. Am. Chem. Soc. **115**, 8706–8715 (1993)
117. Coe, S., Woo, W.-K., Bawendi, M., Bulovic, V.: Electroluminescence from single monolayers of nanocrystals in molecular organic devices. Nature **420**, 800–803 (2002)
118. Colvin, V., Schlamp, M., Alivisatos, A.: Light-emitting diodes made from cadmium selenide nanocrystals and a semiconducting polymer. Nature **370**, 354–357 (1994)
119. Dabbousi, B.O., Bawendi, M.G., Onitsuka, O., Rubner, M.F.: Electroluminescence from CdSe quantum – dot/polymer composites. Appl. Phys. Lett. **66**, 1316–1318 (1995)
120. Gao, M., Richter, B., Kirstein, S.: White-light electroluminescence from self-assembled Q-CdSe/PPV multilayer structures. Adv. Mater. **9**, 802–805 (1997)
121. Greenham, N.C., Peng, X., Alivisatos, A.P.: Charge separation and transport in conjugated-polymer/semiconductor-nanocrystal composites studied by photoluminescence quenching and photoconductivity. Phys. Rev. B **54**, 17628–17637 (1996)
122. Huynh, W.U., Dittmer, J.J., Alivisatos, A.P.: Hybrid nanorod-polymer solar cells. Science **295**, 2425–2427 (2002)
123. Milliron, D.J., Gur, I., Alivisatos, A.P.: Hybrid organic–nanocrystal solar cells. MRS. Bull. **30**, 41–44 (2005)
124. Chan, W.C.W., Nie, S.: Quantum dot bioconjugates for ultrasensitive nonisotopic detection. Science **281**, 2016–2018 (1998)
125. Ishii, D., Kinbara, K., Ishida, Y., Ishii, N., Okochi, M., Yohda, M., Aida, T.: Chaperonin-mediated stabilization and ATP-triggered release of semiconductor nanoparticles. Nature **423**, 628–632 (2003)
126. Mattoussi, H., Mauro, J.M., Goldman, E.R., Anderson, G.P., Sundar, V.C., Mikulec, F.V., Bawendi, M.G.: Self-assembly of CdSe–ZnS quantum dot bioconjugates using an engineered recombinant protein. J. Am. Chem. Soc. **122**, 12142–12150 (2000)

127. Medintz, I.L., Uyeda, H.T., Goldman, E.R., Mattoussi, H.: Quantum dot bioconjugates for imaging, labelling and sensing. Nat. Mater. **4**, 435–446 (2005)
128. Alivisatos, P.: The use of nanocrystals in biological detection. Nat Biotechnol **22**, 47–52 (2004)
129. Larson, D.R., Zipfel, W.R., Williams, R.M., Clark, S.W., Bruchez, M.P., Wise, F.W., Webb, W.W.: Water-soluble quantum dots for multiphoton fluorescence imaging in vivo. Science **300**, 1434–1436 (2003)
130. Santra, S., Yang, H., Holloway, P.H., Stanley, J.T., Mericle, R.A.: Synthesis of water-dispersible fluorescent, radio-opaque, and paramagnetic CdS:Mn/ZnS quantum dots: a multifunctional probe for bioimaging. J. Am. Chem. Soc. **127**, 1656–1657 (2005)
131. Dabbousi, B.O., Rodriguez-Viejo, J., Mikulec, F.V., Heine, J.R., Mattoussi, H., Ober, R., Jensen, K.F., Bawendi, M.G.: (CdSe)ZnS core−shell quantum dots: synthesis and characterization of a size series of highly luminescent nanocrystallites. J. Phys. Chem. B **101**, 9463–9475 (1997)
132. Hines, M.A., Guyot-Sionnest, P.: Synthesis and characterization of strongly luminescing ZnS-Capped CdSe nanocrystals. J. Phys. Chem. **100**, 468–471 (1996)
133. Kortan, A.R., Hull, R., Opila, R.L., Bawendi, M.G., Steigerwald, M.L., Carroll, P.J., Brus, L. E.: Nucleation and growth of cadmium selendie on zinc sulfide quantum crystallite seeds, and vice versa, in inverse micelle media. J. Am. Chem. Soc. **112**, 1327–1332 (1990)
134. Leizerson, I., Lipson, S.G., Lyushnin, A.V.: Finger Instability in wetting–dewetting phenomena. Langmuir **20**, 291–294 (2003)
135. Lyushnin, A.V., Golovin, A.A., Pismen, L.M.: Fingering instability of thin evaporating liquid films. Phys. Rev. E **65**, 021602 (2002)
136. Cazabat, A., Heslot, F., Troian, S., Carles, P.: Fingering instability of thin spreading films driven by temperature gradients. Nature **346**, 824–826 (1990)
137. Byun, M., Han, W., Li, B., Xin, X., Lin, Z.: An unconventional route to hierarchically ordered block copolymers on a gradient patterned surface through controlled evaporative self-assembly. Angew. Chem. Int. Ed. **52**, 1122–1127 (2013)
138. Han, W., He, M., Byun, M., Li, B., Lin, Z.: Large-scale hierarchically structured conjugated polymer assemblies with enhanced electrical conductivity. Angew. Chem. Int. Ed. **52**, 2564–2568 (2013)
139. Han, W., Byun, M., Li, B., Pang, X., Lin, Z.: A simple route to hierarchically assembled micelles and inorganic nanoparticles. Angew. Chem. Int. Ed. **124**, 12756–12760 (2012)
140. Li, B., Han, W., Byun, M., Zhu, L., Zou, Q., Lin, Z.: Macroscopic highly aligned DNA nanowires created by controlled evaporative self-assembly. ACS Nano **7**, 4326–4333 (2013)
141. Carvalho, M.S., Kheshgi, H.S.: Low-flow limit in slot coating: theory and experiments. AIChE J. **46**, 1907–1917 (2000)
142. Higgins, B.G., Scriven, L.E.: Capillary pressure and viscous pressure drop set bounds on coating bead operability. Chem. Eng. Sci. **35**, 673–682 (1980)
143. Ruschak, K.J.: Limiting flow in a pre-metered coating device. Chem. Eng. Sci. **31**, 1057–1060 (1976)
144. Lee, K.-Y., Liu, L.-D., Ta-Jo, L.: Minimum wet thickness in extrusion slot coating. Chem. Eng. Sci. **47**, 1703–1713 (1992)
145. Sartor, L.: Slot Coating: Fluid Mechanics and Die Design. University of Minnesota, Minneapolis, MN (1990)
146. Fasolka, M., Stafford, C., Beers, K.: Gradient and microfluidic library approaches to polymer interfaces. Adv Polym Sci **225**, 63–105 (2010)
147. Stafford, C.M., Roskov, K.E., Epps, T.H., Fasolka, M.J.: Generating thickness gradients of thin polymer films via flow coating. Rev. Sci. Instrum. **77**, 023908-7 (2006)
148. Meredith, J.C., Karim, A., Amis, E.J.: High-throughput measurement of polymer blend phase behavior. Macromolecules **33**, 5760–5762 (2000)
149. Carson Meredith, J., Karim, A., Amis, E.J.: Combinatorial methods for investigations in polymer materials science. MRS Bull **27**, 330–335 (2002)

150. Tsuruma, A., Tanaka, M., Yamamoto, S., Fukushima, N., Yabu, H., Shimomura, M.: Topographical control of neurite extension on stripe-patterned polymer films. Colloid Surf. A **284–285**, 470–474 (2006)
151. Yabu, H., Inoue, K., Shimomura, M.: Multiple-periodic structures of self-organized honeycomb-patterned films and polymer nanoparticles hybrids. Colloid Surf. A **284–285**, 301–304 (2006)
152. Kim, B.H., Shin, D.O., Jeong, S.-J., Koo, C.M., Jeon, S.C., Hwang, W.J., Lee, S., Lee, M.G., Kim, S.O.: Hierarchical self-assembly of block copolymers for lithography-free nanopatterning. Adv. Mater. **20**, 2303–2307 (2008)

Chapter 4
Nanopatterns Produced by Directed Self-Assembly in Block Copolymer Thin Films

Virginie Ponsinet

4.1 Introduction

For the last 30 years, block copolymers (BCP) have held the promises of new functional nanomaterials due to their fascinating spontaneous spatial organization as well as their complex properties, combining those of the individual blocks. The pathway is long, starting at the first synthesis mastered in the 1950s [1], travelling through full thermodynamics descriptions [2], the discovery of industrially viable synthesis technologies [3], and the first patents in one of the expected most appealing applicative fields [4]. This path still has obstacles and challenges ahead before block copolymer nanostructures really come out of the research laboratories. Until then, and for many more years, BCPs will keep researchers busy and amazed by their capacities to spontaneously produce chemically and topographically patterned surfaces [5]. Block copolymers nanopatterning is only one of the many opportunities offered by macromolecules at interfaces as detailed in the other chapters of this book, as well as previous enlightening literature [6]. But they indisputably present many advantages compared to alternative methods for the production of nanopatterned surfaces, including the tunability of size and morphologies of the domains, the ease of processing on many types of surfaces and the possibility to smarten the systems up with chemical functionalities. Theoretical and experimental tools have been readily derived from that developed for interfacial phenomena in mixed polymer systems [7]. The block copolymer thin film morphologies therefore offer a genuine technological platform for the generation of surface nanopatterns as well as ordered thin films, for a wide range of possible applications (cf. Fig. 4.1) mostly in optics, or based on their templating possibilities [8–10]. Two specific thin

V. Ponsinet (✉)
Centre de Recherche Paul Pascal, UPR 8641, CNRS - University of Bordeaux, Pessac 33600, France
e-mail: ponsinet@crpp-bordeaux.cnrs.fr

J. Rodríguez-Hernández, C. Drummond (eds.), *Polymer Surfaces in Motion*,
DOI 10.1007/978-3-319-17431-0_4

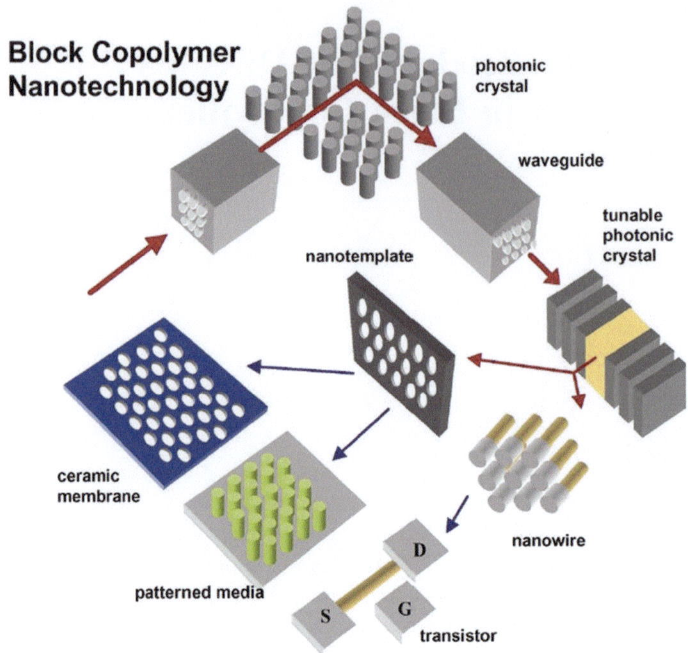

Fig. 4.1 Schematic of various applications of BCPs in nanotechnologies. Reprinted from ref. [8], Polymer 2003, **44**, 6725, Copyright 2003, with permission from Elsevier

film morphologies are particularly attractive, presenting high degree of order and uniaxial symmetry: the lamellar phase in parallel (homeotropic) alignment, which constitutes easily produced well-ordered and periodic multilayered slabs, and the phase of hexagonally packed cylinders in perpendicular orientation, from which can be obtained patterns of cylindrical nanodots or nanopores. The latter structures have attracted a lot of attention, with applications including high-density storage media [11, 12], lithographic masks [13–17], nanofluidics [18], separation membranes [19, 20], templates for nanorods synthesis [21, 22], enhanced patterned LEDs [23]. As it becomes obvious below, many research efforts have been devoted to the control of this orientation.

Extensive and excellent literature [24, 25], can be found on block copolymers and their microphase separation. After a brief reminder of some necessary bases, we will focus here on the conditions allowing to access specific patterns at the free surface of a thin film. This concerns mostly *really thin* films, a regime in which the film structure will be strongly affected by its interfaces on the one hand, and by the commensurability of the film thickness T with a (small) integer number of d_o, the size of the microphase separation-induced period. This natural nanoscopic dimension d_o constitutes a strong, almost inflexible constraint on the nanostructure of block copolymer materials, whatever their shaping, which will have specific consequences in the case of thin films.

4.2 Microphase Separation of Block Copolymers

Block copolymers are the result of covalently linking two or more polymer chains, each called a block. A diblock copolymer is A-A-...-A-A-B-B-...-B-B or A_n-B_m, where the polymerization degrees n and m are typically between 10 and 10^5. The nature of the monomers (A or B) can be chosen within a very wide range of chemical functions, made available by the development of many different synthetic techniques [26]. Thanks to the high degree of control of the copolymer synthesis techniques, block copolymers containing two distinct monomers A and B can also have various architectures, such as linear diblock (A_n-B_m), triblock (A_n-B_m-A_n), pentablock (A_n-B_m-A_p-B_m-A_n), multiblock $(A_n\text{-}B_m)_p$, and star diblocks $(A_n\text{-}B_m)_pX$. With a third ingredient C, linear A_n-B_m-C_p, A_n-C_m-B_p, and B_n-A_m-C_p triblocks, multiblocks (A_n-B_m-C_p-B_m-A_n, etc...) and three-armed stars can be obtained, for instance. For the sake of simplicity, we will focus here on the cases of diblock copolymers, composed of two linear blocks of different chemical nature and linked covalently at one of their extremities. The properties of block copolymers with alternative architectures (multiblocks, branched copolymers, etc.) can be to some extent extrapolated from those of diblocks, although they are not nearly as well understood [24].

In the Flory–Huggins description of polymer melt thermodynamics, the interactions between segments, depending only on the local concentration of the different types of chemical functions, constitute the enthalpy contribution, whereas the entropy is a non-local contribution, depending on the number of allowed conformation of the entire copolymer chains [27, 28]. Entropy is related to elasticity and will be optimized when all tensions are released within the chains.

Most polymers are incompatible with one another and phase separate in a blend. This is also true for the distinct blocks of a block copolymer, but because they are covalently linked, they can segregate only as far as the size of the macromolecule itself: this is called the microphase separation and results in the formation of domains of each block [2, 29, 30], in the melt. They are called *microdomains*, although they actually present nanoscale dimensions: from a few nanometers to a few hundred nanometers depending mostly on the length of the copolymer chain. This segregation at equilibrium occurs when the enthalpic gain of avoiding contact between incompatible blocks overcomes the associated entropic loss. For a diblock A_n-B_m, the two terms in the free energy depend only on the dimensionless Flory–Huggins interaction parameter χ specific of each monomer pair (A,B) and characterizing their incompatibility, the total polymerization degree $N = n + m$ and the volume fractions of the blocks f_A and $f_B = 1 - f_A$. The balance of these two terms in the free energy defines a transition line, similar to the binodal in a blend, called the order–disorder transition (ODT). "Disorder" refers to the homogeneous solid in which all blocks are mixed, when χN is small and the entropy wins. Low values of χN can be obtained for any (A,B) pair when N is small; or at high temperature (since χ decreases when the temperature increases). Inversely, in order to reach high values of χN with short polymer chains, strong incompatibility (high χ) is needed. "Order" refers to the segregated systems, in which the macrodomains form equilibrium organized structures.

In the case of the lamellar phase, the free energy can be approximated to two contributions, the stretching energy and the interfacial energy, respectively:

$$\frac{F_L}{kT} = \frac{3}{8}\left(\frac{d}{a\,N^{1/2}}\right)^2 + \frac{\gamma_{AB}}{kT}\Sigma \tag{4.1}$$

where the interfacial energy is

$$\gamma_{AB} = kT\left(\frac{\chi}{6}\right)^{1/2} a^{-2} \tag{4.2}$$

and the total interface area

$$\Sigma = 2\,\frac{N\,a^3}{d} \tag{4.3}$$

with a the size of one segment of the chain, d, the layer thickness. Equating F_L with the disordered copolymer melt free energy $F_{dis} = f_A\,(1 - f_A)\chi N$, at the order–disorder transition (ODT), leads to $\chi N \sim 10$, independently of the chemical details of the considered system. Near this value lies the weak segregation limit, while far above $\chi N = 10$ is the strong segregation limit. In this latter regime, the interfaces between microdomains are sharp and well-defined [31].

The morphological phase diagrams of diblock copolymers present four equilibrium symmetries: body-centered cubic array of spherical cores (S-bcc), bicontinuous gyroid (G), hexagonally packed cylinders (C), and lamellae (L), which are selected mostly in relation with the volume fractions of the blocks f_A and $f_B = 1 - f_A$, due to interfacial curvature effects [2, 29]. These four phases exhibit long range order of dimension 3, 2, 3 and 1 respectively. With ABC triblocks, a much richer (more than 30) variety of phases [24] can be obtained in the bulk, as for example the striking 2D "knitting pattern" [32] (Fig. 4.2).

When the diblocks are close to being symmetrical ($f_A \sim f_B \sim 0.5$), the volumes occupied by both blocks on either sides of the interface are comparable, and the interface tends to be flat. This leads to a lamellar phase composed of a periodic stack of alternating slab-shaped microdomains extending infinitely in two directions. The period d_o of this stack, of the order of twice the extended length of the copolymer chain, correspond to the equilibrium conformation of the macromolecules and can hardly be modified for a given copolymer. This equilibrium period d_o is simply the result of the minimization of Eqs. (1–3), reading [33, 34],

$$d_o = 2N^{2/3} a \left[\frac{1}{3}\left(\frac{\chi}{6}\right)^{1/2}\right]^{1/3} \tag{4.4}$$

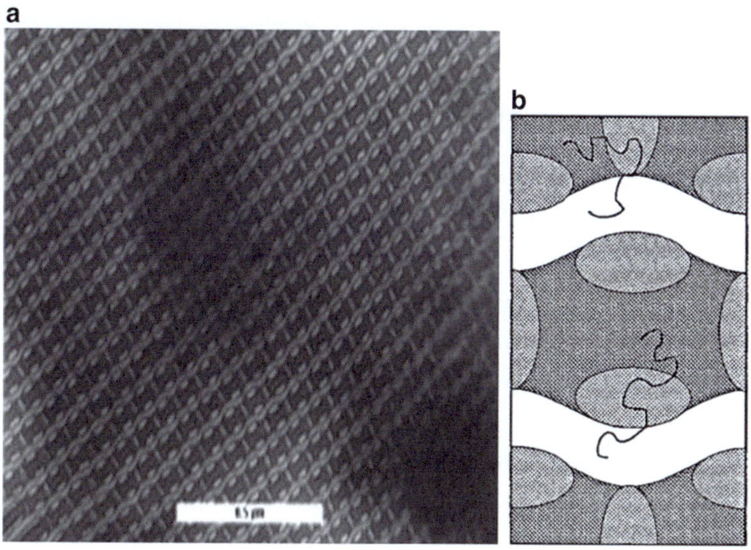

Fig. 4.2 (**a**) Transmission electron micrograph of a polystyrene-block-poly(ethylene-co-butylene)-block-poly(methyl methacrylate) (SEBM) triblock copolymer, stained with RuO_4, presenting the morphology called the "knitting pattern." Bar = 0.5 μm. (**b**) Schematic description of the triblock chain conformation within the unit cell. Adapted with permission from ref. [32], Macromolecules 1998, **31**, 135. Copyright 1998 American Chemical Society

The easiest way to vary d_o is certainly by changing the total polymerization degree N of the copolymer [35]. But playing with the Flory–Huggins parameter χ, either through temperature [34], or initial monomer choices, or diblock chemical composition is also considered [36]. Typically, d_o is easily obtained between 30 and 120 nm with commercially available BCPs, while going below 30 nm or beyond 120 nm may require tailor-made products.

4.3 Block Copolymer Thin Films

As was described before, the equilibrium nanostructures of BCP melt (dry) state are completely controlled by thermodynamics and do not depend on the processing or shaping of a sample. However, the pattern actually exhibited at the surface of a BCP thin film, and the nanostructure organization throughout the thickness of the film, can depend on the shaping and boundary conditions, even at thermodynamic equilibrium. Therefore, the control of orientation of ordered copolymer phases is an important part of the research effort aimed at developing functional surfaces or thin films [37]. This part of the research has benefitted from methodologies adapted from liquid crystal studies, in which the alignment of anisotropic liquids in thin films is required for both fundamental experimental work and applicative developments [38].

The equilibrium nanostructures, while presenting a high degree of order, typically extend over the size of grains, like in polycrystalline materials, leading to macroscopic isotropy. This grain size can be enlarged by annealing processes, but the application of external constraints is usually required to access macroscopic samples with no grain boundaries (monocrystals, or fully aligned samples).

In the case of thin films, meaning for thicknesses ranging from one time until typically ten times the equilibrium characteristic size of the nanostructure, surface effects act as efficient external constraints. It is indeed obvious that boundary conditions can influence the orientation of anisotropic materials in the vicinity of the boundaries and this is likely to affect the whole structure when the thickness of a film is small.

Thin films are usually produced by spin-casting from a relatively dilute solution ($<\sim5$ wt%), or sometimes other casting techniques, and are rarely at equilibrium, due to the fast evaporation of the solvent leaving no time for the macromolecules to adjust their conformations. The copolymer chains are then either fully mixed (the film is said disordered or amorphous) or presenting some kinetically arrested metastable nanostructure [39], depending on the casting solvent (selectivity, saturation vapor pressure) and conditions (concentration, speed, etc). Bringing the film to equilibrium is usually done by thermal or solvent annealing. Elevated temperatures for several hours (above the glass transition temperatures of the constituent blocks and preferably in an inert atmosphere in order to avoid oxidative degradation), or solvent vapor swelling give the chains the necessary mobility for the system to evolve towards the equilibrium nanostructure. This evolution follows the minimization of the system energy, including the terms discussed in Sect. 4.1, leading to the formation of the microphase separated nanostructure, as well as the surface energy at the substrate and atmosphere surfaces. The two blocks, of different chemical nature, usually present a significant contrast in their interfacial energy at the contact of both the substrate and the atmosphere, which will induce a significant energy gain if the right domain is exposed at each interface. In the case of the lamellar structure, this selective alignment at the boundary produces a parallel alignment of the multilayered structure (cf. Fig. 4.3a), which then propagates throughout the thickness of the film.

4.4 Forcing the Perpendicular Alignment

As was said before, a strong motivation for reaching perpendicular alignment of nanostructured thin BCP films is the production of nanolithography masks [8, 11, 40] because they can reach sub-20 nm dimensions, at which classical lithography techniques cannot be applied and sophisticated variations including extreme UV technologies [41] are necessary. BCP are considered an interesting alternative, which motivates many of the research efforts in the search for alignment control.

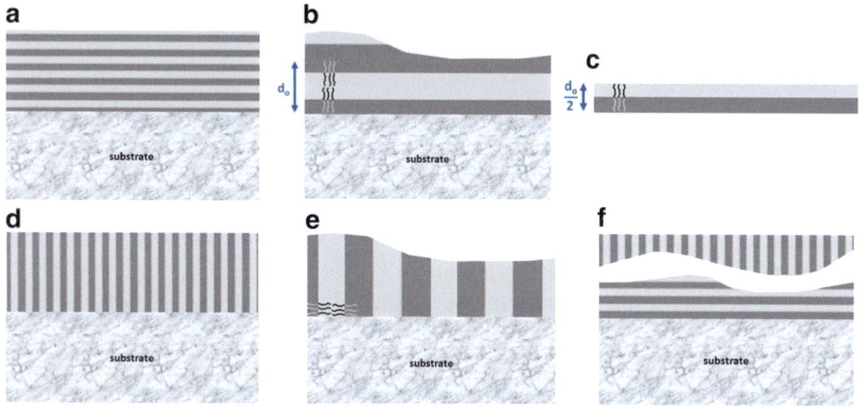

Fig. 4.3 Schematic representation of various alignment situations: (**a**) and (**b**) show parallel alignment, (**c**) represents the elementary brick constituting parallel aligned films, (**d**) and (**e**) show perpendicular alignment, (**f**) shows a mixed alignment

4.4.1 *Surface Energy Tuning*

Rarely occurring naturally, the situation in which the contrast in interfacial energy at the contact of the substrate for the two blocks vanishes, can be induced by a proper surface treatment. This treatment needs to remain unaltered by the thermal or solvent annealing process, which will be used for the alignment step. A clever surface treatment insuring a neutral surface energy towards a given diblock copolymer consists in grafting the substrate with a *random* copolymer of same chemical composition than the diblock [42, 43], although this may induce interactions between the random brush and the block copolymer film, which go beyond a simple surface energy effect [44]. Other surface treatments can be employed, provided that they can be finely tuned, in order to access the proper balanced energy [45].

When the surface energy is neutral for a given diblock, entropic effects will lead to a perpendicular alignment of the nanostructure, due to a preferential conformation of the macromolecules lying along the substrate surface [46]. The Fig. 4.3d, e sketch successful achievements of perpendicular alignment at the substrate interface. Getting perpendicular alignment at the free surface of the film is more difficult, because most ambient conditions will present a preferential surface energy towards one of the blocks of any diblock. It is therefore common to obtain a hybrid alignment [47] (like shown in the Fig. 4.3f), unless the thickness of the film is restricted to a very small value, in which case geometrical constraints will prevent the deformation of the nanostructure to adjust from one orientation to the other. Hybrid orientation films, however, can be processed, by some etching techniques, in order to reveal the perpendicular pattern.

4.4.2 Other Surface Effects

4.4.2.1 Roughness

The effects of the substrate roughness on the orientation of BCPs in thin films have been studied experimentally [48, 49], and theoretically [50]. Typically, if the roughness characteristic lateral length is larger than the period of the copolymer nanostructure, the favored orientation will always be perpendicular, unless the roughness characteristic height is very small. Provided that the substrate roughness is not detrimental to the final use of the supported film, roughness can be an easy control knob for modifying the alignment of a nanostructured film. For instance, ITO-covered surfaces can be used as electrodes in organic electronic devices and double as conveniently rough surfaces.

4.4.2.2 Graphoepitaxy

Graphoepitaxy consists in using substrates with designed topographic features and was developed as a way to better control the effects of the substrate topography, rather than counting on sometimes ill-defined roughness. Substrates with templating grooves could, for instance, constrain a spherical BCP structure to adopt a highly ordered nanopattern [51, 52], when the template dimension is commensurate with the nanostructure characteristic size. The templates can be produced by "standard" lithography techniques because its dimensions are typically one order of magnitude larger than that of the BCP nanostructure. This combination of top-down (lithographic) technology at ~500 nm scale and bottom-up (BCP nanostructures) at ~50 nm scale is promising for well controlled nanosurfaces, if somewhat tedious. Other substrates with specific topographies were successfully used to produce perpendicular alignment of phases of hexagonally packed cylinders: faceted crystals [11] and interferometrically produced sinusoidal polymer gratings [53], are among the interesting examples. The former case proved appropriate for very thin films, whereas the latter requires a thickness above a threshold value of several times the pattern sinusoidal amplitude and the nanostructure period. Provided these structural features are compatible with a given desired structure, and the presence of the groove template is not problematic for the considered applications, these methodologies are interesting.

4.4.2.3 Chemically Patterned Surfaces

Besides topographically patterning the substrate, it can be useful to orient and possibly align PCB thin films with chemically patterned substrates. The propagation of surface patterns into the bulk was predicted to be possible [54], and some of these situations were demonstrated experimentally. For instance, Kim et al. [55]

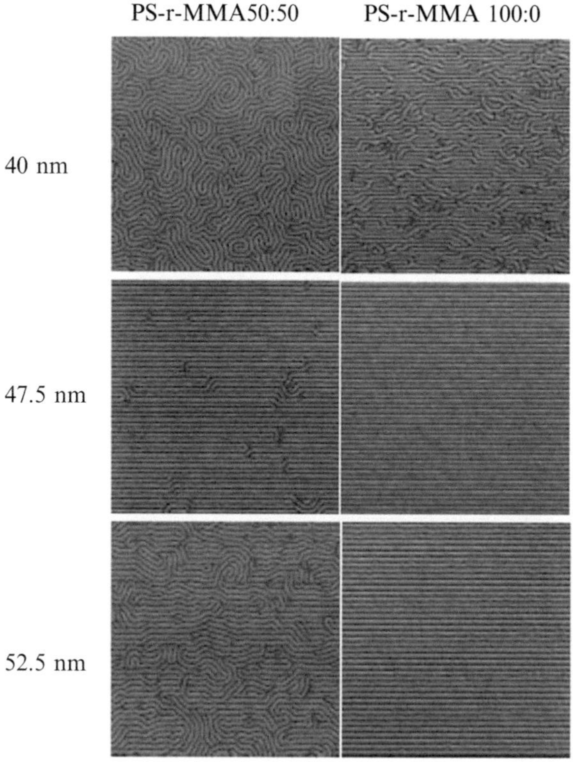

Fig. 4.4 Effect of a chemical pattern on the alignment and azimuthal orientation of lamellar block copolymer nanostructure: top-view SEM images (2 μm × 2 μm) of the nanostructure morphologies of thin films of lamellar copolymer with period $d_o = 48$ nm onto chemically patterned substrates as a function of the pattern pitch and the composition of the random copolymer used to create the pattern. Adapted with permission from ref. [56], Adv. Mater. 2004, **16**, 1315. Copyright 2004 John Wiley and Sons

patterned a substrate using extreme ultraviolet interferometric lithography [41] to transfer alternating lines and spaces to a self-assembled monolayer (SAM) and studied the configurations of thin PS-PMMA films on top of such substrates. For this technique, sometimes referred to as epitaxy, commensurability constraints appear severe since the perpendicular alignment of lamellar microdomains by striped patterns can be obtained only with a pattern pitch equal to the BCP lamellar period within less than 5 % mismatch. The registry can have a slightly larger mismatch tolerance if the chemical pattern is produced with a random copolymer brush [56], supposedly because the interpenetration of the brush chains in the BCP film allows some variation of the lamellar period around the equilibrium value d_o (cf. Fig. 4.4). Contrary to the graphoepitaxy patterning, the chemical template must have the same lengthscale than the structure to be aligned, meaning molecular scale, and therefore requires high-tech lithography techniques. This is why, although it is of fundamental interest to show the mechanisms of such alignment, it cannot claim the finality of an easy production of nanoscale patterns.

Real crystalline epitaxy has also been used to align BCP thin films, while using specific interactions between semicrystalline BCPs and crystalline substrates [57, 58], or directional crystallization [59].

4.4.3 Solvent Annealing

The effect of solvent swelling has become increasingly used in the last decade for the manipulation of block copolymer thin film textures. If a block copolymer thin film is swollen with a solvent, which is a true neutral solvent for the copolymer (meaning the affinities of the two blocks are exactly equal), the presence of the solvent will provide mobility to the copolymer chains and allow them to reach the equilibrium structure, then defined, as before, by the balance between conformation, incompatibility and surface energy effects. In that sense, it would usually produce the same effect as a thermal annealing, but specific effects can also be observed due to the evaporation process, like the presence of a solvent concentration gradient normal to the substrate plane and a subsequent ordering front from the film surface to the substrate, which will on the one hand induce a highly directional nanostructure, and on the other hand increase the impact of the interfacial energy conditions at the free surface (where the solvent concentration is the lowest).

In general, however, the solvent used is not genuinely neutral, which would usually lead to kinetically trapped nanostructures and alignments, especially if one or both blocks are glassy at room temperature [60]. Such metastability makes it possible to force the alignment of the nanostructures without tuning the energy balance, and the glassy nature of the final film insures the long-term absence of evolution of the pattern. Solvent swelling was proven to be a useful process to reach perpendicular alignment of cylinders in different systems [61–64] (cf. Fig. 4.5). This process is however still considered difficult to reproduce, mainly because reproducibility requires a precise mastering and controlling of many experimental

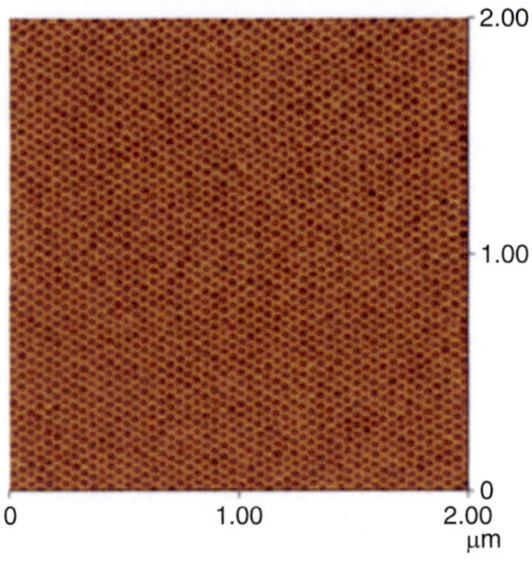

Fig. 4.5 Atomic force microscopy (AFM) phase image of the top surface of a thin film of poly (styrene)-*b*-poly(oxyethylene) diblock copolymer presenting a phase of hexagonally packed cylinders aligned perpendicular to the substrate after annealing in benzene vapor. The image contrast comes from the difference of elastic modulus between the domains. Adapted with permission from ref. [62], Adv. Mater. 2004, **16**, 226. Copyright 2004 John Wiley and Sons

details besides the choice of the solvent: film thickness [65], annealing time with characteristic times sometimes shorter than 1 min [63], vapor pressure and the rate of vapor removal [39, 66], glass transition temperature. All these can affect, often in coupled ways, the resulting metastable morphologies. The effects indeed involve different mechanisms including the modification of the interfacial energies of the swollen film as a function of the solvent loading, as well as hydrodynamic effects [62]. These non-equilibrium processes have not been easily considered by theory, but phase behavior and film patterns could be modeled in great detail by simulations [65] based on dynamic density functional theory. Nevertheless, solvent annealing is one of the most promising method, at this time, for the easy production of large-scale well-aligned nanopatterned thin films of block copolymers, while offering at the same time interesting scientific questions related to these non-equilibrium systems.

4.4.4 *Other Fields*

4.4.4.1 Electrical

The application of an external electric field, adapted from methodologies well optimized for liquid crystal systems, has been used with success [67–69] for large-scale alignment, although it often results in hybrid alignment situations (cf. Fig. 4.6). The electric field effect indeed competes with the surface effects, as was well documented theoretically [70, 71], while also taking into account the effects of mobile charged impurities in the films [72]. The nanostructure development and alignment could be followed by time-resolved X-ray scattering in the cases of initially disordered and ordered films [69, 73]. The reorientation of ordered films occurs by the fragmentation of the initial nanostructure into small domains, all eventually oriented along the field. The application of the field while the film is swollen by a neutral solvent proved beneficial for large scale alignment, and was satisfactorily described by simulations [74, 75].

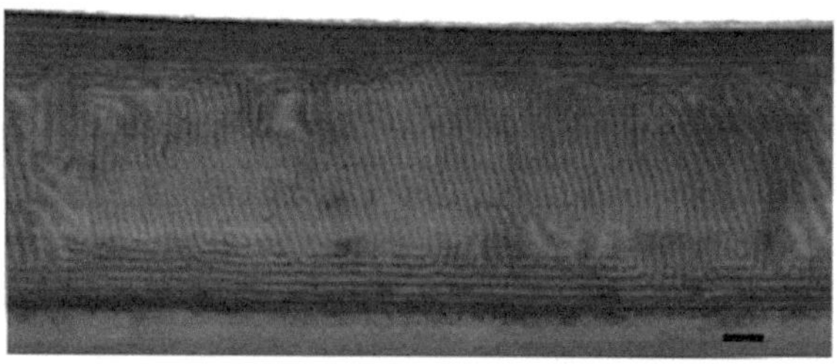

Fig. 4.6 Cross-sectional TEM image of a ~700 nm PS-*b*-PMMA film annealed under ~40 V/μm electric field for 16 h. Scale bar: 100 nm. Reprinted with permission from ref. [47], Macromolecules, 2004, **37**, 2625. Copyright 2011 American Chemical Society

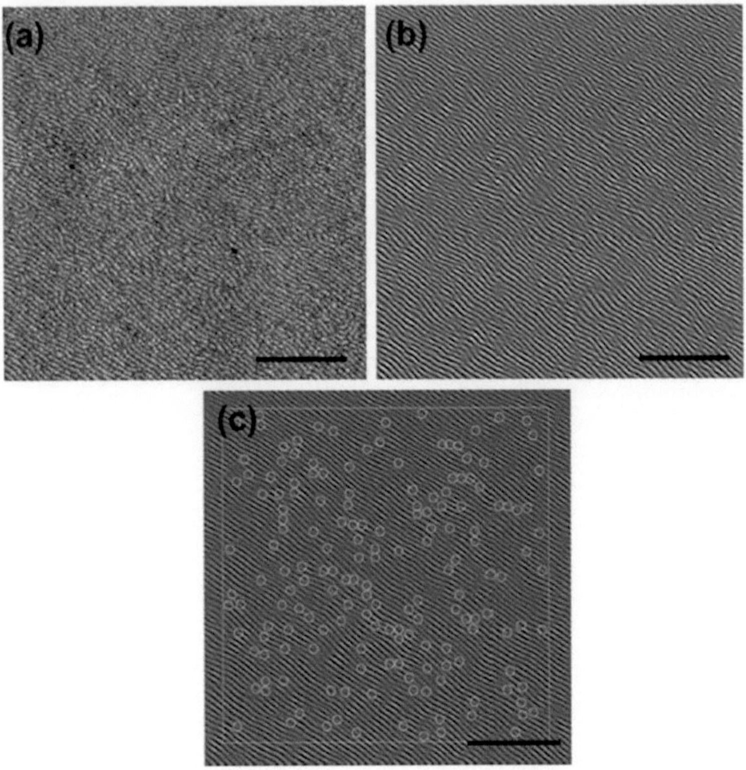

Fig. 4.7 (**a**) AFM image of a PS-PMMA film, 18 nm thick, shear-aligned at 40 kPa. (**b**) The same image after being Fourier-filtered. (**c**) Same image as in (**b**), with *small circles* added to mark the dislocation cores identified by an appropriate image treatment algorithm. Scale bars represent 500 nm, and the shear direction is oriented roughly diagonal (*upper right* to *lower left*) in all images. Adapted from ref. [81] with permission of The Royal Society of Chemistry

4.4.4.2 Flow

In bulk systems, shear induced alignment has proven the most successful, and has focused large research efforts, both experimentally [76, 77], and theoretically [78], which go beyond the scope of this chapter and are reviewed elsewhere [79]. Shear alignment of thin films was shown to be very efficient on spherical and cylindrical block copolymer phases, if it is performed at a temperature between the highest of the two glass transition temperature and the order–disorder transition temperature of the diblock [80]: $T_g < T < T_{ODT}$. In other cases, however, a good alignment is obtained with a high defect density [81] (Fig. 4.7).

Other alignment methodologies have been cited [82], including optical alignment of liquid crystalline block copolymers, and temperature gradients [83, 84], but are less general or practical.

4.5 Parallel Alignment

As was explained before, the parallel alignment is most of the time obtained spontaneously. For cylindrical nanostructures, this situation usually exposes at the free surface a homogeneous layer of the preferentially wetting block [85], meaning that here again, an etching step needs to be developed for nanopatterning purposes. For lamellar nanostructures, this alignment does not induce any exposed nanostructure-based pattern at the free surface. However, a complete alignment of the copolymer film (shown in Fig. 4.3a) is only possible if the total thickness of the film is commensurate with the lamellar period: $T = nd_o$ or $T = (n + 1/2)d_o$, where n is usually a small integer ($1 \leq n \leq 8$). The reason for that is that due to the free energy constraints on both the nanostructure alignment and the copolymer chain conformations, the nanostructured film has to be built as an assembly of the elementary brick (shown in Fig. 4.3c) composed of a copolymer layer of thickness $d_o/2$, corresponding to half of a bilayer of size d_o, the lamellar period. When deposited by a given casting technique, the film has a thickness controlled by the parameters of the casting (initial concentration and rotation speed and acceleration, for example in the case of spin-coating). While the fine tuning of these parameters, and therefore of the final film thickness, is in principle possible, it is more common to obtain a film with a thickness not exactly equal to nd_o or $(n + 1/2)d_o$. The layered structure of the copolymer then forces the film to present regions of different thicknesses, with the different regions satisfying one of the commensurability conditions [86, 87]. The film is therefore terraced, with a quantized surface topography: the step height of the terraces is d_o, so that the free surface exposes the same block domain on each terrace.

When the film thickness is close to and below the commensurability condition, the film lacks a little matter to reach a flat surface, and the terraces are shaped as holes, whereas when the film thickness is close to and above the commensurability condition, the film has a small excess of matter to reach a flat surface, and the terraces are shaped as islands [88]. This evolution is very obvious when a thickness gradient is produced on a lamellar copolymer thin film, as shown in Fig. 4.8. Interestingly, the film is flat on a finite range of thickness around the commensurability condition, showing that the copolymer layers are able to slightly adjust their surface chain density, so that the film can accommodate small discrepancies in thickness and avoid the cost of building a terrace edge.

The edges of the terraces concentrate the excess free energy related to a structural defect in the lamellar organization. These defect lines were shown to have a specific dynamics [89] and to present apparently stable structures [90], for some conditions of film T and bilayer d_o thicknesses. This allows for the relatively easy achievement of specific patterns, with a length scale one or two orders of magnitude larger than that of the copolymer nanostructure. Moreover, Li et al. have shown that these patterns can be somehow manipulated [91] as they present registry or anti-registry with underlying topographically patterned surfaces.

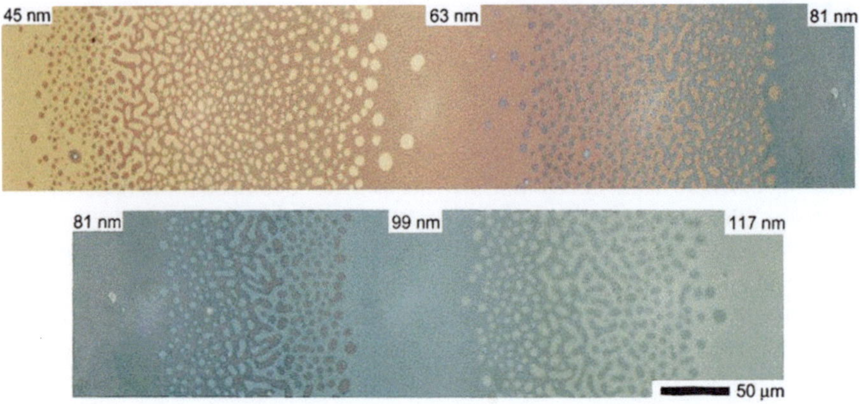

Fig. 4.8 True color optical micrograph of a continuous 26k PS-b-PMMA film with thickness gradient. The lower section is a continuation of the upper section. Film was annealed for 6 h at 170 °C, and the image shows the addition of four successive lamellae to the block copolymer film with increasing thickness, and the corresponding terraced patterns. *Labels* indicate when the film thickness verifies $h_s = (n + 1/2)d_o$, where $n = 2$–6. Reprinted with permission from ref. [88], J. Polym. Sci.: Part B: Polym. Phys. 2001, **39**, 2141. Copyright 2001 John Wiley and Sons

4.6 In-Plane Degeneracy and Single-Grain Formation

Beyond the formation of a specific alignment with regards to the substrate plane, ensuring the formation of a single-grain BCP lattice on large areas is critical in many cases. Also, in the cases of parallel oriented cylinders and perpendicular oriented lamellae, the orientation of the domains is azimuthally degenerate, and a complete in-plane alignment of the structure, usually desired in view of applications such as interconnects or light polarizers, requires some additional action.

An additional field can be applied orthogonally, which can be again of different nature. For instance, nicely ordered lamellar microdomains have been obtained by the combined application of a electric field and a flow field in a rather thick film [92]. Temperature treatment can also be combined with shear [93]. A combination of static and dynamic effects was successfully applied by Berry et al. [94], who used a graphoepitaxy and dynamic thermal annealing.

An intrinsically directional single field can align BCP patterns. As was mentioned before, chemically patterned surfaces, having the same length scale as the BCP nanostructure, apply a strong constrain on BCP thin films. It was shown to actually induce unidirectional orientation of perpendicular aligned lamellar phases [55, 56]. Graphoepitaxy alone can also unidirectionally align line patterns and organize spherical domains. Use of faceted surfaces [95] and 1D groove templates [52], due to entropic effects, as well as nanoimprint lithography [96, 97] (NIL), due to surface effects on the vertical groove walls, are efficient techniques to orient BCP nanostructures without requiring macromolecule-scale prior patterning. In the case of reconstructed sapphire surfaces, the atomic crystalline ordering of the

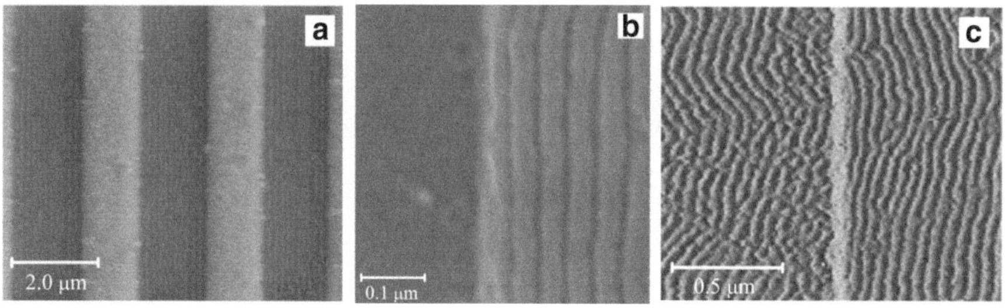

Fig. 4.9 (**a**) Top view of a SEM image of the BCP film after nanoimprinting with a NIL mold with main pitch 1.5 μm. The groove height is 50 nm. The perpendicularly lying BCP lamellae are further oriented along the groove long axis. (**b**) An enlargement of (**a**) where several lamellae (*left*) are well ordered in-plane. (**c**) An AFM top view of an enlarged section of the BCP film close to thin-thick boundary (the *middle vertical line*). To the *right*, the BCP film is thicker and is ordered by the groove vertical wall, whereas to the *left*, the BCP is thinner and less ordered. Reprinted with permission from ref. [96], Macromolecules, 2011, **44**, 2206. Copyright 2011 American Chemical Society

substrate can be transferred, over multiple length scales, to the block copolymer microdomains, via facets regular pattern. In the latter case, a micrometer-sized carved mold is imprinted onto the film, in order to guide the self-assembly of the BCP at the nanometer scale (cf. Fig. 4.9). Self-consistent field theory successfully accounted for the experimental results on NIL azimuthal alignment of lamellar domains, as shown in ref. [96].

Recent progress in the generation of large single-grain BCP thin films include zone processing, which induces sharp ordering front able to guide self-assembly (cf. Fig. 4.10). This can be found in a number of variations, like zone casting, using a scanning polymer deposition process [98], raster solvent vapor annealing [99], generating a localized solvent annealing zone with a scanning solvent vapor delivery nozzle, or thermal gradient zone processing [93, 100].

Device-oriented structures like sharp corners and curved patterns can be transferred into the BCP pattern [101]. In these studies, a small quantity of compatible homopolymer was added to the copolymer, which can redistribute in locations where it will compensate the mismatch between the pattern pitch and the natural lamellar domain d_o, thus allowing the copolymer pattern to follow even the sharp features of the underlying guide (cf. Fig. 4.11).

4.7 Experimental Techniques for the Structural Studies of the Nanopatterns

Atomic force microscopy (AFM) is a very useful technique to study the patterns exhibited at the surface of thin films. Used in taping mode, which exploits the interaction of the tip with the surface during intermittent contacts between

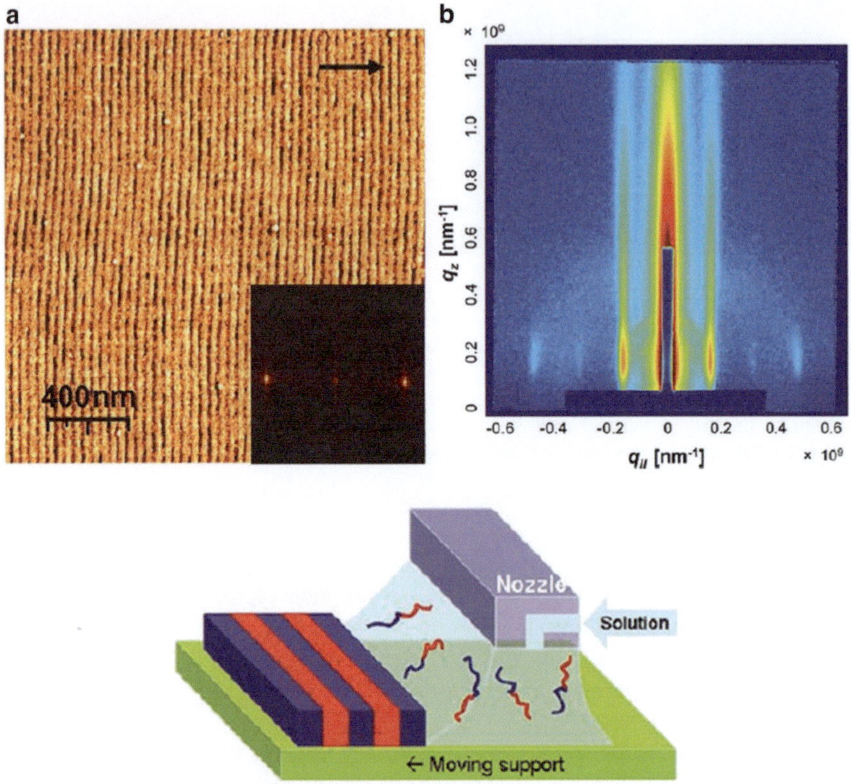

Fig. 4.10 Zone casting of a lamellar BCP thin film. (**a**) Taping mode AFM phase image (*inset*: FT transform), and (**b**) GISAXS pattern of PODMA-*b*-PtBA-*b*-PODMA block copolymer films prepared by zone casting at 40 °C. *Arrow*: substrate withdrawal direction. (**c**) Scheme of the zone casting technique. Adapted with permission from ref. [98], J. Am. Chem. Soc. 2011, **133**, 11802. Copyright 2011 American Chemical Society

them, this scanning microscopy technique can provide topographic images of the free surface with a resolution allowing the observation of 10 nm features or sometimes less, but it can also provide images of chemical contrast between microdomains. Phase images, in particular, relate to the damping of the tip oscillations, which varies with the local elasticity and more generally with the type of tip-polymer interactions at play. This is very complementary to topographic images, specifically because the nanopatterns of block copolymers may sometimes coexist with relatively flat and smooth surfaces. AFM cannot report on in-depth nanostructure but is well suited for studies of top surface patterns and their evolution upon annealing. Defect mobility and annihilation can be imaged, for instance [102] as is shown on Fig. 4.12.

As for other organized soft matter systems with nanoscale characteristic sizes, *X-ray and neutrons scattering techniques* are very appropriate for the study of block

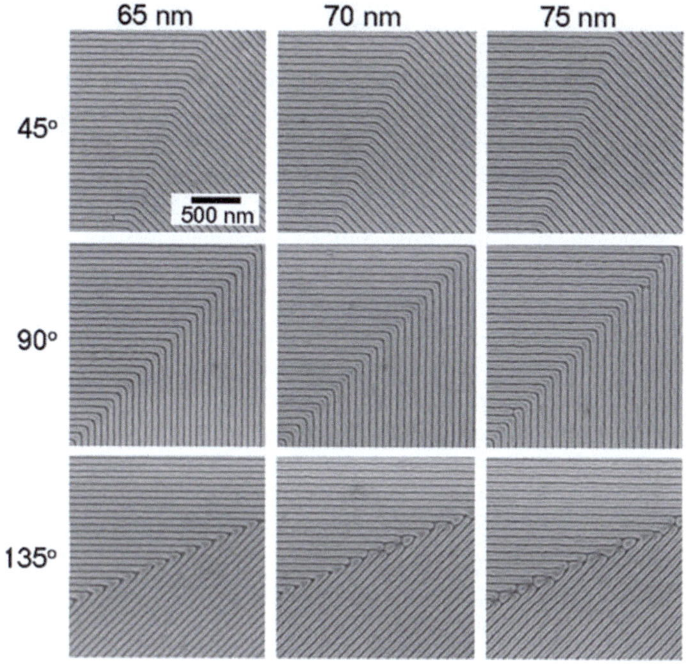

Fig. 4.11 Effect of a chemical pattern on the azimuthal orientation of lamellar block copolymer lamellar nanostructure: top-view SEM images (2 μm × 2 μm) of the nanostructure morphologies of thin films of lamellar copolymer with period $d_o = 70$ nm onto chemically patterned substrates with a bent of 45°, 90°, and 135° as a function of the pattern pitch. Adapted from ref. [101]

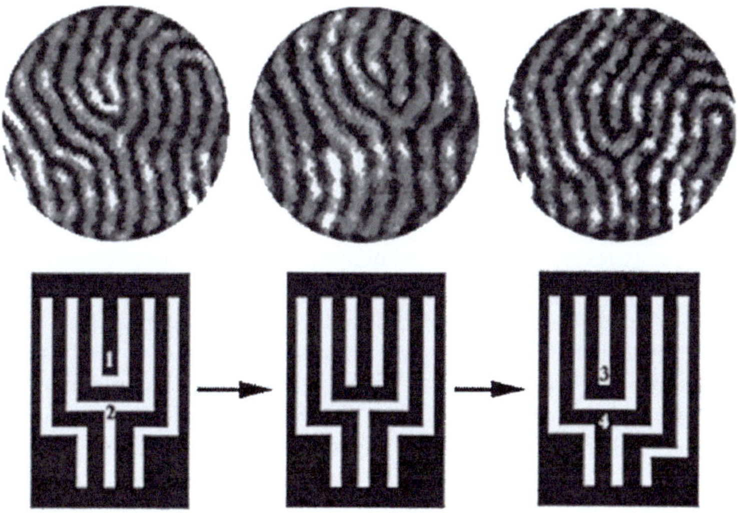

Fig. 4.12 AFM images taken between repeated annealing treatments of a further hour at 523 K showing the evolution of a disclination pair. In each image, the repeat spacing of the microdomains is 50 nm. Adapted with permission from ref. [102], J. Chem. Phys. 2001, **114**, 4730. Copyright 2001 AIP Publishing LLC

copolymer self-assembled nanostructures. In the case of thin films, two main experimental configurations are of interest: the reflectivity [103] and the small-angle scattering at grazing incidence (GISAXS) [104–106]. These techniques provide a large foot-print of the beam on the samples, allowing the collection of data in spite of the sometimes minute thickness of the films. Reflectivity is sensitive to the ordering along the film normal, and is well adapted for layered films. It can also efficiently quantify roughness. GISAXS measures the off-specular signal and records both in-plane and out-of-plane order, giving access to the in-depth nanostructure. It presents a very good temporal resolution, allowing following in detail the nanostructure evolutions during solvent or thermal treatments.

Spectroscopic *ellipsometry* is an elaborate technique, which has been used as a robust and extremely sensitive tool for material characterization [107]. During the last decade, its domain of application has been extended to the study of nanostructured samples, although it is mostly appropriate for the determination of order along the film normal. Small refractive index contrast multilayer structures and birefringence can be experimentally accessed [108] in some conditions, but may require sophisticated models.

Emerging *new methods* include resonant X-ray scattering [109], for enhanced contrast, critical-dimension SAXS, giving access to a 3D reconstruction [110] of in-depth structural features in the films, and scattering infrared near-field microscopy [111], for nanoscale chemical analysis at surfaces.

4.8 Perspectives

Macroscopically aligned BCP thin films provide invaluable robust, versatile, chemically functionalizable nanopatterned surfaces. Due to the interplay between film thickness, nanostructure and alignment, careful and complex processes are required to reach large scale alignment monodomains. Moreover, with the multiplication of the actual possibilities in term of copolymer architecture, monomer choices, advanced substrates, high-tech deposition and treatment processes, as well as new time- or spatially resolved analysis techniques, the field of nanopatterned BCP thin films is still ready to provide remarkable model study subjects as well as routes for defect-free or tailor-made nanomaterials for technological applications.

Device-oriented structures with complex patterns will be obtained by the design and juxtaposition of guiding tiles on the substrates, as has been done by Chang et al. [112], who achieved patterns including dense bends, junctions, and lines with a simple and predictable topographic template (cf. Fig. 4.13).

Search for ever-smaller characteristic sizes, in order to reach the sub-20 nm limits desired for lithography applications, has involved the choices of new monomer combinations [13, 113], salt complexation [114], and exploration of the weak segregation regime [115]. However, this goal is conflicting with the need for defect-free patterns, since the energy cost of a defect formation typically scales as the third

Fig. 4.13 From *left* to *right*: template layout to fabricate complex patterns consisting of design patterns of posts. Each color represents the pattern templated by each arrangement; SEM images of the templates of posts to fabricate dense patterns of bends and terminations; SEM images of the PDMS patterns formed on the template shown on the left; SCFT simulation result showing the constant 50 % density surface of PDMS cylinders assembled with the template on the *left*. Scale bars are 50 nm. Adapted by permission from Macmillan Publishers Ltd: Nature Comm. 2014, **5**, 3305 (ref. [112]), copyright 2014

power of the nanostructure characteristic size [116]. Future research effort will be necessary to reach beyond this irreconcilability.

Other expected further studies will likely assess the nanopatterns and the mechanisms of their formation, in thin films of block copolymers with more complex architectures. For instance, the phase behavior of star copolymers is increasingly well described [117]. For enlarging the application openings, more advanced substrates will likely be implemented, like the curved surfaces theoretically considered in recent studies [118, 119]. Efforts will be devoted to the generation of reconfigurable and responsive surfaces, which can adapt to surrounding environments [120], as for instance thermoresponsive nanopatterned surfaces [121].

References

1. Swarc, M., Levy, M., Milkovich, R.: Polymerization initiated by electron transfer to monomer. A new method of formation of block polymers. J. Am. Chem. Soc. **78**, 2656 (1956)
2. Matsen, M.W., Bates, F.S.: Unifying weak- and strong-segregation block copolymer theories. Macromolecules **29**, 1091 (1996)
3. Taton, D., Gnanou, Y.: Guidelines for synthesizing block copolymers. In: Lazzari, M., Liu, G., Lecommandoux, S. (eds.) Block Copolymers in Nanoscience. Wiley, Weinheim (2006)
4. Harrison, C., Park, M., Register, R., Adamson, D., Mansky, P., Chaikin, P.: Method of nanoscale patterning and products made thereby. US Patent 5,948,470, 7 Sept 1999
5. Russell, T.P.: Copolymers at surfaces and interfaces. Curr. Opin. Colloid Interface Sci. **1**, 107 (1996)

6. Granick, S., Kumar, S.K., Amis, E.J., Antonietti, M., Balazs, A.C., Chakraborty, A.K., Grest, G.S., Hawker, C., Janmey, P., Kramer, E.J., Nuzzo, R., Russell, T.P., Safinya, C.R.: Macromolecules at surfaces: research challenges and opportunities from tribology to biology. J. Polym. Sci. B Polym. Phys. **41**, 2755 (2003)
7. Bucknall, D.G.: Influence of interfaces on thin polymer film behaviour. Prog. Mater. Sci. **49**, 713 (2004)
8. Park, C., Yoon, J., Thomas, E.L.: Enabling nanotechnology with self assembled block copolymer patterns. Polymer **44**, 6725 (2003)
9. Kim, J.K., Yang, S.Y., Lee, Y., Kim, Y.: Functional nanomaterials based on block copolymer self-assembly. Prog. Polym. Sci. **35**, 1325 (2010)
10. Crossland, E.J.W., Nedelcu, M., Ducati, C., Ludwigs, S., Hillmyer, M.A., Steiner, U., Snaith, H.J.: Block copolymer morphologies in dye sensitized solar cells: probing the photovoltaic structure–function relation. Nano Lett. **9**, 2813 (2009)
11. Park, S., Lee, D.H., Xu, J., Kim, B., Hong, S.W., Jeong, U., Xu, U.T., Russell, T.P.: Macroscopic 10-terabit-per-square-inch arrays from block copolymers with lateral order. Science **323**, 1030 (2009)
12. Ross, C.A., Cheng, J.Y.: Patterned magnetic media made by self-assembled block-copolymer lithography. MRS Bull. **33**, 838 (2008)
13. Kennemur, J.G., Yao, L., Bates, F.S., Hillmyer, M.A.: Sub-5 nm domains in ordered poly(cyclo-hexylethylene)-blockpoly(methyl methacrylate) block polymers for lithography. Macromolecules **47**, 1411 (2014)
14. Tang, C., Hur, S., Stahl, B.C., Sivanandan, K., Dimitriou, M., Pressly, E., Fredrickson, G.H., Kramer, E.J., Hawker, C.J.: Thin film morphology of block copolymer blends with tunable supramolecular interactions for lithographic applications. Macromolecules **43**, 2880 (2010)
15. Park, M., Harrison, C., Chaikin, P.M., Register, R.A., Adamson, D.H.: Block copolymer lithography: periodic arrays of similar to 10^{11} holes in 1 square centimeter. Science **276**, 1401 (1997)
16. Harrison, C., Park, M., Chaikin, P.M., Register, R.A., Adamson, D.H.: Lithography with a mask of block copolymer microstructures. J. Vac. Sci. Technol. B **16**, 544 (1998)
17. Lammertink, R.G.H., Hempenius, M.A., van den Enk, J.E., Chan, V.Z.H., Thomas, E.L., Vancso, G.J.: Nanostructured thin films of organic–organometallic block copolymers: one-step lithography with poly(ferrocenylsilanes) by reactive ion etching. Adv. Mater. **12**, 98 (2000)
18. Jiang, Y., Liu, N., Guo, W., Xia, F., Jiang, L.: Highly-efficient gating of solid-state nanochannels by DNA supersandwich structure containing ATP aptamers: a nanofluidic IMPLICATION logic device. J. Am. Chem. Soc. **134**, 15395 (2012)
19. Yang, S.Y., Ryu, I., Kim, H.Y., Kim, J.K., Jang, S.K., Russell, T.P.: Nanoporous membranes with ultrahigh selectivity and flux for the filtration of viruses. Adv. Mater. **18**, 709 (2006)
20. Phillip, W.A., O'Neill, B., Rodwogin, M., Hillmyer, M.A., Cussler, E.L.: Self-assembled block copolymer thin films as water filtration membranes. ACS Appl. Mater. Interfaces **2**, 847 (2010)
21. Kwon, S., Shim, M., Lee, J.I., Lee, T.-W., Cho, K., Kim, J.K.: Ultrahigh density array of CdSe nanorods for CdSe/polymer hybrid solar cells: enhancement in short-circuit current density. J. Mater. Chem. **21**, 12449 (2011)
22. Lee, J.I., Cho, S.H., Park, S.-M., Kim, J.K., Kim, J.K., Yu, J.-W., Kim, Y.C., Russell, T.P.: Highly aligned ultrahigh density arrays of conducting polymer nanorods using block copolymer templates. Nano Lett. **8**, 2315 (2008)
23. Alam, M.M., Lee, Y.-R., Kim, J.-Y., Jung, W.-G.: Variety of nanopatterns on different substrates using PS-b-PMMA and their applications. J. Nanosci. Nanotechnol. **12**, 1634 (2012)
24. Bates, F.S., Fredrickson, G.H.: Block copolymers—designer soft materials. Phys. Today **52**(2), 32 (1999)

25. Hamley, I.W.: The Physics of Block Copolymers. Oxford Science, Oxford (1998)
26. Matyjaszewski, K.: Advances in controlled/living radical polymerization. ACS Series 854, K. Matyjaszewski Ed., 2003, 1 and special issue "Frontiers in Polymer Chemistry". Chem. Rev. **101**, 3579 (2001)
27. Matsen, M.W.: The standard gaussian model for block copolymer melts. J. Phys. Condens. Matter. **14**, 21 (2002)
28. de Gennes, P.-G.: Scaling Concepts in Polymer Physics. Cornell University Press, Ithaca, NY (1979)
29. Bates, F.S., Fredrickson, G.H.: Block copolymer thermodynamics: theory and experiment. Annu. Rev. Phys. Chem. **41**, 525 (1990)
30. Lodge, T.P.: Block copolymers: past successes and future challenges. Macromol. Chem. Phys. **204**, 265 (2003)
31. Broseta, D., Fredrickson, G.H., Helfand, E., Leibler, L.: Molecular weight and polydispersity effects at polymer-polymer interfaces. Macromolecules **23**, 132 (1990)
32. Breiner, U., Krappe, U., Thomas, E.L., Stadler, R.: Structural characterization of the "knitting pattern" in polystyrene-block-poly(ethylene-co-butylene)-block-poly(methyl methacrylate) triblock copolymers. Macromolecules **31**, 135 (1998)
33. Semenov, A.N.: Contribution to the theory of microphase layering in block-copolymer melts. Sov. Phys. JETP **61**, 733 (1985)
34. Ren, Y., Lodge, T.P., Hillmyer, M.A.: Effect of selective perfluoroalkylation on the segregation strength of polystyrene−1,2-polybutadiene block copolymers. Macromolecules **35**, 3889 (2002)
35. Hashimoto, T.: Generalized view of molecular weight dependence of microdomain size of block polymers. Appraisal of Hadziioannou-Skoulios' data on binary mixtures of block polymers. Macromolecules **15**, 1548 (1982)
36. Bendejacq, D., Ponsinet, V., Joanicot, M., Vacher, A., Airiau, M.: Chemical tuning of the microphase separation in diblock copolymers from controlled radical polymerization. Macromolecules **36**, 7289 (2003)
37. Lazzari, M., DeRosa, C.: Methods for the alignment and the large-scale ordering of block copolymer morphologies. In: Lazzari, M., Liu, G., Lecommandoux, S. (eds.) Block Copolymers in Nanoscience. Wiley, Weinheim (2006)
38. de Gennes, P.-G., Prost, J.: The Physics of Liquids Crystals. Oxford University Press, New York, NY (1993)
39. Kim, G., Libera, M.: Kinetic constraints on the development of surface microstructure in SBS thin films. Macromolecules **31**, 2670 (1998)
40. Black, C.T., Ruiz, R., Breyta, G., Cheng, J.Y., Colburn, M.E., Guarini, K.W., Kim, H.-C., Zhang, Y.: Polymer self assembly in semiconductor microelectronics. IBM J. Res. Dev. **51**, 605 (2007)
41. Solak, H.H., David, C., Gobrecht, J., Golovkina, V., Cerrina, F., Kim, S.O., Nealey, P.F.: Sub-50 nm period patterns with EUV interference lithography. Microelectron. Eng. **67–8**, 56 (2003)
42. Mansky, P., Liu, Y., Huang, E., Russell, T.P., Hawker, C.J.: Controlling polymer–surface interactions with random copolymer brushes. Science **275**, 1458 (1997)
43. Huang, E., Russell, T.P., Harrison, C., Chaikin, P.M., Register, R., Hawker, C.J., Mays, J.W.: Using surface active random copolymers to control the domain orientation in diblock copolymer thin films. Macromolecules **31**, 7641 (1998)
44. Han, E., Stuen, K.O., La, Y.-H., Nealey, P.F., Gopalan, P.: Effect of composition of substrate-modifying random copolymers on the orientation of symmetric and asymmetric diblock copolymer domains. Macromolecules **41**, 9090 (2008)
45. Liu, P.-H., Thebault, P., Guenoun, P., Daillant, J.: Easy orientation of diblock copolymers on self-assembled monolayers using UV irradiation. Macromolecules **42**, 9609 (2009)
46. Khanna, V., Cochran, E.W., Hexemer, A., Stein, G.E., Fredrickson, G.H., Kramer, E.J., Li, X., Wang, J., Hahn, S.F.: Effect of chain architecture and surface energies on the ordering

behavior of lamellar and cylinder forming block copolymers. Macromolecules **39**, 9346 (2006)
47. Xu, T., Zhu, Y., Gido, S.P., Russell, T.P.: Electric field alignment of symmetric diblock copolymer thin films. Macromolecules **37**, 2625 (2004)
48. Sivaniah, E., Hayashi, Y., Iino, M., Hashimoto, T., Fukunaga, K.: Observation of perpendicular orientation in symmetric diblock copolymer thin films on rough substrates. Macromolecules **36**, 5894 (2003)
49. Sivaniah, E., Matsubara, S., Zhao, Y., Hashimoto, T., Fukunaga, K., Kramer, E.J., Mates, T. E.: Symmetric diblock copolymer thin films on rough substrates: microdomain periodicity in pure and blended films. Macromolecules **41**, 2484 (2008)
50. Tsori, Y., Andelman, D.: Parallel and perpendicular lamellae on corrugated surfaces. Macromolecules **36**, 8560 (2003)
51. Segalman, R.A., Yokoyama, H., Kramer, E.J.: Graphoepitaxy of spherical domain block copolymer films. Adv. Mater. **13**, 1152 (2001)
52. Cheng, J.Y., Mayes, A.M., Ross, C.A.: Nanostructure engineering by templated self-assembly of block copolymers. Nat. Mater. **3**, 823 (2004)
53. Aissou, K., Shaver, J., Fleury, G., Pécastaings, G., Brochon, C., Navarro, C., Grauby, S., Rampnoux, J.-M., Dilhaire, S., Hadziioannou, G.: Nanoscale block copolymer ordering induced by visible interferometric micropatterning: a route towards large scale block copolymer 2D crystals. Adv. Mater. **25**, 213 (2013)
54. Tsori, Y., Andelman, D.: Diblock copolymer ordering induced by patterned surfaces above the order-disorder transition. Macromolecules **34**, 2719 (2001)
55. Kim, S.O., Solak, H.H., Stoykovich, M.P., Ferrier, N.J., de Pablo, J.J., Nealey, P.F.: Epitaxial self-assembly of block copolymers on lithographically defined nanopatterned substrates. Nature **424**, 411 (2003)
56. Edwards, E.W., Montague, M.F., Solak, H.H., Hawker, C.J., Nealey, P.F.: Precise control over molecular dimensions of block-copolymer domains using the interfacial energy of chemically nanopatterned substrates. Adv. Mater. **16**, 1315 (2004)
57. De Rosa, C., Park, C., Lotz, B., Fetters, L.J., Wittman, J.C., Thomas, E.L.: Control of molecular and microdomain orientation in a semicrystalline block copolymer thin film by epitaxy. Macromolecules **33**, 4871 (2000)
58. De Rosa, C., Park, C., Lotz, B., Thomas, E.L.: Microdomain patterns from directional eutectic solidification and epitaxy. Nature **405**, 433 (2000)
59. Park, C., De Rosa, C., Thomas, E.L.: Large area orientation of block copolymer microdomains in thin films via directional crystallization of a solvent. Macromolecules **34**, 2602 (2001)
60. Xuan, Y., Peng, J., Cui, L., Wang, H., Li, B., Han, Y.: Morphology development of ultrathin symmetric diblock copolymer film via solvent vapor treatment. Macromolecules **37**, 7301 (2004)
61. Fukunaga, K., Elbs, H., Magerle, R., Krausch, G.: Large-scale alignment of ABC block copolymer microdomains via solvent vapor treatment. Macromolecules **33**, 947 (2000)
62. Kim, S.H., Misner, M.J., Xu, T., Kimura, M., Russell, T.P.: Highly oriented and ordered arrays from block copolymers via solvent evaporation. Adv. Mater. **16**, 226 (2004)
63. Yin, J., Yao, X., Liou, J.-Y., Sun, W., Sun, Y.-S., Wang, Y.: Membranes with highly ordered straight nanopores by selective swelling of fast perpendicularly aligned block copolymers. ACS Nano **7**, 9961 (2013)
64. Bang, J., Kim, S.H., Drockenmuller, E., Misner, M.J., Russell, T.P., Hawker, C.J.: Defect-free nanoporous thin films from ABC triblock copolymers. J. Am. Chem. Soc. **128**, 7622 (2006)
65. Knoll, A., Horvat, A., Lyakhova, K.S., Krausch, G., Sevink, G.J.A., Zvelindovsky, A.V., Margele, R.: Phase behavior in thin films of cylinder-forming block copolymers. Phys. Rev. Lett. **89**, 035501 (2002)

66. Albert, J.N.L., Young, W.-S., Lewis, R.L., Bogart, T.D., Smith, J.R., Epps, T.: Systematic study on the effect of solvent removal rate on the morphology of solvent vapor annealed ABA triblock copolymer thin films. ACS Nano **6**, 459 (2012)
67. Thurn-Albrecht, T., Schotter, J., Kastle, G.A., Emley, N., Shibauchi, T., Krusin-Elbaum, L., Guarini, K., Black, C.T., Tuominen, M.T., Russell, T.P.: Ultrahigh-density nanowire arrays grown in self-assembled diblock copolymer templates. Science **290**, 2126 (2000)
68. Xu, T., Hawker, C.J., Russell, T.P.: Interfacial energy effects on the electric field alignment of symmetric diblock copolymers. Macromolecules **36**, 6178 (2003)
69. DeRouchey, J., Thurn-Albrecht, T., Russell, T.P., Kolb, R.: Block copolymer domain reorientation in an electric field: an in-situ small-angle x-ray scattering study. Macromolecules **37**, 2538 (2004)
70. Pereira, G.G., Williams, D.R.M.: Diblock copolymer melts in electric fields: the transition from parallel to perpendicular alignment using a capacitor analogy. Macromolecules **32**, 8115 (1999)
71. Tsori, Y., Andelman, D.: Thin film diblock copolymers in electric field: transition from perpendicular to parallel lamellae. Macromolecules **35**, 5161 (2002)
72. Tsori, Y., Tournilhac, F., Andelman, D., Leibler, L.: Structural changes in block copolymers: coupling of electric field and mobile ions. Phys. Rev. Lett. **90**, 145504 (2003)
73. Thurn-Albrecht, T., DeRouchey, J., Russell, T.P., Kolb, R.: Pathways toward electric field induced alignment of block copolymers. Macromolecules **35**, 8106 (2002)
74. Böker, A., Elbs, H., Hänsel, H., Knoll, A., Ludwigs, S., Zettl, H., Zvelindovsky, A.V., Sevink, G.J.A., Urban, V., Abetz, V., Müller, A.H.E., Krausch, G.: Electric field induced alignment of concentrated block copolymer solutions. Macromolecules **36**, 8078 (2003)
75. Böker, A., Knoll, A., Elbs, H., Abetz, V., Müller, A.H.E., Krausch, G.: Large scale domain alignment of a block copolymer from solution using electric fields. Macromolecules **35**, 1319 (2002)
76. Chen, Z.R., Kornfield, J.A., Smith, S.D., Grothaus, J.T., Satkowski, M.M.:Pathways to macroscale order in nanostructured block copolymers Science **277**, 1248 (1997)
77. Polis, D.L., Smith, S.D., Terrill, N.J., Ryan, A.J., Morse, D.C., Winey, K.I.: Shear-induced lamellar rotation observed in a diblock copolymer by in situ small-angle x-ray scattering. Macromolecules **32**, 4668 (1999)
78. Chen, P.L.: Shear alignments in three-dimensional simulations of lamellar phases. Phys. Rev. E **71**, 061503 (2005)
79. Fasolka, M.J., Mayes, A.M.: Block copolymer thin films: physics and applications. Annu. Rev. Mater. Res. **31**, 323 (2001)
80. Angelescu, D.E., Waller, J.H., Register, R.A., Chaikin, P.M.: Shear-induced alignment in thin films of spherical nanodomains. Adv. Mater. **17**, 1878 (2005)
81. Pujari, S., Keaton, M.A., Chaikin, P.M., Register, R.A.: Alignment of perpendicular lamellae in block copolymer thin films by shearing. Soft Matter **8**, 5358 (2012)
82. Darling, S.B.: Directing the self-assembly of block copolymers. Prog. Polym. Sci. **32**, 1152 (2007)
83. Hashimoto, T., Bodycomb, J., Funaki, Y., Kimishima, K.: The effect of temperature gradient on the microdomain orientation of diblock copolymers undergoing an order–disorder transition. Macromolecules **32**, 952 (1999)
84. Mita, K., Tanaka, H., Saijo, K., Takenaka, M., Hashimoto, T.: Cylindrical domains of block copolymers developed via ordering under moving temperature gradient. Macromolecules **40**, 5923 (2007)
85. Turner, M.S., Rubinstein, M., Marques, C.M.: Surface-induced lamellar ordering in a hexagonal phase of diblock copolymers. Macromolecules **27**, 4986 (1994)
86. Green, P.F.: Wetting and dynamics of structured liquid films. J. Polym. Sci. B Polym. Phys. **41**, 2219 (2003)
87. Maaloum, M., Chatenay, D., Coulon, G., Gallot, Y.: Edge profile of relief 2D domains at the free surface of smectic copolymer thin films. Phys. Rev. Lett. **68**, 1575 (1992)

88. Smith, A.P., Douglas, J.F., Meredith, J.C., Amis, E.J., Karim, A.: High-throughput characterization of pattern formation in symmetric diblock copolymer films. J. Polym. Sci. B Polym. Phys. **39**, 2141 (2001)
89. Coulon, G., Collin, B., Chatenay, D., Gallot, Y.: Kinetics of growth of islands and holes on the free surface of thin diblock copolymer films. J. Phys. **3**, 697 (1993). II France
90. Smith, A.P., Douglas, J.F., Meredith, J.C., Amis, E.J., Karim, A.: Combinatorial study of surface pattern formation in thin block copolymer films. Phys. Rev. Lett. **87**, 015503 (2001)
91. Li, Z., Qu, S., Rafailovich, M.H., Sokolov, J., Tolan, M., Turner, M.S., Wang, J., Schwarz, S. A., Lorenz, H., Kotthaus, J.P.: Confinement of block copolymers on patterned surfaces. Macromolecules **30**, 8410 (1997)
92. Xu, T., Goldbach, J.T., Russell, T.P.: Sequential, orthogonal fields: a path to long-range, 3-D order in block copolymer thin films. Macromolecules **36**, 7296 (2003)
93. Ye, C.H., Singh, G., Wadley, M.L., Karim, A., Cavicchi, K.A., Vogt, B.D.: Anisotropic mechanical properties of aligned polystyrene-block-polydimethylsiloxane thin films. Macromolecules **46**, 21 (2013)
94. Berry, B.C., Singh, G., Kim, H.-C., Karim, A.: Highly aligned block copolymer thin films by synergistic coupling of static graphoepitaxy and dynamic thermal annealing fields. ACS Macro Lett. **2**, 346 (2013)
95. Hong, S.W., Huh, J., Gu, X.D., Lee, D.H., Jo, W.H., Park, S., Xu, T., Russell, T.P.: Unidirectionally aligned line patterns driven by entropic effects on faceted surfaces. Proc. Natl. Acad. Sci. U. S. A. **109**, 1402 (2012)
96. Man, X., Andelman, D., Orland, H., Thébault, P., Liu, P.H., Guenoun, P., Daillant, J., Landis, S.: Organisation of block copolymers using nanoimprint lithography: comparison of theory and experiments. Macromolecules **44**, 2206 (2011)
97. Park, S.-M., Liang, X., Harteneck, B.D., Pick, T.E., Hiroshiba, N., Wu, Y., Helms, B.A., Olynick, D.L.: Sub-10 nm nanofabrication via nanoimprint directed self-assembly of block copolymers. ACS Nano **5**, 8523 (2011)
98. Tang, C., Wu, W., Smilgies, D.-M., Matyjaszewski, K., Kowalewski, T.: Robust control of microdomain orientation in thin films of block copolymers by zone casting. J. Am. Chem. Soc. **133**, 11802 (2011)
99. Seppala, J.E., Lewis, R.L., Epps, T.H.: Spatial and orientation control of cylindrical nanostructures in ABA triblock copolymer thin films by raster solvent vapor annealing. ACS Nano **6**, 9855 (2012)
100. Yager, K.G., Fredin, N.J., Zhang, X., Berry, B.C., Karim, A., Jones, R.L.: Evolution of block-copolymer order through a moving thermal zone. Soft Matter **6**, 92 (2010)
101. Stoykovich, M.P., Muller, M., Kim, S.O., Solak, H.H., Edwards, E.W., Pablo, J.J., Nealey, P.F.: Directed assembly of block copolymer blends into nonregular device-oriented structures. Science **308**, 1442 (2005)
102. Hahm, J., Sibener, S.J.: Time-resolved atomic force microscopy imaging studies of asymmetric PS-b-PMMA ultrathin films: dislocation and disclination transformations, defect mobility, and evolution of nanoscale morphology. J. Chem. Phys. **114**, 4730 (2001)
103. Anastasiadis, S.H., Russell, T.P., Satija, S.K., Majkrzak, C.F.: Neutron reflectivity studies of the surface-induced ordering of diblock copolymer films. Phys. Rev. Lett. **62**, 1852 (1989)
104. Ree, M.: Probing the self- assembled nanostructures of functional polymers with synchrotron grazing incidence x-ray scattering. Macromol. Rapid Commun. **35**, 930 (2014)
105. Renaud, G., Lazzari, R., Leroy, F.: Probing surface and interface morphology with grazing incidence small-angle Y-ray scattering. Surf. Sci. Rep. **64**, 255 (2009)
106. Lu, X., Yager, K.G., Johnston, D., Black, C.T., Ocko, B.M.: Grazing-incidence transmission X-ray scattering: surface scattering in the Born approximation. J. Appl. Crystal. **46**, 165 (2013)
107. Azzam, R.M.A., Bashara, N.M.: Ellipsometry and Polarized Light. Elsevier, Amsterdam (1987)

108. Johannsmann, D.: Investigation of soft organic films with ellipsometry. In: Knoll, W., Advincula, R.C. (eds.) Functional Polymer Films, Characterization and Applications, vol. 2. Wiley, Weinheim (2011)
109. Virgili, J.M., Tao, Y., Kortright, J.B., Balsara, N.P., Segalman, R.A.: Analysis of order formation in block copolymer thin films using resonant soft x-ray scattering. Macromolecules **40**, 2092 (2007)
110. Sunday, D.F., Hammond, M.R., Wang, C.Q., Wu, W.-L., Delongchamp, D.M., Tjio, M., Cheng, J.Y., Pitera, J.W., Kline, R.J.: Determination of the internal morphology of nanostructures patterned by directed self-assembly. ACS Nano **8**, 8426 (2014)
111. Filimon, M., Kopf, I., Ballout, F., Schmidt, D.A., Bründermann, E., Rühe, J., Santer, S., Havenith, M.: Smart polymer surfaces: mapping chemical landscapes on the nanometre scale. Soft Matter **6**, 3764 (2010)
112. Chang, J.B., Choi, H.K., Hannon, A.F., Alexander-Katz, A., Ross, C.A., Berggren, K.K.: Design rules for self-assembled block copolymer patterns using tiled templates. Nat Commun. **5**, 3305 (2014)
113. Cushen, J.D., Otsuka, I., Bates, C.M., Halila, S., Fort, S., Rochas, C., Easley, J.A., Rausch, E.L., Thio, A., Borsali, R., Willson, C.G., Ellison, C.J.: Oligosaccharide/silicon-containing block copolymers with 5 nm features for lithographic applications. ACS Nano **6**, 3424 (2012)
114. Kim, S.H., Misner, M.J., Yang, L., Gang, O., Ocko, B.M., Russell, T.P.: Salt complexation in block copolymer thin films. Macromolecules **39**, 8473 (2006)
115. Kim, S., Nealey, P.F., Bates, F.S.: Directed assembly of lamellae forming block copolymer thin films near the order–disorder transition. Nano Lett. **14**, 148 (2014)
116. Mishra, V., Fredrickson, G.H., Kramer, E.J.: Effect of film thickness and domain spacing on defect densities in directed self-assembly of cylindrical morphology block copolymers. ACS Nano **6**, 2629 (2012)
117. Jong, S., Moon, H.C., Kwak, J., Bae, D., Lee, Y., Kim, J.K., Lee, W.B.: Phase behavior of star-shaped polystyrene-block-poly(methyl methacrylate) copolymers. Macromolecules **47**, 5295 (2014)
118. Hexemer, A., Vitelli, V., Kramer, E.J., Fredrickson, G.H.: Monte Carlo study of crystalline order and defects on weakly curved surfaces. Phys. Rev. E **76**, 051604 (2007)
119. Vitelli, V., Lucks, J.B., Nelson, D.R.: Crystallography on curved surfaces. Proc. Natl. Acad. Sci. U. S. A. **103**, 12323 (2006)
120. Stuart, M.A.C., Huck, W.T.S., Genzer, J., Müller, M., Ober, C., Stamm, M., Sukhorukov, G.B., Szleifer, I., Tsukruk, V.V., Urban, M., Winnik, F., Zauscher, S.I., Luzinov, I., Minko, S.: Emerging applications of stimuli-responsive polymer materials. Nat. Mater. **9**, 101 (2010)
121. Kelly, J.Y., Albert, J.N.L., Howarter, J.A., Kang, S., Stafford, C.M., Epps, T.H., Fasolka, M. J.: Investigation of thermally responsive block copolymer thin film morphologies using gradients. ACS Appl. Mater. Interfaces **2**, 3241 (2010)

Chapter 5
Nanostructured Interfaces by Surface Segregation of Block Copolymers

Antoine Bousquet and Juan Rodríguez-Hernández

5.1 Introduction

The modification of polymer surfaces both in terms of structure and functionality is a subject that has been evidenced to be of theoretical and practical interest [1, 2]. In this sense, a large amount of literature has been published describing a wide variety of surface chemical modification approaches such as flame or corona treatments, chemical reactions (in solution), plasma or UV treatments or the application of polymer coatings among others [3–11].

Equally, the generation of nano/microstructured polymer film surfaces has been a challenge during the last decades. The advances in the fabrication of structured surfaces to obtain micro and nano patterns have been accomplished following two different approaches, i.e., either by adapting techniques, such as molding (embossing) or nano/microimprinting or by developing novel techniques including laser ablation or soft lithography [12]. Thus, higher resolution capabilities are directly related with technological advances. In contrast to the use of highly sophisticated tools required by the above mentioned techniques, surface instabilities produced by different mechanisms take advantage of the inherent properties of polymers to induce particular surface patterns. Some of the surface

A. Bousquet (✉)
Institut des Sciences Analytiques et de Physico-chimie pour l'Environnement et les Matériaux (IPREM/EPCP), Technopole Helioparc 2, av. Pdt Angot 64 053 PAU, Cedex 09, France
e-mail: antoine.bousquet@univ-pau.fr

J. Rodríguez-Hernández
Instituto de Ciencia y Tecnología de Polímeros (ICTP), Consejo Superior de Investigaciones Científicas (CSIC), C/Juan de la Cierva 3, 28006 Madrid, Spain
e-mail: jrodriguez@ictp.csic.es

J. Rodríguez-Hernández, C. Drummond (eds.), *Polymer Surfaces in Motion*,
DOI 10.1007/978-3-319-17431-0_5

instabilities are well known since decades but have been only recently extended their use to pattern polymer surfaces. This recent interest relies on the rich and complex patterns obtained as a result of self-organizing processes that are rather difficult if not impossible to fabricate by using traditional patterning techniques.

Within this context, this chapter discusses the possibilities to produce surface patterns by surface segregation of an additive towards the interface in polymer blends. Surface segregation is the result of the preferential migration of one blend component to the interface thereby inducing selective enrichment at the near-surface level. As will be discussed below in detail, this effect is directed by the surface thermodynamics that favors the presence at the interface of the component of a mixture lowering the surface tension. As a consequence, this phenomenon is the cause of having large differences between the surface and the bulk composition.

Surface segregation, which is also a common phenomenon in other materials, has received limited attention and most of the studies applied this concept for the control of the surface chemical composition. Moreover, surface segregation has been typically considered as a non-desirable effect. This is particularly true in the case of materials with precise bulk properties provided by the presence of different additives such as plasticizers or UV-absorbers. The segregation of these additives towards the interface modifies the bulk properties and can provoke large variations on their mechanical behavior. Several contributions have explored this phenomenon in order to reduce the amount of oligomers/polymeric additives that bloom to polymer surfaces [13]. However, in many other applications it is desirable to have surface properties that vary to a large extent from those found in the bulk. For example, to favor adhesion or increase the wettability, or the opposite, i.e., increase the hydrophobicity, to improve the biocompatibility of commercial polymers or to enhance the chemical resistance.

In this chapter, we aim to provide an overview of the possibilities of using surface segregation not only to functionalize and but also to nanostructure polymer surfaces. For this purpose, we first discuss the factors that may favor or reduce the presence of a particular additive at the interface. As depicted here, the molecular structure, the functional groups contained within the polymer, and other factors such as hydrophilicity of the environment or the temperature will play a key role on the migration of an additive towards the surface. More interestingly, the use of block copolymers within the blends permits the formation, by self-assembly, of different structures at the nanometer scale, thus providing an interesting way to fabricate different nanometer scale structures at the interface.

It has to be mentioned at this point that this chapter focuses on the segregation and eventually interfacial self-assembly of nonmiscible blends. These blends are susceptible to form microstructured and nanostructured domains at the polymer interface. Readers interested in surface segregation in homopolymer miscible blends should refer to the following references [14–22].

5.2 From Surface Rearrangement to Surface Segregation

Pioneer studies of several groups including Whitesides [23, 24], McCarthy [25, 26], Ter-Minassian-Saraga [27], Donaruma [28], or Bergbreiter [29, 30], just to mention few of them, evidenced a reorganization of functional groups present at the surface of treated polymer films exposed to a particular treatment. In particular, these initial works concluded that when considering polymer–air interfaces a surface reorganization occurs in order to decrease the surface energy. This reorganization process may be eventually improved by annealing. In effect, a general trend found in their studies was that surface reorganization forms less polar surfaces with only few exceptions [30]. More precisely, depending on the nature of the groups introduced during the surface treatment they will face two different situations: on the one hand, interfacial reconstruction is observed if the functional groups introduced are of higher surface energy than the pristine material. On the other hand, we will not observe this phenomenon when the surface functionalization leads to a decrease in the surface energy, for instance, by introducing low surface energy fluorocarbon groups (CF).

Generally speaking, surface reorganization is related to materials formed by one single component because it just involves a movement of a part of the molecules presents in the surface first layer. However, the concepts describe above can be applied to multicomponent materials. In this case, we will need to define the concept of surface segregation. Surface segregation is a thermodynamical phenomenon that directs the interfacial migration of certain components within a particular blend (Fig. 5.1). In general, provided an appreciable difference in energy between the components of the blend, the lower energy component is preferred at the interface in order to minimize the surface free energy. One of the most extensively studied

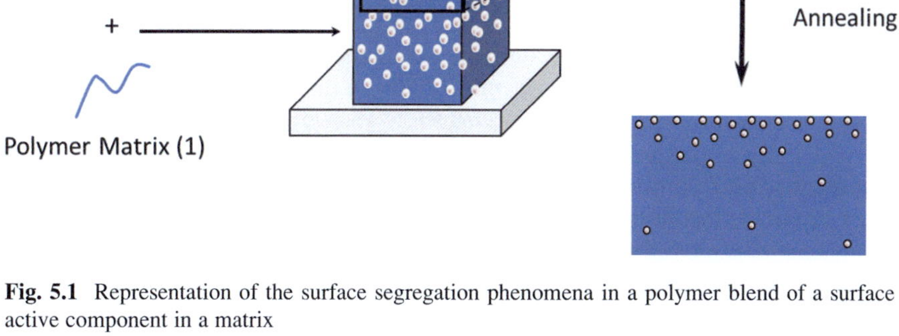

Fig. 5.1 Representation of the surface segregation phenomena in a polymer blend of a surface active component in a matrix

system is the compatible blend of polystyrene (PS) with poly(vinylmethylether). Many studies showed by different techniques that the surface of such a blend was largely enriched with PVME moieties. Indeed PS has a higher surface tension than PVME inducing the latter to migrate towards the surface to lower the interfacial free energy [31–36]. Incompatible polymer blends are also affected by the surface segregation phenomenon. For example a PS/Poly(methyl methacrylate) (PMMA) film is at thermodynamic equilibrium enriched at the film/air interface with PS which monomer unity present a lower surface tension than the PMMA one [36].

This analysis is valid for end functional polymers or block copolymers. The latter is discussed in detail throughout this chapter since block copolymers are able to form nanostructured domains.

The surface segregation can therefore be explained considering a thermodynamical equilibrium condition for a multicomponent polymer at a precise interface [37]. As depicted by the Gibbs–Duhem equation (5.1), surface segregation is controlled by the equilibrium between the bulk free energy penalty that is incurred if a constituent is removed from the bulk phase, and the change in the interfacial free energy that occurs when the low surface energy component substitutes the high energy component.

$$\text{Gibbs} - \text{Duhem Equation}\; d(n_1\mu_1) + d(n_2\mu_2) = -Ad\gamma = A(\gamma_1 - \gamma_2)d\phi_{s,f} \quad (5.1)$$

The two terms on the left side of the equation are related to the energy required when one of the components is extracted from the bulk as a consequence of the surface segregation. In the simplest case (we consider here a binary blend) the 1 and 2 correspond to each component. In addition n refers to the number of moles and μ is the chemical potential. On the right hand of the equation are included the terms related to the variation of the free surface energy produced in a determined surface area (A) when the surface concentration of the low surface energy component increases and the high energy component is embedded within the material. Whereas the surface concentration of the components appears in the equation with the term $d\phi_{s,f}$, the term γ designs the surface tension of each component.

The force for the surface segregation to occur is directly related to the term $A(\gamma_1\text{–}\gamma_2)$. Thus, a positive value of this term indicates that the additive possess lower surface energy than the matrix and therefore will migrate to the interface. On the contrary, a negative value would be related to an additive with a high surface energy. In the latter, the polymer matrix will be favored at the polymer–air interface.

Other groups have developed alternative models in order to explain the surface segregation in multicomponent systems, briefly discussed in the next part. However, it is outside the scope of this chapter to thoroughly describe the theoretical models developed in this topic but rather provide a simple overview of the main aspects involved in the surface segregation of polymer blend. Those readers interested in the theoretical approaches reported to understand the surface segregation phenomena are referred to the following references [18, 38–41].

As mentioned above, surface segregation is a spontaneous phenomenon. However, the kinetics of this process depends not only on the different surface tension of the components but also on the material employed. For instance, the use of polystyrene based blends would require weeks or even months in order to observe the surface segregation of one of the components at room temperature. In this concern, some authors suggested the employment of annealing in order to enhance the kinetics towards the equilibrium state [42]. In some cases, an increase of the temperature may induce, at least to some extent, the degradation of the material. In this particular case, surface segregation can be achieved by modifying the experimental conditions employed for the preparation of the films. For instance, the nature of the solvent employed may favor the formation of a compositional gradient towards the interface. As described by Chen et al. [43], varying the solvent, both the evaporation rate and the chain mobility during the film formation from a polymer solution will play a key role and improve the migration of one of the components.

From this preliminary definition, we can conclude that, by taking advantage of this spontaneous phenomenon and considering the aspects that rule the migration of a particular component in a polymer blend we will be able to vary the surface behavior. In the next paragraphs of this chapter we illustrate how this phenomenon has been employed both to modify the chemical composition of the polymer surface and to produce nanostructured interfaces. Moreover, the adaptive/responsive behavior of the polymer blend surfaces as a function among others of the environment, temperature or pH will be also described. But first of all we have to review the various parameters that require a consideration in order to obtain functional surfaces by segregation.

5.3 Factors Involved in the Surface Segregation of Polymer Blends

There are some general rules one should take into account to control the surface segregation phenomenon. From a macroscopic point of view, and as previously mentioned, the surface of a binary polymer blend will generally be enriched in the component with the smaller pure-component free energy. The difference in surface tension between the blend components, i.e., $\gamma_1 - \gamma_2$ serves as a "surface field" that directs the migration of the component with lower surface tension. This field is opposed by the osmotic forces that act for a mixing in the polymer blend. From a microscopic perspective, both enthalpic and entropic factors contribute to the "surface field" and to counter osmotic forces. Enthalpic factors refer to those functional groups included the additive introduced within the blend. Entropic factors include the topological characteristics of the polymers employed such as chain length or the degree of branching.

In this part we summarize the different factors driving surface segregation and how to control them to obtain desired surface properties.

5.3.1 Enthalpically Driven Segregation

5.3.1.1 An Illustrative Case of Surface Segregation: Fluorine End-Functionalized Polystyrene

Surface segregation of end-functional polymers requires additional considerations since, in comparison with polymer blends, high and low surface tension functional groups are covalently bonded within the same macromolecular chain. Functional polymers have drawn attention for their potential application as additive for surface modification of a homologue but non-functional matrix. Indeed, their inherent compatibility with the homopolymer matrix does not vary the mechanical properties of the host material. Probably one of the most studied systems is the blend of a polystyrene end-fluorinated (PS-F) with a non-functional polystyrene (PS) matrix. The fluorine end group has a lower surface tension than the polystyrene backbone resulting in its preferential segregation to the air–polymer interface in order to reduce the overall surface energy. As a result of the fluorine surface enrichment very hydrophobic materials were obtained.

In the late 1980s, research groups observed that mixture of hydrogenous and deuterated polystyrene had unfavorable thermodynamics and a positive Flory–Huggins interaction parameter χ [44]. This effect arises from the differences in zero point energy of H and D atoms leading to C–D bonds being slightly more polar than C–H bonds. Additionally, this difference in polarity influences the surface tension so that the deuterated polymer has a slightly lower surface tension and thus would segregate at the air–polymer interface (entry 1 Table 5.1) [14, 45]. Based on these findings, in 1994, Affrossman et al. realized an extensive study on the surface segregation of a fluoro end-terminated PS [46]. For the preparation of the functional polymer, they used anionic polymerization to terminate deuterated polystyrene with a perfluorooctyl group (*d*PS-F). When this polymer is blended with protonated polystyrene (*h*PS) (15/85 wt%, *d*PS-F/*h*PS), the authors observed a superficial enrichment of 36 wt% in *d*PS-F (entry 2 Table 5.1). This value increased astonishingly to 62 wt% after a thermal annealing at 130 °C for 24 h. They attributed this behavior to the presence of the fluorinated group in *d*PS-F which has a lower surface energy than PS repetitive unit. For comparison PS has a surface free energy of 40.7 mN m^{-1} when polytetrafluoroethylene one is 20 mN m^{-1}. Hence, the fluorine end-functional PS migrate toward the surface to reduce the interfacial tension. This segregation

Table 5.1 Blends and chemical composition at the surface before and after annealing (Data extracted from [46])

Entry	Polymer 1	Polymer 2	ϕ_b	ϕ_s Before annealing	ϕ_s After annealing
1	*d*PS	*h*PS	0.14	0.17	0.18
2	*d*PS-F	*h*PS	0.15	0.36	0.62
3	*d*PS	*h*PS-F	0.85	0.65	0.59

ϕ_b Deuterated polystyrene bulk molar fraction
ϕ_S Deuterated polystyrene surface molar fraction

happens during the spin-coating process [43] but is amplified through a thermal annealing above the glass-transition temperature (T_g) of the materials. In both cases enough mobility is given to the chains to reorganize and minimize surface tension. In polymer *d*PS-F, the deuterium and fluorine both act towards surface segregation. On the contrary, by analyzing a mixture of deuterated polymer and fluoro end-capped hydrogenous polystyrene (entry 3, Table 5.1) relative contribution of the two moieties can be evaluated. The surface is enriched with the fluoropolymer with a value of 41 wt% after annealing. This illustrated the dominant effect of a small number of fluorine atoms over the fully deuterated styrene repetitive unit.

Isotopic polymer blends have been used as model systems for studying the behavior of polymer blends since isotopic substitution, which is expected to only cause a relatively minor perturbation of properties, provides sufficient contrast for experimental probes such as neutron reflectivity (NR) or secondary ion mass spectrometry (SMIS). For the above mentioned study the mass contrast between deuterated and hydrogenated polymers has been extensively employed to prove the surface segregation. Elman et al. [47] used the nuclear contrast to evidence the segregation of end-functional deuterated PS in a non-functional hydrogenated matrix by neutron reflectivity. Three terminal groups were investigated: a "neutral" control specimen prepared with hydrogen as a terminal group, an "attractive" end group containing low surface energy fluorine atoms, and a "repulsive" high surface energy group composed of a carboxylic acid moiety. The authors demonstrated, based on the NR curves and X-ray photoelectron spectroscopy (XPS) data, that the fluoroterminated polymer preferentially migrated at the air–polymer interface and that the carboxylic acid terminated polymer was repelled from the surface.

These pioneered reports were followed by many others using PS-F [48–50], all supporting an air–polymer surface enrichment in fluorine. Theoretical models have then been developed to predict this migration behavior [51, 52]. The Lattice Model is one of particular interest because it provides end-group concentration depth profiles [18, 53]. The basis of the model is a cubic lattice confined between two impenetrable surfaces that are separated to form a bulk-like region in the center of the film. In this lattice a chemical group, being the repetitive unit or the end-group, has to be chosen to be the unity reference volume (Fig. 5.2). Quantitative prediction of end-group concentration depth profiles must also consider:

- The surface interaction parameter, describing the preference of the end-group for the surface compared to that of the macromolecule backbone.
- The bulk interaction parameters, representing the bulk interaction between the end group and the polymer main chain.
- The chain length of both the matrix and the additive.
- The amount of additive in the blend.

This lattice model was found to provide an excellent representation of experimental XPS data over a wide range of blend compositions and molecular weights when an end-fluorinated polystyrene was used as additive [53]. Thus, a surface enrichment of the low-energy end groups is confined to the first lattice layer. Interestingly, the second layer shows maximum depletion of functional group

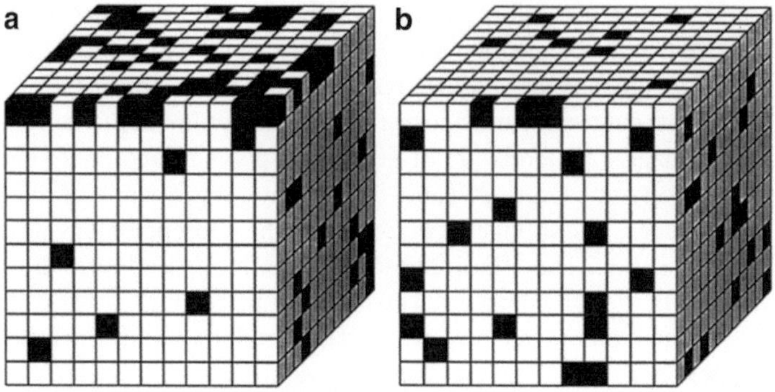

Fig. 5.2 Schematic representation of functional polymer chains configured on a cubic lattice. The darker cubes indicate a lattice site occupied by a functional end group, and the lighter cubes are occupied by polymer chain segments: (**a**) illustrates a chain with a low-energy attractive end group, (**b**) depicts a nonfunctional polymer with neutral end groups. Reproduced with permission from [54]

while the concentration in the following layers gradually recover bulk mean concentration over a depth comparable to the chain dimensions.

These lattice calculations were also employed to explore the surface segregation of end-chain polymers functionalized by one or several groups with either higher or lower surface tension than the backbone repetitive unit [54]. The authors concluded that adjacency of low surface energy group is the most effective way for enhancing surface segregation. Moreover, having two groups with similar surface energy character (low or high) is always preferable to having two groups of opposite character, the so-called Pushmi-Pullyu architecture, where one group prefers the surface while the other is repelled from it (Fig. 5.3). From this inspiring study grew the idea of using diblock copolymer, containing one block surface active and the other matrix compatibilizer, as additive for surface segregation (see Sect. 5).

It is then relatively easy to achieve very hydrophobic PS surfaces using polymer additives that are end-functionalized with low-surface energy groups such as fluorine or even siloxanes [55], because they spontaneously segregate to film surfaces, reducing their surface tension. However, it is less trivial to enrich a surface with polar group and solutions developed are discussed in the next part.

5.3.1.2 Interfacial Attraction of Functional Polymers: Hydrophilic vs. Hydrophobic

The problem in obtaining hydrophilic surfaces arises from the fact that polar groups have a significantly higher surface energy than most of the polymer repetitive units (including obviously styrene); therefore, their surface segregation is thermodynamically unfavorable [56]. However, Elman et al. demonstrated by neutron reflectivity that the carboxylic acid end-functional polystyrene (PS-COOH) was migrating towards the polymer–native oxide hydrophilic interface (the polymer films were

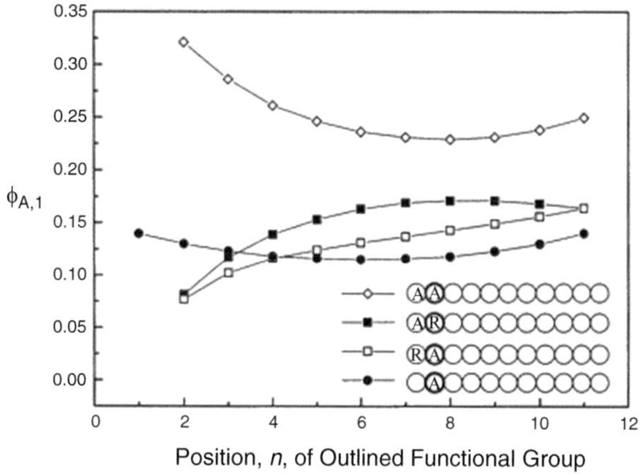

Fig. 5.3 Surface volume fraction of attractive functional groups as a function of the position, *n*, of a mid-chain group (denoted by *bold circles* on the figure *inset*) for four different architectures: functional polymer with two attractive functional groups (*open diamonds*); functional Pushmi-Pullyu polymer with an attractive functional end group and a repulsive functional group at position *n* (*filled squares*); functional inverse Pushmi-Pullyu polymer with a repulsive functional end group and an attractive functional group at position n (*open squares*); functional polymer with one attractive functional group at position n (*filled circles*). The *inset* illustrates each chain structure with *A* representing an attractive group and *R* representing a repulsive group. Reproduced with permission from [54]

spin-coated onto silicon wafers) [47]. Kawagushi et al. reported the same behavior using XPS and SIMS analysis [57]. After a thermal annealing of 36 h at 120 °C the PS-COOH additive was located at the PS–silicon oxide substrate interface. These two studies proves that the tension decrease of all kind of interfaces is driving component segregation; if the interface is hydrophilic then hydrophilic moieties will migrate towards it, on the contrary a hydrophobic interface (e.g., air) will attract hydrophobic moieties.

From these inspiring results, researcher's efforts were directed towards the elaboration of hydrophilic polymer surfaces using additive segregation. Koberstein et al. proved that, by immersing the polymer film in a polar environment, it was possible to induce the adsorption of polar functional groups at the surface [37]. Figure 5.4 shows that annealed blend of a PS matrix with PS-F or PS-COOH in saturated water vapor at 55 °C, resulted in the increase of the surface hydrophilic behavior. Surface reorganization occurred when the polymer environment is changed even in the glassy state (glass-transition temperature of polystyrene is around 100 °C). The fluorine terminated polymer adsorbed initially to the surface moved away from the interface with water vapor and is replaced by PS repetitive units [58]. On the contrary, for PS-COOH the contact angle is initially the same of a pure PS film indicating the absence of superficial carboxylic acid functions. When exposed to water vapor, the acid functions are dragged to the new interface to reduce the surface tension. By this way hydrophilic PS film were created. Later Wong et al. modeled that this sub-Tg reorganization shows diffusion kinetic [59].

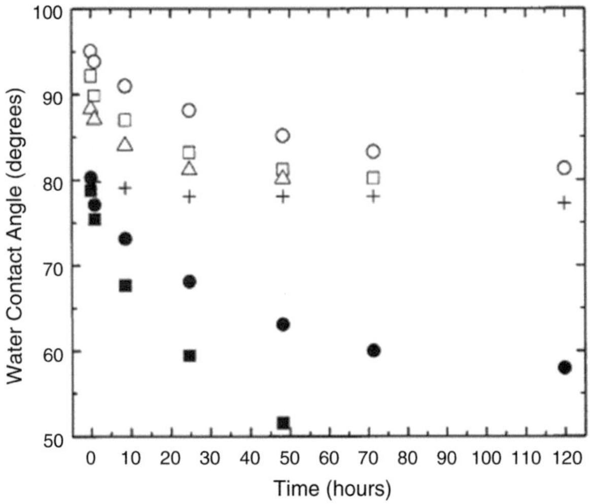

Fig. 5.4 Water contact angles for ω-functional PS as a function of its time in contact with saturated water vapor at 55 °C: (*open symbols*) fluorosilane-terminated PS [molecular weight (*circle*) 5,000, (*square*) 10,000, or (*triangle*) 25,000 g.mol^{-1}], (+) PS control, and (*filled symbols*) carboxylic acid terminated PS [molecular weight (*circles*) 6,500 or (*squares*) 10,000 g.mol^{-1}]. Reproduced with permission from [37]

The limitation of this approach is that polar groups that are not directly adjacent to the surface cannot contribute to the wetting modification due to the glassy nature of the polymer matrix. To overcome this drawback, Narrainen et al. showed that surface segregation of polar end-functionalized polymers was possible in blends by thermal annealing above the matrix T_g in a polar non-solvent [42]. They explored the use of glycerol as a high boiling point solvent to anneal polystyrene at 150 °C in a hydrophilic environment that favored the end-functional *d*PS-COOH additive (Fig. 5.5). They proved by nuclear reaction analysis that a blend containing 20 v% (volume %) of additive could be surface enriched at 60 v% after a 30 min treatment.

In 2009, the same group developed the first selective method in surface segregation of polar groups. In an attempt to discriminate the migration of polystyrene end-functionalized with carboxylic acid group or primary amine function, Thompson et al. studied the influence of pH during the annealing of PS film. Indeed the polymer film were immersed in pH-buffered aqueous solutions and then annealed at 120 °C for 2 h [60]. The samples were allowed to cool down to 70 °C before removing them from the solution, to avoid any surface rearrangement. Neutron reflectivity and nuclear reaction analysis were used to quantify the concentration of each functional polymer at the surface after annealing at controlled pH ranging from 1.9 to 9.4. With increasing pH, there was a systematic increase in the surface content of carboxylic acid groups from 0.05 to 0.34 functions/nm^2. On the contrary when pH decreased the amine surface concentration increased from 0.11 to 0.39 functions/nm^2.

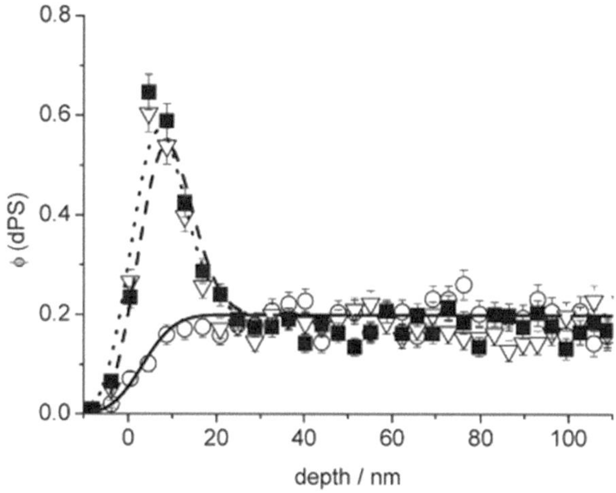

Fig. 5.5 Depth concentration profiles of deuterated α-ω-end functional *d*PS-COOH after annealing in glycerol at 150 °C for 0 min (*open circles*), 10 min (*filled squares*) and 30 min (*open triangles*). Reproduced with permission from [42]

5.3.2 *Entropically Driven Segregation*

5.3.2.1 Effect of the Molecular Weight

In 1993, Hariharan et al. studied, by neutron reflectivity, the influence of molecular weight on the surface segregation of a blend *d*PS and *h*PS [19]. The results, exposed in Table 5.2, first confirmed that when isotopic blends of approximately equal molecular weights were annealed at 150 °C in air, the deuterated component always partitioned preferentially to the surface (entries 1, 2 and 7). Moreover, the relative surface enrichment increased with increasing molecular weight because the χN factor increases (N being the average number degree of polymerization DPn), and thus the miscibility of the components diminishes. This result was in good qualitative agreement with the predictions of the constant density mean-field lattice theories [18].

In contrast, the surface segregation of asymmetric blends is controlled by the disparity in molecular weights between the additive and the matrix. The authors proved that either isotopic species can partition to the free surface depending on the specific combinations of chain lengths employed. In Table 5.2 (entries 3–7) it can be seen that deuterated polystyrene of DPn 4,808 units only segregate to the air–polymer interface when the protonated matrix has a molecular weight higher than 50,000 g mol^{-1} (styrene molar mass is 100 g mol^{-1}). Below this critical value the surface is enriched with the smaller protonated chains. In these situations, the surface segregation is balanced between the enthalpic effect that partitions the deuterated species to the surface and the entropic contribution that induced the superficial migration of the shortest chains.

Theoretical arguments suggested that the low molecular weight component in a bimodal blend tend to locate preferentially to the surface in order to minimize loss of conformational entropy at the material boundary [52]. Indeed, the surface tension

Table 5.2 Surface segregation in a polystyrene isotopic blend, effect of the molecular weight (Data extracted from [19])

Entry	DPn *d*PS	DPn *h*PS	ϕ_b	ϕ_s	$R_S=(\phi_s-\phi_b)/\phi_b$
1	288	288	0.45	0.48	0.067
2	480	480	0.41	0.53	0.293
3	4,808	112	0.48	0.1	−0.792
4		291	0.48	0.32	−0,333
5		500	0.48	0.49	0.021
6		923	0.51	0.60	0.176
7		4,808	0.40	0.67	0.675

ϕ_b deuterated polystyrene bulk volume fraction
ϕ_s deuterated polystyrene surface volume fraction
R_s is the relative surface enhancement of the deuterated polystyrene

of polystyrene is empirically known to have significant molecular weight dependence, as noted in Eq. (5.2) [61].

$$\gamma_{A,i} = \gamma_{A,\infty} - k_e N_i^{-2/3} \tag{5.2}$$

where $\gamma_{A,i}$ and $\gamma_{A,\infty}$ are respectively the surface tension of the examined polystyrene and that of an infinite molecular weight polymer. k_e is a constant related with and N_i is the degree of polymerization.

Thus, the lower molecular weight component was segregated at the surface, and this tendency became more pronounced with an increasing molecular weight difference. This behavior can be explained in term of the existence of a conformational or translational entropic penalty for a longer chain at the surface in addition to a preferential surface localization of end groups. With decreasing chain length, the entropic penalty at the surface drops, and the number of end groups at the surface increases.

As a consequence it is obvious that polymer dispersity will have an influence on surface segregation. Smaller chains in the samples will migrate at the interfaces [62]. Tanaka et al. used scanning force microscopy in order to investigate the surface molecular motion of PS films. It was revealed that the surface was in a glass/rubber transition state at 293 K due to the surface segregation of the lower molecular weight chains of a polydisperse blend (compared with 373 K in the bulk) [63].

The effect of the molecular weight is also affecting the surface chemical composition in a blend of symmetrical component. Kawagushi et al. showed by XPS that the number density of chain ends at the surface is inversely proportional to the molecular weight [57]. Figure 5.6 shows the integral intensity ratio of signals $F_{(1s)}$ to $C_{(1s)}$ for three mass symmetrical blends of *h*PS-F/*d*PS of 93,000, 48,000 and 25,000 g mol^{-1}. Since chain-end groups contain F atoms, the value of I_F/I_C reflects the chain-end concentration in the surface region. The emission angle of the photoelectrons corresponds to a certain analytical depth, which increases with sin θ. For all the films, the ratio I_F/I_C increased with decreasing analytical depth, but also increased with decreasing molecular weight at a given analytical depth. Thus, it is

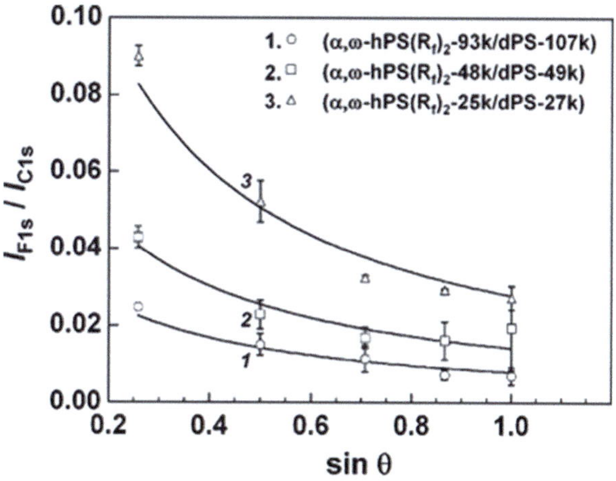

Fig. 5.6 F/C intensity ratio in function of depth in symmetrical blend of fluoro α,ω-end-functional protonated polystyrene (α,ω-*h*PS(Rf)$_2$) and deuterated PS, measured by angle dependent XPS. Reproduced with authorization from [57]

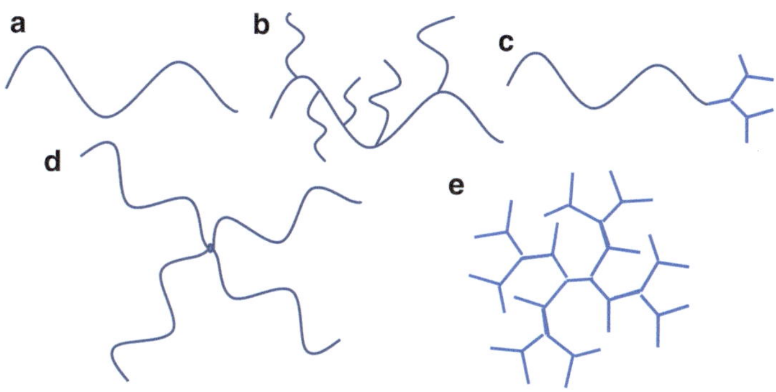

Fig. 5.7 Polymer topologies studied for surface segregation, (**a**) linear, (**b**) comb, (**c**) dendritic end-chain/pompom, (**d**) star, (**e**) dendritic/branched

clear that fluoro chain ends were partitioned to the surface, and their concentration increased with decreasing molecular weight. It as to be mentioned that this effect is the inverse than the one above mentioned when Hariharan mixed *h*PS and *d*PS of the same molecular weight in which segregation was exalted by increasing molar masses [19]. The difference is that the enthalpic gain due to surface enrichment of high energy end-chains has a greater influence than the entropic penalty resulting from mixing of high molecular weight components.

5.3.2.2 Effect of the Chain Architecture

Theoretical studies have predicted that branched polymers such as stars, combs or dendrimers (Fig. 5.7) should be preferred at the surface in the absence of other differences between the two components of a blend [13, 64, 65]. Three explanations have been suggested to explain this behavior: the favoring of multiple chain ends at the polymer–air interface, the connectivity points of these structures, and a diminution of conformational entropic penalty and/or augmentation translational entropic gain when the branched structures are localized at the surface [66, 67]. Indeed, to explain this last point, a common surface definition is that the material density should drop from its bulk value to zero. As a consequence this depletion region leads to a favorable coupling between the surface and the component with the lower value of the pure-component parameter $\beta^2 = R_g{}^2/V$ (R_g being the chain radius of gyration and V its molecular volume) [65]. Stars, comb, dendritic and all multi-branched macromolecules are known to have a lower conformational entropy related parameter β [68–70].

The first experimental verification of these models came in 2000 when Foster et al. synthesized a six-arm star polystyrene using anionic polymerization [71]. They demonstrated the surface enrichment of the star in isotopic blend with a linear polystyrene counterpart by dynamic secondary ion mass spectroscopy. However, in this study the star polymer was the deuterated component and had a lower molar mass than the linear matrix. Therefore, one could question if these two factors aren't themselves driving the surface segregation.

In 2008, Qian et al. synthesized two series of four-arm and eleven-arm star polystyrene with six different molar masses for each case. They determined surface tension of single component polymer film using a Wilhelmy plate technique [72]. Their findings were that star polymers presented a lower surface tension that linear polystyrene and that this value decreased with increasing number of arms. This behavior was explained by a lattice model considering finite compressibility of the chain and density gradient in the polymer blend. Thus this study implied that when mixed with a linear polymer star-shaped macromolecules would segregate to reduce the surface tension.

Hyperbranched and comb polymers have also been used as surface active additive. Ariura et al. synthesized by combination of anionic and cationic polymerization a monodispersed hyperbranched polystyrene [73]. The authors proved by combination of DSIMS and neutron reflectivity the preferential surface enrichment of the branched protonated macromolecules when blended with its deuterated linear polystyrene counterparts with the same molar mass. Other systems involving the segregation of the branched macromolecules in binary blends were demonstrated such as in polyamide [74] or poly(methylmethacrylate) [75].

In all these cases is reported the importance of the end-chain on the superficial migration of the additive. On the contrary, diffusion of cyclic polymers having no chain ends would be of great interest in the study of polymer surface segregation. In 2006, Tanaka compared time evolution of isotopic interfaces in two bilayers films, one made of cyclic polystyrene (c-*h*PS/c-*d*PS) and the linear counterpart (l-*h*PS/l-*d*PS) [76]. He showed that interdiffusion of the cyclic macromolecules was faster than the linear polymers one. Therefore, it would be very interesting to examine the surface segregation behavior of a binary film composed of a cyclic additive in a linear matrix.

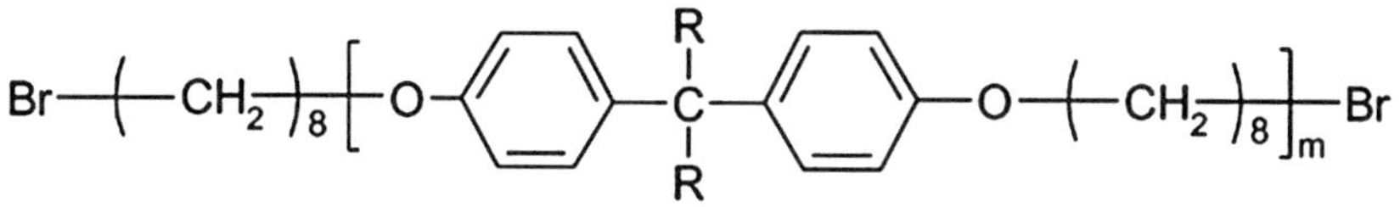

Fig. 5.8 Chemical structure of the copolymers BA-C8 and 6FBA-C8 [77]

5.3.2.3 Effect of the Crystallinity

The use of crystalline or semicrystalline polymers blends also affects the surface segregation. An illustrative example of the use of how semicrystalline polymers induce variations on the migration phenomena was reported by Li et al. [77]. This group studied blends of two different polymers BA-C8 (matrix) and 6FBA-C8 (additive) that are respectively semi-crystalline and amorphous polymers (Fig. 5.8). Even though the polymers are miscible in the bulk as a result of the similar chemical structure, surface segregation of the 6FBA-C8 polymer still occurs due to its lower surface energy. Moreover, this effect is largely observed at the crystalline regions. An additional driving force for 6FBAC8 to segregate to the surface was provided by the crystallization of BA-C8. The authors showed an increase in the surface energy of pure BA-C8 as it crystallizes. This would increase the thermodynamic driving force for the migration of the low surface energy component of the blend. As BA-C8 crystallized, more 6FBA-C8 migrated to the crystalline regions, causing the surface of the crystalline regions to rise above the amorphous regions.

5.3.3 *Surface Segregation kinetics*

The segregation of one of the additives towards the interface is (as mentioned above) a thermodynamic process and will occur until the two terms of equation (1) achieve the equilibrium state. However, depending on the polymer blend (type of components, ratio between the components, molecular characteristics of the components) the kinetic of this process can vary significantly. It has to be mentioned that an annealing process is required in order to accelerate the surface migration. Clarke et al. studied the segregation of end-functional carboxylic acid polystyrene in a PS matrix by neutron reflectivity and helium (3) nuclear reaction analysis [78, 79]. Figure 5.9 shows the concentration/depth profile of the *d*PS-COOH additive (80,000 g mol^{-1}) in the *h*PS matrix (500,000 g mol^{-1}) during an annealing at 150 °C in vacuum. They observed an additive depletion from the surface and an enrichment at the polymer–SiO_2 substrate interface that continue up to 7 days.

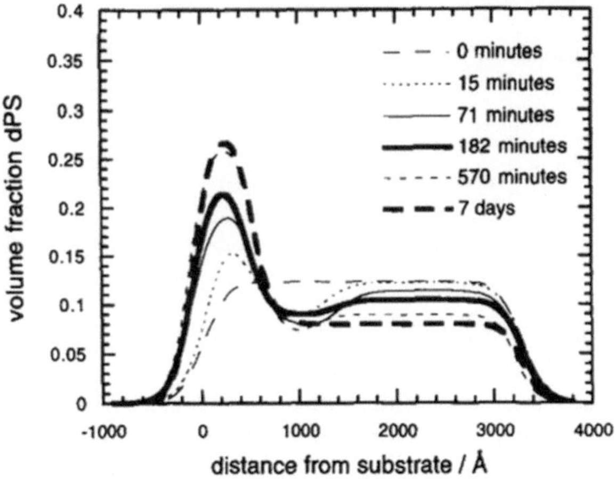

Fig. 5.9 Concentration/depth profile of a *d*PS-COOH additive (80,000 g mol^{-1}) in the *h*PS matrix (500,000 g mol^{-1}) during an annealing at 150 °C in vacuum. Reproduced with permission from [78]

The shape of this profile is particularly interesting; it shows a substantial depletion region behind the enriched zone, where the volume fraction of the functionalized polymer is less than the bulk one. This behavior gradually disappeared as the annealing time increased. The authors described this phenomenon as a two steps mechanism; first the additive in the first 150 nm close to the interface will migrate to lower the interfacial tension, and secondly the "bulk" additive will segregate towards the depletion zone to reach an osmotic equilibrium between the components in the film. When thermodynamic equilibrium is reached (in this case after 10 h) the depletion layer does not exist anymore. Diffusion coefficient of the functionalized chains has been estimated to be around 3.10^{-14} cm^2 s^{-1}. This value cannot be compared to other systems because it strongly depends of the glass transition temperature of the polymer matrix and the annealing temperature [59, 80].

As a last example of the migration kinetic and the power of the interface driven segregation, Schulze et al. showed by Forward recoil spectrometry that amino end-group polystyrene could go through more than 500 nm of non-functional PS to reach the interface with PMMA during a vacuum annealing at 174 °C (Fig. 5.10) [81].

5.3.4 Surface Segregation of a Functional Polymer: Other Systems

Most of the cases discussed above are related to the use of polystyrene, but surface segregation is a universal phenomenon and can be transposed to other systems. Table 5.3 reports a non-exhaustive list of binary blends in which segregation occurred and in which the surfaces were decorated with functional groups. End-functional fluoro polymers are commonly developed via different chemistries (ring opening, controlled radical or anionic polymerization) and used as surface

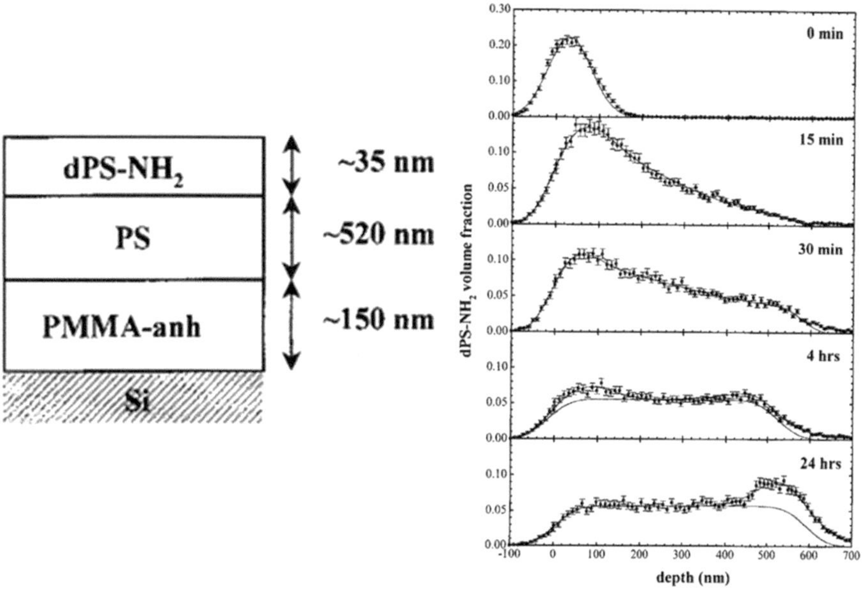

Fig. 5.10 Diffusion of deuterated PS-NH$_2$ through PS to a PPMA interface at 174 °C. Reproduced with permission from ref. [81]

Table 5.3 Surface segregation in polymer blends

Matrix[a]	Additive	Annealing conditions	ϕ_b^b	ϕ_s^b	Remark	Ref.
PDMS	PDMS-NH$_2$	Vaccum, RT, 24 h	–	$\phi_s < \phi_b$	NH$_2$ depletion	[82–84]
	PDMS-COOH			$\phi_s < \phi_b$	COOH depletion	
	PDMS-OH			$\phi_s = \phi_b$	No effect	
	PDMS-CH$_3$			$\phi_s > \phi_b$	CH$_3$ enrichment	
PE	PE-F	Vaccum, 120 °C, 1 h	0.04	0.3	Semi-crystalline polymer matrix	[85]
PVP	PVP-F	Vaccum, RT,	–	$\phi_s > \phi_b$	Influence of end-group on surface Tg	[86]
		Vaccum, 150 °C, 48 h	–	$\phi_s > \phi_b$		[87]
PEP	*d*PEP	Vaccum, 75 °C, 5 h	–	$\phi_s > \phi_b$	Influence of annealing temperature and time	[88]
PLA	PLA-F	Vaccum, 65 °C, 16 h	–	$\phi_s > \phi_b$	The polymer has a diffusion coefficient of 160 nm^2 s^{-1}	[89]
		Vaccum, 90 °C, 1 h	0.05	0.5		[90]
PEO	PEO-F	Vaccum, 90 °C, 12 h	–	$\phi_s > \phi_b$	Influence of polymer conformation	[91]

[a]*PDMS* polydimethylsiloxane, *PE* polyethylene, *PVP* poly(*N*-vinylpyrrolidone), *PEP* poly(ethylene propylene), *dPEP* deuterated poly(ethylene propylene), *PLA* poly(ᴅ,ʟ-lactide), *PEO* poly (ethylene oxide)

[b]ϕ_b and ϕ_s are the mass fraction of the additive respectively in the bulk and on surface

active additives. Hydrophobic and hydrophilic polymer films have had their surface tension lowered by the surface segregation of these fluorine end-chains. Deuterated polymer and high energy end-functional macromolecules have been shown to respectively enrich and deplete various polymer surfaces. Different annealing conditions have been used but they have all in common the fact that the temperature has to be raised above the glass-transition temperature of the used polymer, to ensure an efficient migration process.

However, some examples reported the surface segregation driven by surface freezing. Surface freezing is an exception in nature where almost all the materials exhibit the opposite behavior in surface melting. This particular feature was reported by Dhinojwala et al. [92] that employed linear alkanes and alkane analogues which are unique materials to exhibit surface freezing. More precisely, they determined the surface composition and surface ordering temperatures in binary mixtures containing acrylates with two different lengths of side chains. The chemical attachment of alkyl chains to the backbone leads to striking differences from the surface freezing in binary alkane mixtures. Above $T_{s,22}$ (order-to-disorder transition of the polymer constructed with acrylates having an alkyl chain of 22 units), the short disordered side chains are preferred on the surface. However, below $T_{s,22}$ the surfaces change discontinuously from a complete miscible to a complete immiscible surface layer. The long ordered side chains completely cover the surface for bulk composition as small as 2 wt%.

5.3.5 Avoiding Surface Segregation in Polymer Blends

Surface segregation has been depicted in both compatible and incompatible polymer blends. In general, miscibility does not alter the preferential placement of one of the components at the interface. Therefore, preventing surface segregation would require additional considerations. An interesting approach has been reported by Liu et al. [93]. They demonstrated that it is possible to eliminate surface enrichment of the low energy component at the surface when the interaction between the components in a particular blend is strong enough. To prove this concept they studied blends of poly(styrene-*co*-hexafluorohydroxyisopropyl-α-methyl styrene)/poly(4-vinyl pyridine) (PS(OH)/PVP). The interaction between the two components could be controlled by changing the density of hydrogen bonding through the adjustment of the hydroxyl content of the PS (OH) component. When the hydroxyl content was lower than 5 mol%, the surface of these polymer blends was largely enriched with PS(OH) because of the difference in surface free energy between PS(OH) and PVP. On the contrary, when the hydroxyl content was higher than 21 mol%, complexes formed render the surface and bulk compositions very similar.

In the following parts we aim to provide an overview of the possibilities of using surface segregation not only to functionalize and but also to nanostructure polymer surfaces. For this purpose, we will introduced the micro phase separation of block coppolymers first in bulk and then to surfaces.

5.4 Phase Separation in Block Copolymers and Block Copolymer Blends

5.4.1 *Block Copolymer Self-Assembly in Bulk: Few Concepts*

As has been already introduced, block copolymers consist of two distinct polymer chains that are covalently bonded. This unique feature allows them to exhibit particular microphase segregated domains. Effectively, whereas the incompatibility between the two blocks induces the macrophase separation, the covalent bond avoids the latter to occur. Instead, block copolymers self-assemble into ordered micro-phases. This is today a well-established phenomenon, and the phase behavior as well as the domains size both in bulk and thin films has been described in terms of both interfacial tension between the blocks and entropic stretching energies of the two blocks [94]. Thus, block copolymer with a low interaction Flory–Huggins parameter or short chain lengths of the blocks will produce disordered phases. On the other hand, by assuming a large Flory–Huggins interaction parameter [95], the degree of polymerization and relative volume of each block f_X dictates the morphology obtained. Common periodic phases for a diblock copolymer with increasing f_A include lamellae (symmetric block copolymers), to cylinders or spheres (large asymmetric block copolymers) (Fig. 5.11). However, whereas this is particularly true in bulk, thin film with the presence of interfaces (air or substrate) introduces new variables that need to be considered. Some of them have been already introduced and can be equally applied for the case of block copolymers.

5.4.2 *Block Copolymer Composition and Microphase Separated Morphologies at Interfaces*

In addition to the microphase separation phenomenon, in the presence of an interface, the affinity of one of the blocks by the interface influences the final rearrangement of the block copolymer at the outmost surface as has been already reported, for instance by Coulon et al. [96] for the case of polystyrene-*b*-poly (methyl methacrylate) block copolymers (Fig. 5.12). Initially, upon spin coating the block copolymers are rather disordered due to the fast evaporation process. However, upon annealing reorganization occurs and nanometer scale phases rich in each of the components are observed. Finally, the difference in the surface energies of the components forces the orientation of these domains parallel to the surface, with the lower-surface-energy block located at the surface [96–98].

The initially favorable situation where the low surface energy component is in contact with the air interface can be modified if, for instance, the environment of exposure changes. In effect, surface rearrangement occurs when polymeric films of amphiphilic copolymers reorganize at the surface depending on the environment of exposure. As a result of changes in the hydrophilicity of the environment, the

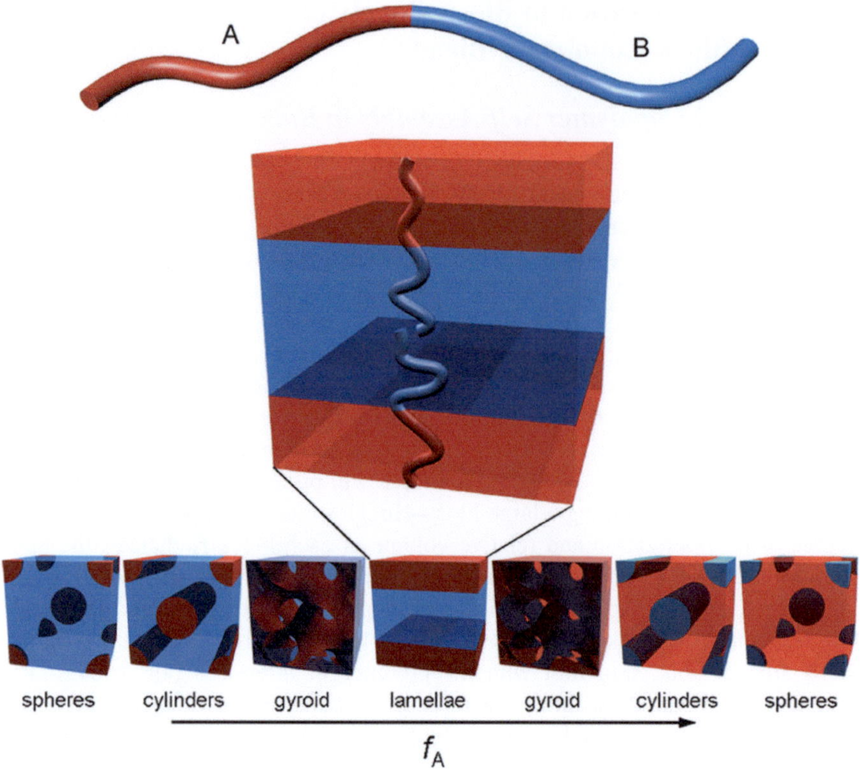

Fig. 5.11 Schematics of thermodynamically stable diblock copolymer phases. The A–B diblock copolymer, such as the PS-*b*-PMMA molecule represented at the *top*, is depicted as a simple two-color chain for simplicity. The chains self-organize such that contact between the immiscible blocks is minimized, with the structure determined primarily by the relative lengths of the two polymer blocks (f_A). Reproduced with permission from ref. [95]

moieties have the capability to exhibit either the low or the high energy functional groups at the interface. Actually, the interaction between the environment and the components as a function of their surface energy has been demonstrated in several works [99–101]. For instance, Nakahama and coworkers [102, 103] used this property to design materials that reconstruct in response to a change in the environment. In particular, the authors described the preparation of block copolymers based on a hydrophobic segment (polystyrene, polyisoprene or poly (4-octylstyrene)) and a hydrophilic poly(2,3-dihydroxypropyl methacrylate) (poly (DIMA)). As evidenced by different techniques (TEM, XPS and contact angle measurements) the as-cast film of poly(styrene-*b*-DIMA) exhibits the enrichment of polystyrene segments in the top surface region and that the polystyrene layer at the surface is substituted with the hydrated poly(DIMA) segment upon exposure of the film to water. The longer the exposure to water, the greater is the reconstruction, until only the hydrophilic block is in contact with the water. The reversal of the

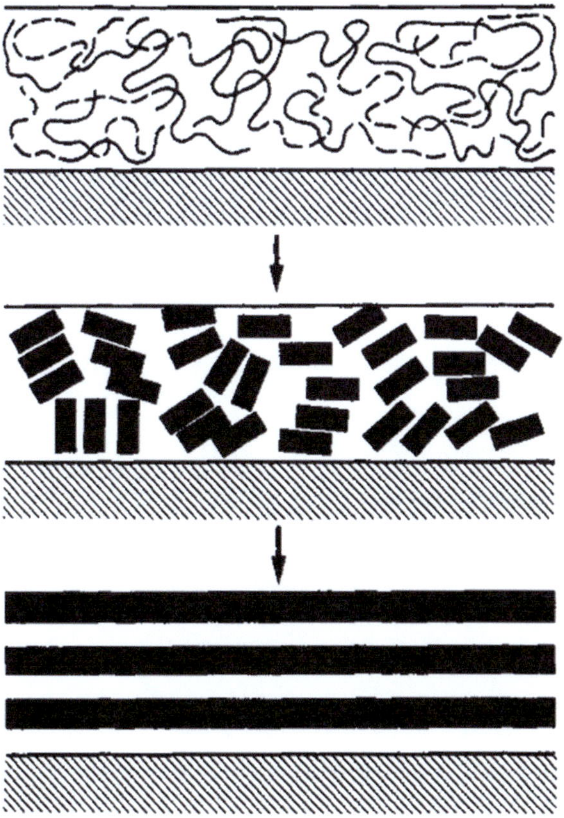

Fig. 5.12 Schematic of the surface-ordering process in a diblock copolymer. The *top* diagram shows the diblock copolymer in a phase mixed state. This is not accessible for PS/PMMA co-polymers via solvent-casting processes. The *center* diagram corresponds to a microphase-separated morphology where the periodic lamellar microdomains are randomly oriented in the specimen. After annealing for 24 h at 170 °C, the copolymer exhibits a lamellae morphology oriented parallel to the surface as shown in the *bottom* picture. Reprinted with permission from ref. [96]

hydrated surface back to the hydrophobic one through annealing treatment was detected by XPS.

Other examples include the use of segmented block copolymers rather and diblock copolymers [100, 104]. For instance, Pike et al. [104] of poly(dimethyl-siloxane-urea-urethanes) by dynamic contact angle. The authors immersed the polymer films in water and studied the time-dependent advancing and receding contact angles. They observed that whereas the advancing contact angles are relatively constant with immersion time, the receding contact angles decrease to some equilibrium value after a few days exposure to water. From these experimental results, they proposed a mechanism in which the hard block urethane-urea domains migrate through the soft block silicone to the polymer–water interface.

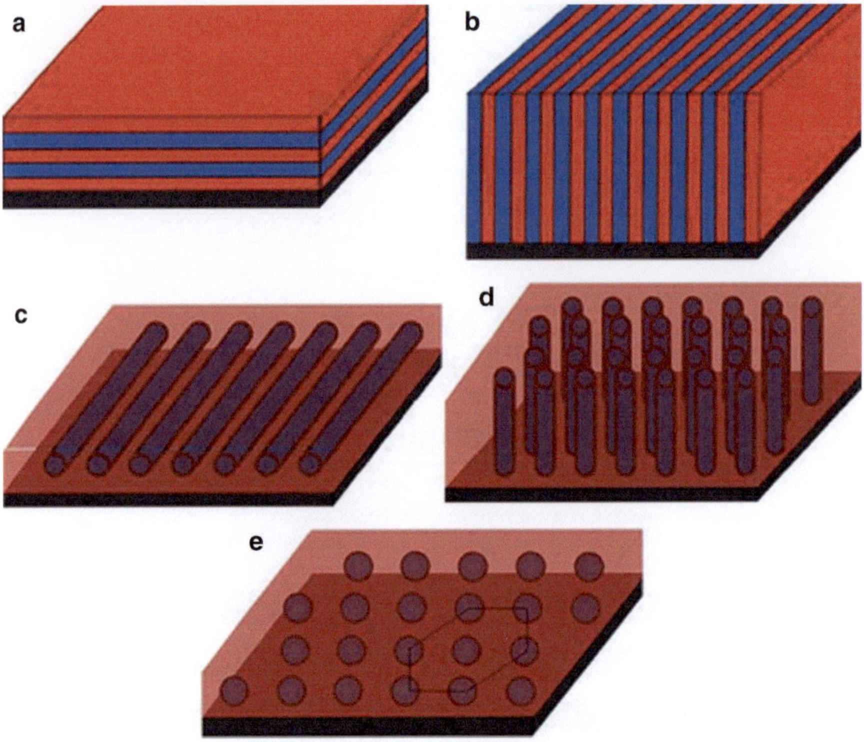

Fig. 5.13 When confined to a thin film, the orientation of block copolymer domains with respect to the substrate surface is crucial for many applications. (**a**) Lamellae lying parallel to the substrate, (**b**) lamellae aligned perpendicular, (**c**) cylinders lying parallel, (**d**) cylinders perpendicular, and (**e**) spheres. In the case of lamellae in the perpendicular orientation and cylinders in parallel, lines can be patterned if the persistence length of the structure can be controlled. In the case of upright cylinders and spheres, the grain size and perfection of the hexagonal array is of primary importance. Reproduced with permission from ref. [105]

Block copolymers in thin films tend to self-assemble in different structures depending on the volume fraction of the blocks (f_X) but also on the interface. In the case of an A-B diblock copolymer and using the volume fraction of the A block (f_A) as reference, we can observe the formation of spheres in a body-centered cubic (BCC) lattice surrounded by a matrix of B when the f_A is small. An increase of f_A up to 0.5 will induce the formation of lamellae. Intermediate values will lead to a hexagonal or double gyroid structure. When considering thin films (fewer than 100 nm) the population of polymer molecules close to the interface is significantly higher than in bulk and therefore interfaces play a central role in the orientation and order of the microdomains. Hence, the nanodomain formation is strongly influenced by the affinity of each block with the surface [105]. As a consequence, similar structures can exhibit different orientation to the substrate surface (Fig. 5.13). Thus, cylinders (hexagonal packing) and lamellae can be either parallel or perpendicular to the surface. The final orientation of these structures will determine the domain of application. As an example, perpendicularly oriented cylinders may be of interest for data storage whereas cylinders lying parallel to the surface can be on interest to prepare nanowires. It is outside the scope of this chapter to provide a thorough revision of this field but rather introduce the key aspects on thin film formation. For a detailed discussion in this field the readers are referred to previous contributions [94, 105–109].

5.4.3 Role of the Film Thickness on the Thin Film Morphology

The film thickness determines the particular arrangement of the block copolymer. As an example of the role of the film thickness on the morphology, Knoll et al. [110] studied the case of polystyrene-*block*-polybutadiene-*block*-polystyrene (PS-*b*-PB-*b*-PS) with tendency to form cylindrical microdomains. This group observed, parallel orientations of PS cylinders in thin films with a thickness of 2–3 domain spacing. On the contrary, perpendicular orientation may be favored under controlled conditions of solvent evaporation in spin cast films. More interestingly, a gradual increase of the thicknesses led to a terrace morphology containing different block copolymer structures (Fig. 5.14). They succeeded in the construction of a diagram having these main features. First of all, perpendicular cylinders were observed in the thinnest films and in transition regions between terraces of parallel cylinders. There is a transition from perpendicular cylinders to a perforated lamellar structure upon increasing polymer concentration in the solvent chloroform. This is driven by the segregation of PB to both film surfaces, leading to its depletion in the center of the film and hence the formation of PB-perforated PS lamellae.

5.4.4 Phase Separation in Block Copolymer–Homopolymer Blends

The theory and experimentally structures exhibited by homopolymer–copolymer (A/A-B) blends were reported by De Gennes [111] and Gallot et al. [112] based on previous studies of Hashimoto et al. [113–115] who proposed a classification in terms of solubility of the copolymer brushes by the homopolymer. They established three different situations depending on the molecular weight of the homopolymers in the blend [113–115] (Fig. 5.15). They defined a situation (I) in which the homopolymer (generally a low molecular weight) can be completely or "uniformly" solubilized into the A block domains of similar chemical nature. In situation (II) the solubilization of the homopolymer in the block copolymer

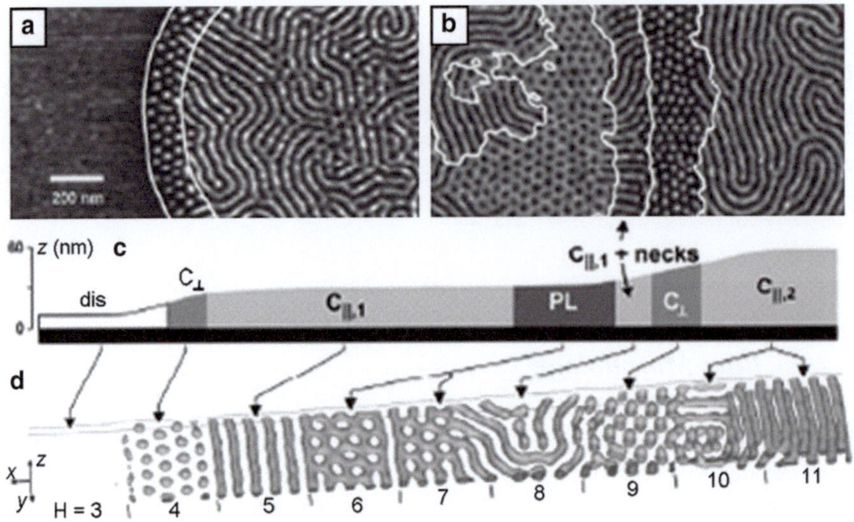

Fig. 5.14 (**a**, **b**) TM-SFM phase images of thin SBS films on Si substrates after annealing in chloroform vapor. The surface is everywhere covered with a ~10-nm-thick PB layer. Bright (*dark*) corresponds to PS (PB) microdomains below this top PB layer. (**c**) Schematic height profile of the phase images shown in (**a**, **b**). (**d**) Simulation of an $A_3B_{12}A_3$ block copolymer film in one large simulation box with increasing film thickness. Reproduced with permission from ref. [110]

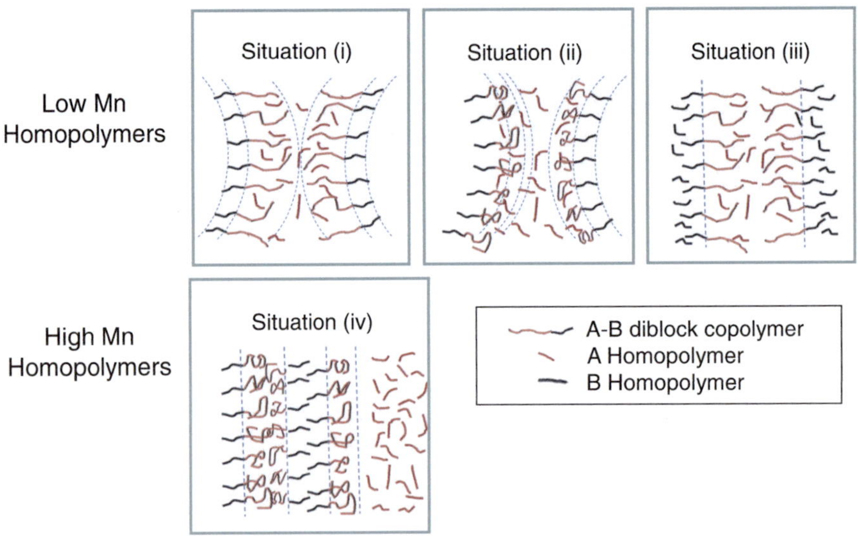

Fig. 5.15 Morphologies obtained in homopolymer–block copolymer blends depending on the molecular weight of the homopolymer employed. Reproduced with permission from ref. [120]

occurs only to a limited extent, i.e., we speak about partial or "localized" solubilization. In these two situations the increase of the volume of the phase A changes the interfacial curvature thus leading to a transition between different morphologies. As a result of the incorporation of a low molecular weight homopolymer we can elaborate materials with morphologies ranging from lamellar phase or the intriguing OBDD (ordered bicontinuous double-diamond) to hexagonal lattices exclusively depending on the % of homopolymer introduced in the initial blend [116]. When A and B homopolymers are mixed with the block copolymer forming a ternary blend both can be solubilized in each A and B blocks. As a consequence, depending on the proportion of homopolymer/block copolymer introduced, the shape of the interface can be either maintained or modified (Situation III). The interfacial distribution of the block copolymer, i.e., the distance between two block copolymer chains will increase as a consequence of the incorporation of homopolymer in both phases. Finally, DeGennes and Gallot described a situation (IV) in which the low solubility of the homopolymer induces the formation of particular structures combining macrophase separated domains rich in diblock copolymer (in which additionally microphase separation may be observed) randomly distributed and areas formed by the homopolymer host. Situation (IV) is generally found in blends with diblock copolymers with large incompatibility between the blocks (given by the Flory–Huggins parameter (χ)) or in blends in which the molar masses of homopolymer and diblock copolymer differs significantly [117–119].

Provided a large difference in molecular weight between the block copolymer and the matrix, we have to take into account that the relative ratio between the segments in the block copolymer plays an important role in the final morphology. As depicted in Fig. 5.16, typically, for AB block copolymers with similar volumes of block A and B we observe the formation of a lamellar phase that turns into a so-called "onion ring" pattern when blended with a A homopolymer. On the contrary, asymmetric block copolymers with larger A block tend to produce micellar aggregates randomly distributed in the homopolymer phase.

5.5 Surface Segregation in Block Copolymer Blends: Self-Assembly and Nanostructuration at the Interface

Block copolymers have been usually incorporated in polymer blends among others to improve the compatibility of different blend materials. Therefore, the studies carried out have been mainly focused on bulk. As a consequence, the study of polymer blends surfaces having block copolymers have received limited attention.

Studies concerning functional surfaces produced by surface segregation of copolymers typically involved the use of either double hydrophobic copolymers [121, 122] or amphiphilic block copolymers [123–126] as additives to be placed at the surface. These copolymers are designed to contain hydrophobic segments

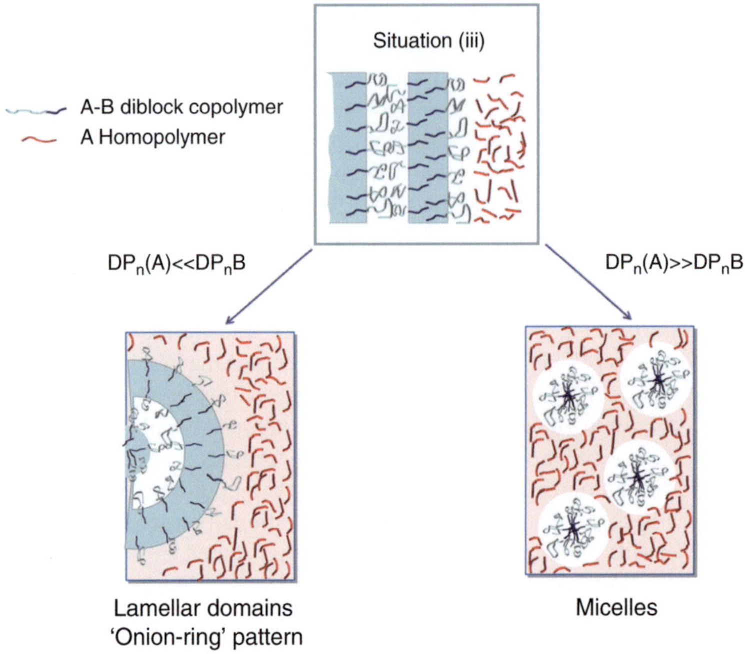

Fig. 5.16 Structures obtained depending on the volume fraction of the blocks. Reproduced with permission from ref. [120]

similar in nature to the polymer matrix employed. Examples of amphiphilic block copolymers employed include polystyrene-*block*-poly(acrylic acid), polystyrene-*block*-poly(L-glutamic acid) or polystyrene-*block*-poly(2-[2-(2-methoxyethoxy) ethoxy]ethyl methacrylate). In the first two examples, the hydrophilic groups were placed at the surface by annealing in water vapor. The latter exhibits a particular feature. Yokohama et al. used a block copolymer of polystyrene (PS) and 2-[2-(2-methoxyethoxy)ethoxy]ethyl methacrylate (PME3MA) that spontaneously exposes the PME3MA block, which was soluble in water, to the surface in a vacuum. PS-*b*-PME3MA mixed with polystyrene (PS) segregated to the PS surface and changed the hydrophobic PS surface into hydrophilic surface. [123, 124] As depicted in Fig. 5.17b, segregation of dPS-*b*-PME3MA to the surface of PS provides a surface which is apparently hydrophobic but becomes hydrophilic upon contact with water. More interestingly, the segregation of dPS-PME3MA occurs even in hydrophobic environment and persists after prolonged annealing in a vacuum or air. The authors claimed that the methyl termini of the side chains that exhibit low surface energy are the cause of this observation. Those methyl termini cover the surface, and the ethylene oxide part of the side chains are screened out from the surface. Once the surface is exposed to water, the surface reconstruction occurs quickly, and the wetting behavior of water changes drastically.

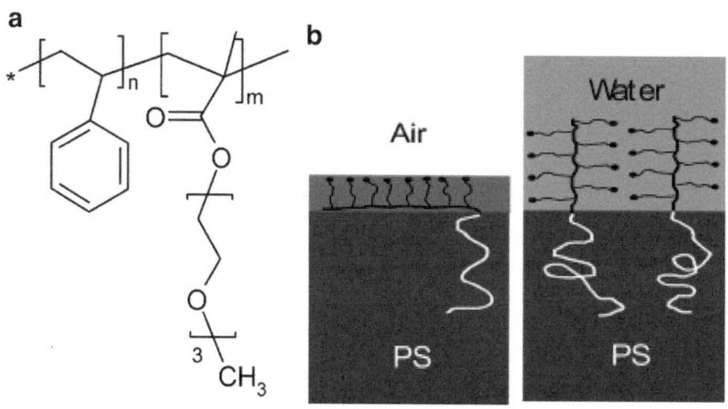

Fig. 5.17 (**a**) Chemical Structure of Poly{(deuterated styrene)-*block*-2-[2-(2-methoxyethoxy) ethoxy]ethyl methacrylate}(dPS-*b*-PME3MA). (**b**) Schematic pictures of the conformations of dPS-*b*-PME3MA in a mixture with PS in air (or in a vacuum) and in water. At the surface in a vacuum or air, the terminal methyl groups of the side chains, *filled circle*, cover the surface. In water, the water-soluble PME3MA blocks are anchored onto the surface of polystyrene and stretching out into water. PME3MA blocks cover the surface in either hydrophobic or hydrophilic environment

The incorporation of block copolymers instead of single functional groups and the use of a polymer matrix with a relative high Tg exhibits two major advantages. First, the surface composition long-term stability is assured by the high glass transition temperature of the polymer (for example polystyrene) used as a matrix with very low chain mobility at room temperature. Second, since we are introducing segments rather than single functions, the system may form a microphase segregated interface between the two incompatible blocks that limits, at least to some extent, further rearrangement of the surface. The latter, i.e., the particular rearrangement of the surface segregated block copolymers at the interface has also more interesting consequences. As mentioned in the previous paragraph, the addition of block copolymers into a polymer matrix modifies the structure of the blend either altering the microstructure or inducing the formation of both microphase and macrophase separated domains. This phenomenon, largely described for polymer blends in bulk has received limited attention in surfaces. In this section we introduce some examples on polymer blends containing block copolymers in which the block copolymer can self-assemble at the interface producing nanostructured and eventually microstructured domains.

A crucial aspect on polymer blends containing block copolymers concerns the concentration of block copolymer both in bulk and at the surface. The former can be controlled in the design of the initial blend and the latter depends not only on the amount initially introduced in the blend but also on the eventual surface segregation. In Fig. 5.18 is schematically illustrated the block copolymer arrangement of binary blends with variable amount of block copolymer. At low concentrations (a, b) the block copolymer can reside either in the bulk or at the interface and the amount of block copolymer segregated at the interface is directly related to the

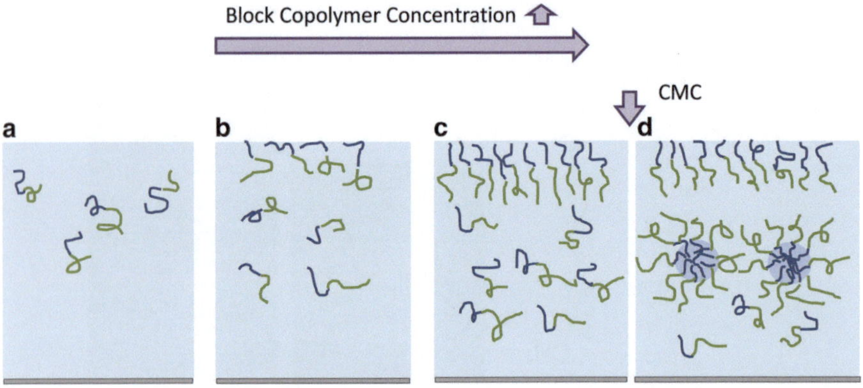

Fig. 5.18 (**a**) Schematic illustration of segregation mechanism as a function of the block copolymer concentration within the blend. At low concentrations (**a**, **b**), the segregation isotherm linearly depends on the concentration in bulk. Further segregation is prevented in (**c**) as a consequence of the brush layer at the surface. Above this concentration (**d**) micelles start to form. Adapted from ref. [123]

block copolymer in the bulk. An increase of the amount of block copolymer in the initial blend will eventually lead to a saturation of the block copolymer concentration at the surface (c). From this point, addition of copolymer in the blend will lead to formation of micelles in the bulk. This effect indicates that self-assembly by reorganization of block copolymers at the surface can produce nanostructures above the so-called critical micelle concentration (CMC).

As a consequence of the surface self-assembly different microdomain morphologies can be formed. These morphologies can be observed directly after film formation or may be formed upon segregation of block copolymers to preexisting interfaces [125, 126]. According to Shull et al. [127] two aspects play a major role on the segregation of block copolymer micelles to the interfaces. On the one hand, segregation is the result of the attractive interactions between the polymer brushes and a free surface. On the other hand, both surface and interface segregation of block copolymer micelles requires high molecular weight homopolymer matrix phases [125]. Taking into account this molecular weight requirements surface segregation has been employed, for instance, by Ibarboure et al. [128] to both functionalize and nanostructure interfaces of homopolymer–*block* copolymer blends. The use of rod-coil block copolymers, i.e., poly(L-glutamic acid)-*block*-polystyrene affords different patterns depending on the block copolymer composition (Fig. 5.19). Short polypeptide blocks are unable to stabilize the formation of α-helical secondary structures and form either β-sheet or coil structures. As a consequence, block copolymers composed by polypeptide chains below 20 units mixed with homopolystyrene tend to form micelles within the homopolystyrene phase. In the case of longer polypeptides that stabilize the α-helical structure, a rod-like structure is formed. As a result of these structures and in spite of the asymmetrical composition of the block copolymers, lamellar structures were observed.

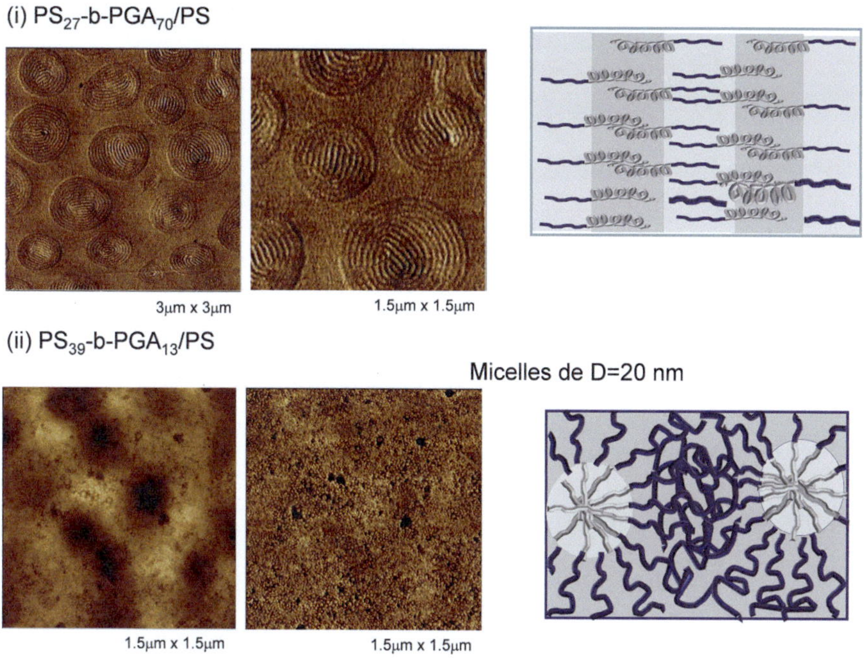

Fig. 5.19 AFM images of polymer blends having (*i*) 30 wt of diblock copolymer PS$_{27}$-*b*-PGA$_{70}$ and 70 % of polystyrene and (*ii*) 20 % of diblock copolymer PS$_{49}$-*b*-PGA$_{17}$ and 80 % of polystyrene. The diblock copolymer employed is asymmetric having either a larger hydrophilic polypeptide block (*i*) or a larger hydrophobic polystyrene block (*ii*). Reproduced with permission from ref. [128]

Similarly, Ansari et al. [121] described the synthesis of polystyrenes functionalized with multiple fluorocarbon groups with are known to be highly surface-active in blends with homopolymer polystyrene (hPS). This additive is able to diffuse to the polymer air interface and the fluorocarbon groups enhanced the hydrophobicity of the surface. The authors established a correlation between the surface properties and the bulk behavior founding parallels between the aggregation and adsorption of these functionalized polymers in blends similarly to the model proposed among others by Yokohama and Shultz. Moreover, the amount of fluorocarbon within the additive plays a key role on the surface activity. More precisely, the surface activity of the additives employed is enhanced by increasing the amount of fluorocarbon within the additive. In addition, depending on the molecular structures the surface segregated additives formed particular nanometer size morphologies. Over the molecular weight range considered, polystyrene functionalized with two C_8F_{17} fluorocarbon groups was fully miscible with *h*PS but appeared to form soluble micelles at concentrations above 12 % (w/w). Polystyrenes functionalized with more than two C_8F_{17} groups were only partially miscible with *h*PS and formed multilamellar vesicles.

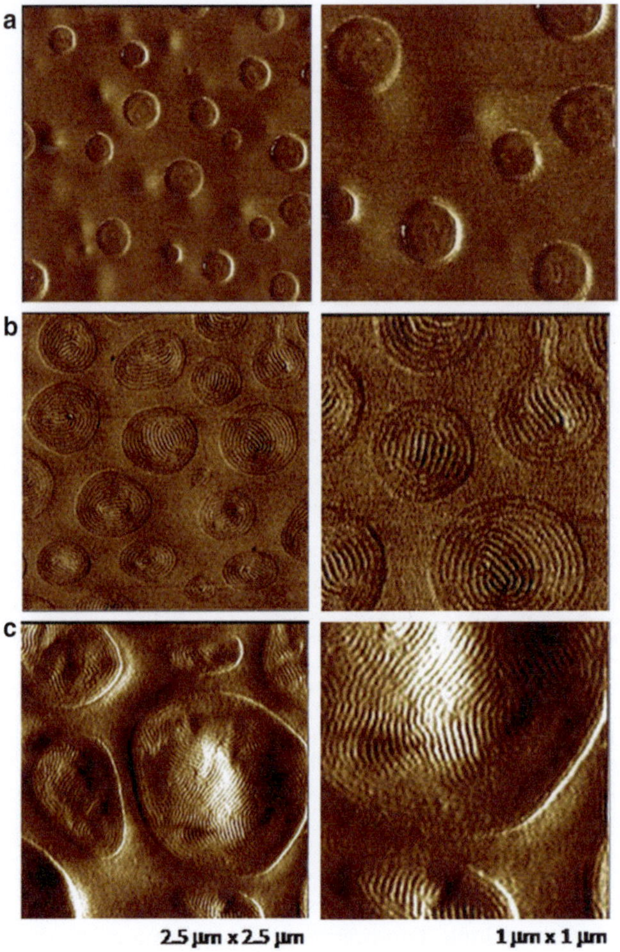

Fig. 5.20 Dependence of the macrophase separation on the block-copolymer concentration within the blend. (**a**) 15 wt%, (**b**) 30 wt%, and (**c**) 60 wt% of diblock copolymer within the blend. Reproduced with permission from ref. [128]

The polymer concentration may not only influence the formation of microphase separated structures but also the dimension of the block copolymer domains at the interface. For instance, Ibarboure et al. [128] elaborated nanostructured surfaces composed of polystyrene-*block*-poly(L-glutamic acid) (PS_{27}-*b*-PGA_{70}) mixed with a high molecular weight linear polystyrene homopolymer (PS_{600}). As depicted in Fig. 5.20, after spin coating from THF solutions, the surface exhibits diblock copolymer rich circular domains in a homopolymer-rich disordered phase in which the diblock copolymer, i.e., PS_{27}-*b*-PGA_{70}, appears to form alternating lamellar microdomains of PS and PGA blocks. The size of the diblock copolymer rich domains has been found to vary with the quantity of diblock copolymer within the blend. While the mixtures having 15 wt% of diblock copolymer lead to microphase separated domains with sizes around 200–350 nm, the mixtures charged with 30 wt% afford 400–600 nm domains. The areas formed with major proportion of diblock copolymer (60 wt%) are rather large (several microns) and polydisperse.

In addition to the nanoscale features produced by interfacial self-assembly of block copolymers other patterning approaches can be additionally combined to induce microscale features. As an example, the elaboration of micrometer size patterned surfaces by UV-light lithography was reported where, in addition, the surface chemical composition can be controlled by surface segregation of a fluorinated copolymer, i.e., poly(methyl methacrylate-*co*-2,2,2-trifluoroethyl methacrylate) (P(MMA-*co*-TFMA)) incorporated in the photopolymerizable mixture: methyl methacrylate (MMA) and ethylene glycol dimethacrylate (EGDMA) [122]. The authors evidenced that the surface composition can be modified depending on such factors as with the environmental conditions or the concentration of copolymer in the blend. Moreover, the surface wettability of the copolymer is enhanced by the surface pattern created. As a consequence, the wettability of the films can be modified depending on the pattern and composition of the blend.

The lateral pattern resolution was affected by the incorporation of P(MMA-*co*-TFMA) and the amount of cross-linking agent (EGDMA) employed. The height difference between the non-irradiated and the irradiated areas versus the amount of cross-linking agent (EGDMA) are represented in Fig. 5.21a. Equally, Fig. 5.21a depicts the image profiles obtained by SEM of the samples with compositions MMA/EGDMA 85:15, 90:10, and 95:05 in which two different amounts of P (MMA-*co*-TFMA) were added (10 and 20 wt%). Moreover, the films obtained from mixtures with increasing amount of EGDMA exhibit better resolved patterns since the depth increases with the amount of EGDMA. The height difference between the irradiated and non-irradiated areas increase with the percentage of EGDMA, as can be observed in the graph of the Fig. 5.21a where in the 10 wt% of copolymer series the average highs are 8.1 μm, 9.5 μm, and 10.8 μm while in the 20 wt% series these highs are 8.7 μm, 10.4 μm, and 13.0 μm, respectively. An increase of the EGDMA amount raises the viscosity of the photopolymerizable mixture thus limiting the diffusion of the monomers and improving the depth of the pattern created.

5.6 Stimuli Responsive Interfaces Obtained by Surface Segregation

Surface segregated block copolymers have been fabricated with responsive segments that allow them upon surface segregation to vary the surface behavior as a function of the external stimuli. Examples of responsive segments incorporated in the block copolymer design include poly(acrylic acid) or synthetic polypeptides

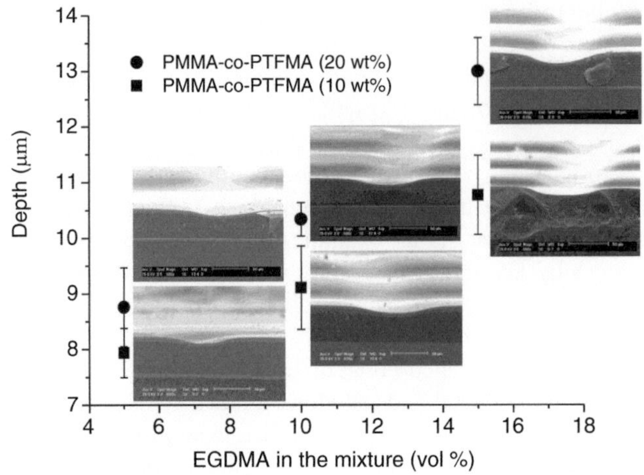

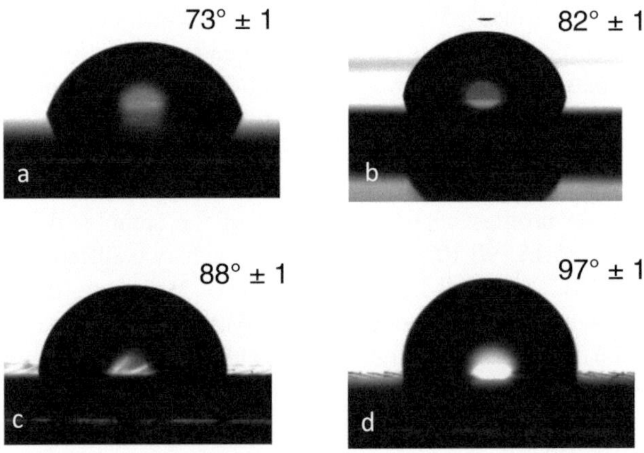

Fig. 5.21 (*A*) Relation between the percentage of EGDMA and the height differences between the exposed and non-exposed areas. *Squares* correspond to a series of mixtures MMA/EGDMA with 10 wt% of fluorinated copolymer and *circles* to the same mixtures but with 20 wt% of fluorinated copolymer. *Insets*: SEM cross section images of each pattern. (*B*) Contact angle measurements of non-structured films of PMMA (**a**), p(MMA-*co*-TFMA) (**b**), a microstructured network from a mixture p(MMA/EGDMA) 85:15 with 10 wt% of p(MMA-*co*-PTFMA) and either a hydrophilic cover (**c**) or a hydrophobic cover (**d**). In both experiments photomask with 100 μm hole has been used. Reproduced with permission form ref. [122]

(poly(L-glutamic acid) or poly(L-lysine)) which are pH responsive and poly (dimethylaminoethyl methacrylate) with is simultaneously pH and temperature responsive.

For instance, surface segregation in water vapor of the blend films prepared with polystyrene-*block*-poly(L-glutamic acid) PS_{27}-*b*-PGA_{20} improved the reorientation of the hydrophilic polypeptide block at the interface (Fig. 5.22).[129] Water contact angle measurements supported the surface enrichment of polypeptide segments as a

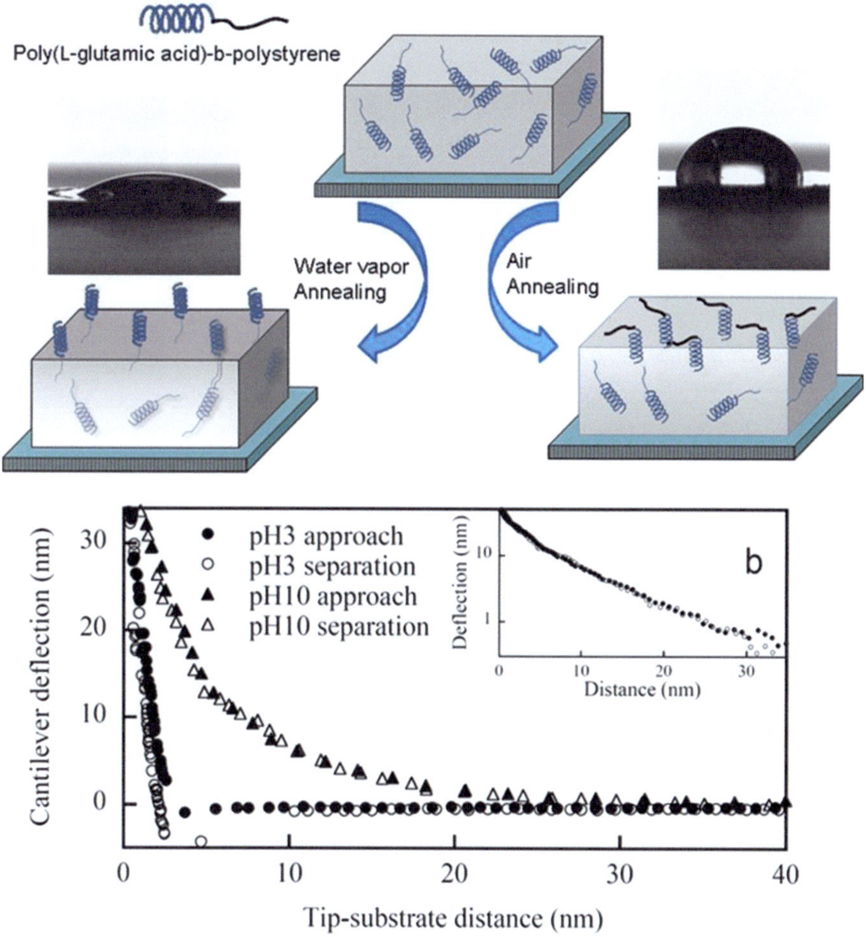

Fig. 5.22 *Above*: Scheme of a polymer blend film containing 20 wt% of diblock copolymer PS_{27}-*b*-PGA_{20} either annealed to air (*right*) or annealed to water vapor (*left*). *Below*: Normal tip–surface interaction force between the Si_3N_4 AFM tip and a water annealed surface at different pH values. The semilog representation of the force measured for the water annealed surface (*inset*) illustrates the long-range exponential decay of the repulsive force. Reproduced with permission from ref. [129]

consequence of the humidity treatment. Hence, whereas the surfaces exposed to air exhibit water contact angles of 90°, the films treated in water vapor are rather hydrophilic, with contact angles below 50°. More interestingly, atomic force microscopy (AFM) by measuring the interaction force between a silicon nitride AFM tip and the substrates provided further insight on the pH responsive character of the surface. Figure 5.21 below represents the interaction force profile between the Si_3N_4 AFM tip and the poly(L-glutamic acid) functionalized surfaces annealed to water vapor (b) measured under water at different pH values. A long-range

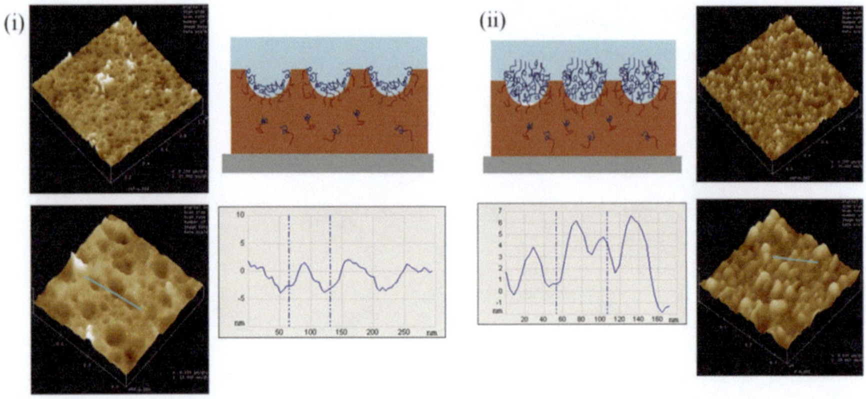

Fig. 5.23 AFM height images of PAA-*b*-PS/PS blends upon annealing to water vapor (during 3 days at 95 °C) obtained in aqueous media at two different pH values: (*i*) pH 3.0 and (*ii*) pH 7.0. *Right* images above are 1×1 μm^2, *below* images are 0.35×0.35 μm^2, and the height scale is 0–10 nm. As depicted in the cartoons, depending on the environmental pH values, the PAA either protonated or dissociated collapse or stretch respectively resulting in the formation of nanometer sized holes or grains. Reproduced with permission from ref. [130]

exponential repulsion, typical of the interaction between charged surfaces was observed at basic pH values as a consequence of the repulsion between the tip and the surface. However, this repulsion disappeared at low pH values below the isoelectric point of Si_3N_4—where the tip is positively charged but the carboxylic acid groups are neutralized—illustrating the responsive character of the surfaces.

Changes in the environmental pH can be also traduced in morphological changes at the surface. Particularly interesting is the case when micellar assemblies migrate towards the interface and modify their structure as a function of the environmental pH. In this sense, PS-*b*-PAA block copolymers have been employed as additives in blends with PS and annealed to water vapor in order to reorient the PAA block towards the interface as is schematically shown in Fig. 5.23 [130]. Poly(acrylic acid), is a soft acid, can exist either as a charged anion at pH values above 4.5 or as a protonated neutral group below 4.5. Thus, in principle depending on the pH of the water droplet, the wettability of the annealed surfaces varied accordingly.

More interestingly, the exposure of the functionalized surface to a different pH changed the surface morphology at the nanoscale dramatically. At low pH values holes with characteristic dimensions of 35–45 nm could be observed (Fig. 5.23i). On the contrary, above the p*K*a, the number of dissociated carboxylic acid groups increased producing a greater repulsion between the acrylic acid monomer units. As a consequence, at pH > 7.0 the chains become fully dissociated which results in the maximum degree of swelling of the core of the micelles and formed nanometer scale grains.

In addition to pH responsive block copolymer, other multifunctional stimuli-responsive additives that, via surface segregation, decorate the interface have been also reported. [131] As an example an amphiphilic diblock copolymer,

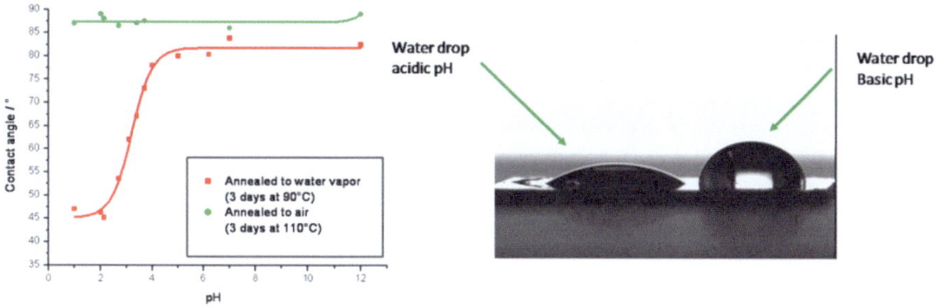

Fig. 5.24 *Left*: Contact angle measurements of a blend containing 20 wt% of (3) and 80 wt% of PS as a function of the pH annealed to air (*green curve*) or to water vapor (*red curve*). *Right*: Water drops at either acidic pH or basic pH on a polymer surface annealed to water vapor. Reproduced with permission from ref. [131]

i.e., PS-*b*-poly(*N*,*N*′-dimethylaminoethylmethacrylate) (PS-*b*-PDMAMEMA) or PS-*b*-poly(*N*,*N*′-diethylaminoethylmethacrylate) (PS-*b*-PDEAEMA) was blended with a homopolystyrene employed as polymer host. The surface prepared using these two block copolymers as additives exhibit both thermal and pH responsive behavior. More precisely, the pH response of the pristine surfaces and the water vapor annealed films was first investigated by contact angle measurements employing water drops at different pH values. Figure 5.24 depicts the water contact angle values obtained as a function of the pH of the drop between pH = 1 and 12. Whereas the pristine films exhibit contact angle values of about 86–90° independently of the pH of the water drop, the contact angle values of the samples exposed to water vapor remain constant above pH 4.5 at values of 83–85° but decreased significantly below this pH to values around 45°. Moreover, the PDMAEMA exhibit a lower critical solubility temperature (LCST). Below the LCST the polymer is soluble in aqueous solution whereas above the LCST the polymer chains prefer to form intramolecular interactions and precipitate in water. The surface wettability of these films was investigated by means of the Wilhelmy technique that provides values about the decrease of the surface tension. For that purpose, spin coating of the polymer blends of diblock copolymer and homopolystyrene on two plates of similar size were bonded by the opposite faces. The bath temperature was varied between 20 and 60 °C, below and above the LCST of PDMAEMA and maintained at high pH values (above the p*K*a). For the non-annealed samples at temperatures above the LCST, the surface tension decreases as a consequence of the enhanced hydrophobicity of the surface. At lower temperatures, the same surface exhibit larger surface tension values, indicating higher surface hydrophilicity.

Variations on the environment of exposure, for instance, between hydrophilic water vapor or dry air will also induce responses at the surface level [100, 132]. The response to environmental changes has been typically followed contact angle measurements during successive annealing treatments either to water or to dry air [133]. Figure 5.25 shows the contact angles obtained as a function of the exposure

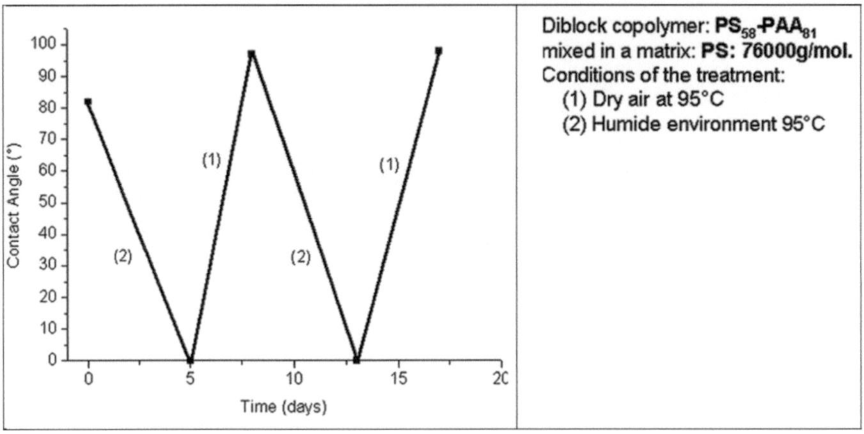

Fig. 5.25 Contact angles measured after successive annealing either to air at 95 °C (1) or to water at 95 °C (2). Reproduced with permission from ref. [133]

time. Whereas water contact angles decrease for material annealed in contact with water, they perfectly recover after annealing in air. Consecutive treatments with dry air/humid vapor produced hydrophilic and hydrophobic surfaces and evidenced the reversibility of the surface rearrangement.

Not only the surface wettability but also the morphology has appeared to be largely influenced by the environment of exposure. As an example of the microstructural changes at polymers blend surfaces, an example is include in which a block copolymer a hydrophilic polypeptide, poly(L-glutamic acid) (PGA) able of changing back and forth the secondary structure between rigid rod-like structures to disordered random-coil. It is well known that although PGA forms preferentially α-helical structures in bulk, films exposed to solvents, such as water or TFA disrupt the structure and a transition is observed to a rather random-coil structure. This ability has been employed to finely tune the structuration within the diblock copolymer rich droplets. Films were exposed to water vapor in a closed vessel and kept at 90 °C during 2 days and rapidly cooled to room temperature to fix the structure and dried under vacuum. As depicted in the AFM images of Fig. 5.26, the structure observed in the microphase separated domains changes from a lamellar phase to either cubic or a perpendicular oriented hexagonal phase. According to the authors, during annealing to water vapor, water (pH 7) is partially condensed at the surface and deprotonates, at least partially, the carboxylic acid functional groups (p*K*a 4.8) of PGA. Repulsion between negatively charged side chain groups leads to monomer repulsion, the α-helical structure is disrupted and the polypeptidic hydrophilic domains change from a rod-like α-helical structure obtained directly after spin coating to a more extended random coil conformation. Hence, deprotonation by water of hydrophilic block involves changes in the secondary conformation but also the hydrophilic/hydrophobic volume ratio is altered which may is turn responsible of the variations on the morphology formed.

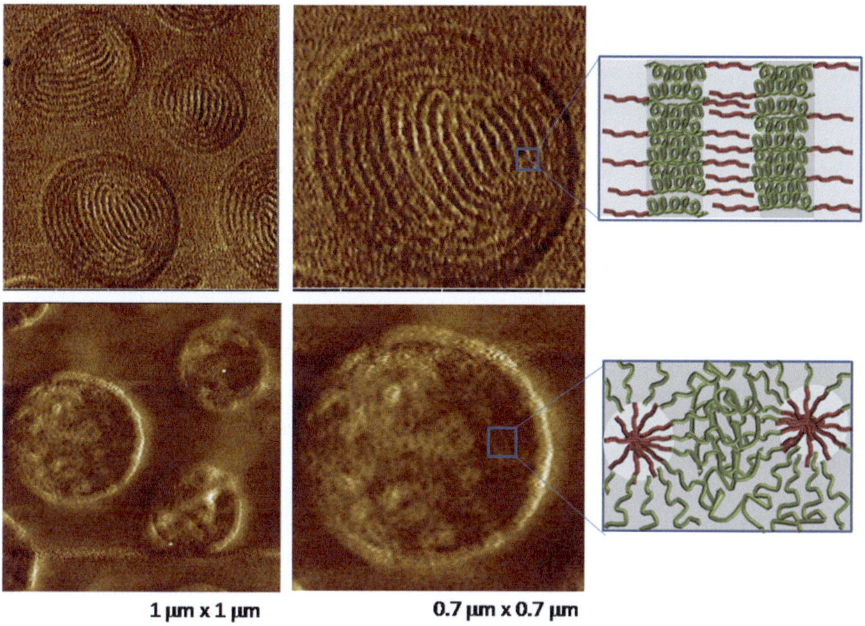

Fig. 5.26 The AFM images (recorded in tapping mode) evidenced the morphological changes on the local nanostructure by annealing the film to water vapor. The lamellar structure (*top*) is transformed into a cubic phase (*bottom*). Reproduced with permission from ref. [128]

5.7 Conclusions and General Remarks

This chapter summarizes the potential of surface segregation in order to vary the surface chemical composition based on thermodynamical parameters. As a result, materials containing at least two constituents, one being of higher surface energy than the other will evolve towards a state where the interfacial tension is minimized. Minimization of the surface energy requires overcoming entropic forces and drives the movement of one of the components towards the material surface. So that, in principle, and as has been reviewed in this chapter one can tailor the surface properties of a polymer film.

This chapter also discusses those factors that play a key role on the segregation process that can be classified as entropic and enthalpic factors. Whereas entropic factors include those related with the macromolecular structure (molecular weight or degree of branching), enthalpic factors involve the chemical functionality of the components in the polymer blend. In addition, the environment of exposure can dramatically modify the segregation depending on the polarity. Exposure of a polymer surface to a moist atmosphere resulted in an interfacial enrichment in the polar component. On the contrary, vacuum or dry environments favor the interfacial segregation of the less polar additive.

Several major advantages of using surface segregation in comparison with other surface modification approaches include the control of type and density of the functional groups at the surface or the presence of a "reservoir" of the functional component remaining in the bulk so that damaged surfaces may be repaired simply by annealing the film.

At the same time, based on the microphase separation occurring in block copolymers blends, including the latter may result in phase separated films exhibiting nano-domains. The formation of nanostructured domains at the interface induced by surface segregation of block copolymers is an interesting strategy to fabricate surfaces with controlled functionality and patterned in one single step.

Since changes on the environment can significantly affect the thermodynamics at the surface level structural reorganization may occur. Hence, reversible changes on the surface properties of these materials can be easily accomplished, thus, creating switchable materials with controlled surface wettability, charge, adhesion, and chemical functionality.

Surface segregation is today a well-known phenomenon, and the factors involved in this thermodynamic process are well established. Nevertheless, their use to functionalize surface or even form nanostructured materials has received limited attention. We believe this approach can be an excellent alternative to modulate the surface properties of polymeric materials in multiple applications requiring a permanent surface chemical composition or to elaborate responsive surfaces in which changes on the surface properties reflect precise variations on the environmental conditions.

Acknowledgment JRH would like to thank financial support from Ministerio de Economía y Competitividad (MINECO) (Projects MAT2010-426 17016, MAT2010-21088-C03-01and MAT2013-47902-C2-1-R).

References

1. Li, D., Zheng, Q., Wang, Y., Chen, H.: Combining surface topography with polymer chemistry: exploring new interfacial biological phenomena. Polym. Chem. **5**, 14–24 (2014)
2. Ross, A.M., Lahann, J.: Surface engineering the cellular microenvironment via patterning and gradients. J. Polym. Sci. B Polym. Phys. **51**, 775–794 (2013)
3. Ward, L.J., Badyal, J.P.S., Goodwin, A.J., Merlin, P.J.: Solventless coupling of perfluoroalkylchlorosilanes to atmospheric plasma activated polymer surfaces. Polymer **46**, 3986–3991 (2005)
4. Cheng, T.S., Lin, H.T., Chuang, M.J.: Surface fluorination of polyethylene terephthalate films with RF plasma. Mater. Lett. **58**, 650–653 (2004)
5. Selli, E., Mazzone, G., Oliva, C., Martini, F., Riccardi, C., Barni, R., et al.: Characterisation of poly(ethylene terephthalate) and cotton fibres after cold SF6 plasma treatment. J. Mater. Chem. **11**, 1985–1991 (2001)
6. Sigurdsson, S., Shishoo, R.: Surface properties of polymers treated with tetrafluoromethane plasma. J. Appl. Polym. Sci. **66**, 1591–1601 (1997)

7. Kawase, T., Sawada, H.: End-capped fluoroalkyl-functional silanes. Part II: modification of polymers and possibility of multifunctional silanes. J. Adhes. Sci. Technol. **16**, 1121–1140 (2002)
8. Kawase, T., Yamane, M., Fuji, T., Minagawa, M., Sawada, H., Moriya, Y.: Fluoroalkylation of polyester by end-capped fluoroalkyl-functional silanes. J. Adhes. Sci. Technol. **11**, 1381–1397 (1997)
9. Biltresse, S., Descamps, D., Boxus, T., Marchand-Brynaert, J.: Attachment of bis-(trifluoromethyl)aryl labels onto the chain ends of poly(ethylene terephthalate) (PET) track-etched membranes and films by surface wet chemistry. J. Polym. Sci. A Polym. Chem. **38**, 3510–3520 (2000)
10. Saidi, S., Guittard, F., Guimon, C., Geribaldi, S.: Fluorinated comblike homopolymers: the effect of spacer lengths on surface properties. J. Polym. Sci. A Polym. Chem. **43**, 3737–3747 (2005)
11. Saidi, S., Guittard, F., Guimon, C., Geribaldi, S.: Low surface energy perfluorooctyalkyl acrylate copolymers for surface modification of PET. Macromol. Chem. Phys. **206**, 1098–1105 (2005)
12. del Campo, A., Arzt, E.: Fabrication approaches for generating complex micro- and nanopatterns on polymeric surfaces. Chem. Rev. **108**, 911–945 (2008)
13. Minnikanti, V.S., Archer, L.A.: Entropic attraction of polymers toward surfaces and its relationship to surface tension. Macromolecules **39**, 7718–7728 (2006)
14. Jones, R.A.L., Kramer, E.J., Rafailovich, M.H., Sokolov, J., Schwarz, S.A.: Surface enrichment in an isotopic polymer blend. Phys. Rev. Lett. **62**, 280–283 (1989)
15. Composto, R.J., Stein, R.S., Kramer, E.J., Jones, R.A.L., Mansour, A., Karim, A., et al.: Surface enrichment in polymer blends—a neutron reflection test. Phys. B Condens. Matter. **156–157**, 434–436 (1989)
16. Sokolov, J., Rafailovich, M.H., Jones, R.A.L., Kramer, E.J.: Enrichment depth profiles in polymer blends measured by forward recoil spectrometry. Appl. Phys. Lett. **54**, 590–592 (1989)
17. Cifra, P., Bruder, F., Brenn, R.: Surface segregation in a polymer blend. Comparison between Monte Carlo simulation and mean-field theory. J. Chem. Phys. **99**, 4121–4127 (1993)
18. Hariharan, A., Kumar, S.K., Russell, T.P.: Surface segregation in binary polymer mixtures: a lattice model. Macromolecules **24**, 4909–4917 (1991)
19. Hariharan, A., Kumar, S.K., Russell, T.P.: Reversal of the isotopic effect in the surface behavior of binary polymer blends. J. Chem. Phys. **98**, 4163–4173 (1993)
20. Hariharan, A., Kumar, S.K., Russell, T.P.: Free surfaces of polymer blends. II. Effects of molecular weight and applications to asymmetric polymer blends. J. Chem. Phys. **99**, 4041–4050 (1993)
21. Hariharan, A., Kumar, S.K., Rafailovich, M.H., Sokolov, J., Zheng, X., Duong, D., et al.: The effect of finite film thickness on the surface segregation in symmetric binary polymer mixtures. J. Chem. Phys. **99**, 656–663 (1993)
22. Genzer, J., Faldi, A., Composto, R.J.: Self-consistent mean-field calculation of surface segregation in a binary polymer blend. Phys. Rev. E Stat. Phys. Plasmas Fluids **50**, 2373–2376 (1994)
23. Whitesides, G.M., Laibinis, P.E.: Wet chemical approaches to the characterization of organic surfaces: self-assembled monolayers, wetting, and the physical-organic chemistry of the solid-liquid interface. Langmuir **6**, 87–96 (1990)
24. Ferguson, G., Whitesides, G.: Thermal reconstruction of the functionalized interface of polyethylene carboxylic acid and its derivatives. In: Schrader, M., Loeb, G. (eds.) Modern Approaches to Wettability, pp. 143–177. Springer, New York (1992)
25. Cross EM, McCarthy TJ. Surface reorganizations in functionalized poly(chlorotrifluoroethylene). Abstracts of Papers of the American Chemical Society. 1988;195:141.
26. Holmesfarley, S.R., Reamey, R.H., Nuzzo, R., McCarthy, T.J., Whitesides, G.M.: Reconstruction of the interface of oxidatively functionalized polyethylene and derivatives on heating. Langmuir **3**, 799–815 (1987)

27. Baszkin, A., Deyme, M., Nishino, M., Ter-Minassian-Saraga, L.: Surface chemistry and wettability of modified polyethylene. In: Horst MÜller, F., Weiss, A., Wolfram, E. (eds.) Colloids and Surfaces, pp. 97–108. Steinkopff, Darmstadt (1976)
28. Donaruma, L.G.: Surface and interfacial aspects of biomedical polymers. Volume 1. Surface chemistry and physics, Joseph D. Andrade, Ed., Plenum, New York, 1985, 479 pp. Price: $69.50. J. Polym. Sci. C Polym. Lett. **24**, 427–428 (1986)
29. Rasmussen, J.R., Bergbreiter, D.E., Whitesides, G.M.: Location and mobility of functional—groups at surface of oxidized, low-density polyethylene film. J. Am. Chem. Soc. **99**, 4746–4756 (1977)
30. Bergbreiter, D.E., Kabza, K.: Annealing and reorganization of sulfonated polyethylene films to produce surface-modified films varying hydrophilicity. J. Am. Chem. Soc. **113**, 1447–1448 (1991)
31. Pan, D.H.K., Prest Jr., W.M.: Surfaces of polymer blends: X-ray photoelectron spectroscopy studies of polystyrene/poly(vinyl methyl ether) blends. J. Appl. Phys. **58**, 2861–2870 (1985)
32. Cowie, J.M.G., Devlin, B.G., McEwen, I.J.: A study of surface enrichment in polystyrene/poly(vinyl methyl ether) blends using attenuated total reflectance infra-red spectroscopy: 1. Polymer **34**, 501–504 (1993)
33. Cowie, J.M.G., Devlin, B.G., McEwen, I.J.: Surface enrichment in polystyrene/poly(vinyl methyl ether) blends. 3. An analysis of the near-surface composition profile. Macromolecules **26**, 5628–5632 (1993)
34. Cowie, J.M.G., Devlin, B.G., McEwen, I.J.: Surface enrichment in PS/PVME blends: 2. The effect of specific interactions in the bulk mixture. Polymer **34**, 4130–4135 (1993)
35. Forrey, C., Koberstein, J.T., Pan, D.H.: Surface segregation in miscible blends of polystyrene and poly(vinylmethyl ether): comparison of theory and experiment. Interface Sci. **11**, 211–223 (2003)
36. Tanaka, K., Takahara, A., Kajiyama, T.: Film thickness dependence of the surface structure of immiscible polystyrene/poly(methyl methacrylate) blends. Macromolecules **29**, 3232–3239 (1996)
37. Koberstein, J.T.: Molecular design of functional polymer surfaces. J. Polym. Sci. B **42**, 2942–2956 (2004)
38. Hariharan, A., Kumar, S.K., Russell, T.P.: A lattice model for the surface segregation of polymer chains due to molecular weight effects. Macromolecules **23**, 3584–3592 (1990)
39. Theodorou, D.N.: Structure and thermodynamics of bulk homopolymer/solid interfaces: a site lattice model approach. Macromolecules **21**, 1400–1410 (1988)
40. Theodorou, D.N.: Lattice models for bulk polymers at interfaces. Macromolecules **21**, 1391–1400 (1988)
41. Scheutjens, J.M.H.M., Fleer, G.J.: Statistical theory of the adsorption of interacting chain molecules. 1. Partition function, segment density distribution, and adsorption isotherms. J. Phys. Chem. **83**, 1619–1635 (1979)
42. Narrainen, A.P., Clarke, N., Eggleston, S.M., Hutchings, L.R., Thompson, R.L.: Surface adsorption of polar end-functionalised polystyrenes. Soft Matter. **2**, 981–985 (2006)
43. Chen, J., Zhuang, H., Zhao, J., Gardella Jr., J.A.: Solvent effects on polymer surface structure. Surf. Interface Anal. **31**, 713–720 (2001)
44. Bates, F.S., Wignall, G.D.: Nonideal mixing in binary blends of perdeuterated and protonated polystyrenes. Macromolecules **19**, 932–934 (1986)
45. Tanaka, K., Kajiyama, T., Takahara, A., Tasaki, S.: A novel method to examine surface composition in mixtures of chemically identical two polymers with different molecular weights. Macromolecules **35**, 4702–4706 (2002)
46. Affrossman, S., Hartshorne, J.M., Pethrick, R.A., Richards, R.W.: Surface segregation in blends of hydrogenous polystyrene and perfluorohexane end-capped deuterated polystyrene, studied by SSIMS and XPS. Macromolecules **27**, 1588–1591 (1994)
47. Elman, J.F., Johs, B.D., Long, T.E., Koberstein, J.T.: A neutron reflectivity investigation of surface and interface segregation of polymer functional end groups. Macromolecules **27**, 5341–5349 (1994)

48. Xie, F., He, T., Zhang, H.F., Lee, F.K., Du, B., Tsui, O.K.C., et al.: Effect of low surface energy chain ends on the glass transition temperature of polymer thin films. Macromolecules **35**, 1491–1492 (2002)
49. Affrossman, S., Bertrand, P., Hartshorne, M., Kiff, T., Leonard, D., Pethrick, R.A., et al.: Surface segregation in blends of polystyrene and perfluorohexane double end capped polystyrene studied by static SIMS, ISS, and XPS. Macromolecules **29**, 5432–5437 (1996)
50. Mason, R., Jalbert, C.A., O'Rourke Muisener, P.A.V., Koberstein, J.T., Elman, J.F., Long, T. E., et al.: Surface energy and surface composition of end-fluorinated polystyrene. Adv. Colloid Interface Sci. **94**, 1–19 (2001)
51. Anastasiadis, S.H., Gancarz, I., Koberstein, J.T.: Interfacial tension of immiscible polymer blends: temperature and molecular weight dependence. Macromolecules **21**, 2980–2987 (1988)
52. Schaub, T.F., Kellogg, G.J., Mayes, A.M., Kulasekere, R., Ankner, J.F., Kaiser, H.: Surface modification via chain end segregation in polymer blends. Macromolecules **29**, 3982–3990 (1996)
53. O'Rourke Muisener, P.A.V., Jalbert, C.A., Yuan, C., Baetzold, J., Mason, R., Wong, D., et al.: Measurement and modeling of end group concentration depth profiles for ω-fluorosilane polystyrene and its blends. Macromolecules **36**, 2956–2966 (2003)
54. O'Rourke-Muisener, P.A.V., Koberstein, J.T., Kumar, S.: Optimal chain architectures for the molecular design of functional polymer surfaces. Macromolecules **36**, 771–781 (2003)
55. Cai, Y., Gardner, D., Caneba, G.T.: Surface properties of silicone-containing block-graft copolymer/polystyrene systems. J. Adhes. Sci. Technol. **13**, 1017–1027 (1999)
56. Tanaka, K., Jiang, X., Nakamura, K., Takahara, A., Kajiyama, T., Ishizone, T., et al.: Effect of chain end chemistry on surface molecular motion of polystyrene films. Macromolecules **31**, 5148–5149 (1998)
57. Kawaguchi, D., Tanaka, K., Torikai, N., Takahara, A., Kajiyama, T.: Surface and interfacial segregation in blends of polystyrene with functional end groups and deuterated polystyrene. Langmuir **23**, 7269–7275 (2007)
58. Tanaka, K., Kawaguchi, D., Yokoe, Y., Kajiyama, T., Takahara, A., Tasaki, S.: Surface segregation of chain ends in α,ω-fluoroalkyl-terminated polystyrenes films. Polymer **44**, 4171–4177 (2003)
59. Wong, D., Jalbert, C.A., Orourke-Muisener, P.A.V., Koberstein, J.T.: Surface dynamics of polymer glasses: sub-Tg surface reorganization in end-functional polymers. Macromolecules **45**, 7973–7984 (2012)
60. Thompson, R.L., Hardman, S.J., Hutchings, L.R., Narrainen, A.P., Dalgliesh, R.M.: pH-Controlled polymer surface segregation. Langmuir **25**, 3184–3188 (2009)
61. Wu S. Polymer Interface and Adhesion. 1982:Marcel Dekker, New York.
62. Minnikanti, V.S., Zhenyu, Q., Archer, L.A.: Surface segregation and surface tension of polydisperse polymer melts. J. Chem. Phys. **126** (2007)
63. Tanaka, K., Takahara, A., Kajiyama, T.: Effect of polydispersity on surface molecular motion of polystyrene films. Macromolecules **30**, 6626–6632 (1997)
64. Yethiraj, A.: Entropic and enthalpic surface segregation from blends of branched and linear polymers. Phys. Rev. Lett. **74**, 2018–2021 (1995)
65. Wu, D.T., Fredrickson, G.H.: Effect of architecture in the surface segregation of polymer blends. Macromolecules **29**, 7919–7930 (1996)
66. Wu, D.T., Fredrickson, G.H., Carton, J.P.: Surface segregation in conformationally asymmetric polymer blends: incompressibility and boundary conditions. J. Chem. Phys. **104**, 6387–6397 (1996)
67. Minnikanti, V.S., Archer, L.A.: Surface migration of branched molecules: analysis of energetic and entropic factors. J. Chem. Phys. **123**, 1–9 (2005)
68. Hongxia, Z., Yang, L., Chunqing, Z., Zhansheng, L., Xin, L., Yurong, W.: Synthesis of dendrigraft star-comb polybutadienes by anionic polymerization and grafting-onto methodology. Macromolecules **42**, 5073–5079 (2009)

69. Steiner, U., Klein, J., Fetters, L.J.: Surface phase inversion in finite-sized binary mixtures. Phys. Rev. Lett. **72**, 1498–1501 (1994)
70. Scheffold, F., Budkowski, A., Steiner, U., Eiser, E., Klein, J., Fetters, L.J.: Surface phase behavior in binary polymer mixtures. II. Surface enrichment from polyolefin blends. J. Chem. Phys. **104**, 8795–8806 (1996)
71. Foster, M.D., Greenberg, C.C., Teale, D.M., Turner, C.M., Corona-Galvan, S., Cloutet, E., et al.: Effective χ and surface segregation in blends of star and linear polystyrene. Macromol. Symp. **149**, 263–268 (2000)
72. Qian, Z., Minnikanti, V.S., Sauer, B.B., Dee, G.T., Archer, L.A.: Surface tension of symmetric star polymer melts. Macromolecules **41**, 5007–5013 (2008)
73. Ariura, F., Tanaka, K., Nagamura, T., Deffieux, A., Schappacher, M., Hino, M., et al.: Surface Aggregation Structure of (Linear Polystyrene/Hyperbranched Polystyrene) Blend Films, 2nd edn, pp. 3928–3929. Polymer Preprints, Japan (2005)
74. Hirai, T., Huan, L., Ohta, Y., Yokozawa, T., Tanaka, K.: Surface segregation of well-defied N-substituted hyperbranched polyamides in linear polymer matrix. Chem. Lett. **40**, 366–367 (2011)
75. Haraguchi, M., Hirai, T., Ozawa, M., Miyaji, K., Tanaka, K.: Hydrophobic acrylic hard coating by surface segregation of hyper-branched polymers. Appl. Surf. Sci. **266**, 235–238 (2013)
76. Kawaguchi, D., Masuoka, K., Takano, A., Tanaka, K., Nagamura, T., Torikai, N., et al.: Comparison of interdiffusion behavior between cyclic and linear polystyrenes with high molecular weights. Macromolecules **39**, 5180–5182 (2006)
77. Lei, Y.-G., Cheung, Z.-L., Ng, K.-M., Li, L., Weng, L.-T., Chan, C.-M.: Surface chemical and morphological properties of a blend containing semi-crystalline and amorphous polymers studied with ToF-SIMS, XPS and AFM. Polymer **44**, 3883–3890 (2003)
78. Clarke, C.J., Jones, R.A.L., Clough, A.S.: The effect of matrix molecular weight on the kinetics of formation of end-adsorbed polystyrene layers from the melt. Polymer **37**, 3813–3817 (1996)
79. Clarke, C.J., Jones, R.A.L., Edwards, J.L., Clough, A.S., Penfold, J.: Kinetics of formation of physically end-adsorbed polystyrene layers from the melt. Polymer **35**, 4065–4071 (1994)
80. Crowe, J.A., Genzer, J.: Creating responsive surfaces with tailored wettability switching kinetics and reconstruction reversibility. J. Am. Chem. Soc. **127**, 17610–17611 (2005)
81. Schulze, J.S., Cernohous, J.J., Hirao, A., Lodge, T.P., Macosko, C.W.: Reaction kinetics of end-functionalized chains at a polystyrene/poly(methyl methacrylate) interface. Macromolecules **33**, 1191–1198 (2000)
82. Jalbert, C., Koberstein, J.T., Hariharan, A., Kumar, S.K.: End group effects on surface properties of polymers: semiempirical calculations and comparison to experimental surface tensions for α, ω-functional poly(dimethylsiloxanes). Macromolecules **30**, 4481–4490 (1997)
83. Jalbert, C.J., Koberstein, J.T., Balaji, R., Bhatia, Q., Salvati Jr., L., Yilgor, I.: Surface depletion of end groups in amine-terminated poly(dimethylsiloxane). Macromolecules **27**, 2409–2413 (1994)
84. Koberstein, J.T., Jalbert, C.: Molecular weight dependence and end-group effects on the surface tension of poly(dimethylsiloxane). Macromolecules **26**, 3069–3074 (1993)
85. Hardman, S.J., Hutchings, L.R., Clarke, N., Kimani, S.M., Mears, L.L.E., Smith, E.F., et al.: Surface modification of polyethylene with multi-end-functional polyethylene additives. Langmuir **28**, 5125–5137 (2012)
86. Bergius, W.N.A., Hutchings, L.R., Sarih, N.M., Thompson, R.L., Jeschke, M., Fisher, R.: Synthesis and characterisation of end-functionalised poly(N-vinylpyrrolidone) additives by reversible addition-fragmentation transfer polymerisation. Polym. Chem. **4**, 2815–2827 (2013)
87. Jiang, X., Yang, C.Z., Tanaka, K., Takahara, A., Kajiyama, T.: Effect of chain end group on surface glass transition temperature of thin polymer film. Phys. Lett. A Gen. At. Solid State Phys. **281**, 363–367 (2001)
88. Jones, R.A.L., Norton, L.J., Kramer, E.J., Bates, F.S., Wiltzius, P.: Surface-directed spinodal decomposition. Phys. Rev. Lett. **66**, 1326–1329 (1991)

89. Wong, D.A., O'Rourke-Muisener, P.A.V., Koberstein, J.T.: Effect of chain architecture on surface segregation in functional polymers: Synthesis and surface properties of end- and center-functional poly(D,L-lactide). Macromolecules **40**, 1604–1614 (2007)
90. Hutchings, L.R., Narrainen, A.P., Eggleston, S.M., Clarke, N., Thompson, R.L.: Surface-active fluorocarbon end-functionalized polylactides. Polymer **47**, 8116–8122 (2006)
91. Su, Z., Wu, D., Hsu, S.L., McCarthy, T.J.: Adsorption of end-functionalized poly(ethylene oxide)s to the poly(ethylene oxide)-air interface. Macromolecules **30**, 840–845 (1997)
92. Prasad, S., Hanne, L., Dhinojwala, A.: Surface segregation in polymer blends driven by surface freezing. Macromolecules **39**, 7467–7470 (2006)
93. Liu, S., Jiang, M., Chan, C.-M., Weng, L.-T.: Elimination of surface enrichment in polymer blends via interpolymer complexation. Macromolecules **34**, 3802–3804 (2001)
94. Matsen, M.W.: Self-assembly of block copolymers in thin films. Curr. Opin. Coll. Interface Sci. **3**, 40–47 (1998)
95. Darling, S.B.: Directing the self-assembly of block copolymers. Prog. Polym. Sci. **32**, 1152–1204 (2007)
96. Coulon, G., Russell, T.P., Deline, V.R., Green, P.F.: Surface-induced orientation of symmetric, diblock copolymers—a secondary ion mass-spectrometry study. Macromolecules **22**, 2581–2589 (1989)
97. Henkee, C.S., Thomas, E.L., Fetters, L.J.: The effect of surface constrains on the ordering of block copolymers domains. J. Mater. Sci. **23**, 1685–1694 (1988)
98. Anastasiadis, S.H., Russell, T.P., Satija, S.K., Majkrzak, C.F.: The morphology of symmetric diblock copolymers as revealed by neutron reflectivity. J. Chem. Phys. **92**, 5677–5691 (1990)
99. Thanawala, S.K., Chaudhury, M.K.: Surface modification of silicone elastomer using perfluorinated ether. Langmuir **16**, 1256–1260 (2000)
100. Vaidya, A., Chaudhury, M.K.: Synthesis and surface properties of environmentally responsive segmented polyurethanes. J. Colloid Interface Sci. **249**, 235–245 (2002)
101. Chaudhury, M.K., Whitesides, G.M.: How to make water run uphill. Science **256**, 1539–1541 (1992)
102. Mori, H., Hirao, A., Nakahama, S., Senshu, K.: Synthesis and surface characterization of hydrophilic-hydrophobic block copolymers containing poly(2,3-dihydroxypropyl methacrylate). Macromolecules **27**, 4093–4100 (1994)
103. Senshu, K., Yamashita, S., Mori, H., Ito, M., Hirao, A., Nakahama, S.: Time-resolved surface rearrangements of poly(2-hydroxyethyl methacrylate-block-isoprene) in response to environmental changes. Langmuir **15**, 1754–1762 (1999)
104. Pike, J.K., Ho, T., Wynne, K.J.: Water-induced surface rearrangements of poly(dimethylsiloxane−urea−urethane) segmented block copolymers. Chem. Mater. **8**, 856–860 (1996)
105. Segalman, R.A.: Patterning with block copolymer thin films. Mater. Sci. Eng. R Rep. **48**, 191–226 (2005)
106. Kim, H.-C., Hinsberg, W.D.: Surface patterns from block copolymer self-assembly. J. Vac. Sci. Technol. A **26**, 1369–1382 (2008)
107. Kim, J.K., Yang, S.Y., Lee, Y., Kim, Y.: Functional nanomaterials based on block copolymer self-assembly. Prog. Polym. Sci. **35**, 1325–1349 (2010)
108. Hamley, I.W.: Ordering in thin films of block copolymers: fundamentals to potential applications. Prog. Polym. Sci. **34**, 1161–1210 (2009)
109. Albert, J.N.L., Epps III, T.H.: Self-assembly of block copolymer thin films. Mater. Today. **13**, 24–33 (2010)
110. Knoll, A., Horvat, A., Lyakhova, K.S., Krausch, G., Sevink, G.J.A., Zvelindovsky, A.V., et al.: Phase behavior in thin films of cylinder-forming block copolymers. Phys. Rev. Lett. **89**, 035501 (2002)
111. De Gennes, P.G.: Conformations of polymers attached to an interface. Macromolecules **13**, 1069–1075 (1980)
112. Lowenhaupt, B., Steurer, A., Hellmann, G.P., Gallot, Y.: Microphases and macrophases in polymer blends with a diblock copolymer. Macromolecules **27**, 908–916 (1994)

113. Tanaka, H., Hasegawa, H., Hashimoto, T.: Ordered structure in mixtures of a block copolymer and homopolymers. 1. Solubilization of low molecular weight homopolymers. Macromolecules **24**, 240–251 (1991)
114. Tanaka, H., Hashimoto, T.: Ordered structures of block polymer/homopolymer mixtures. 3. Temperature dependence. Macromolecules **24**, 5713–5720 (1991)
115. Hashimoto, T., Tanaka, H., Hasegawa, H.: Ordered structure in mixtures of a block copolymer and homopolymers. 2. Effects of molecular weights of homopolymers. Macromolecules **23**, 4378–4386 (1990)
116. Winey, K.I., Thomas, E.L., Fetters, L.J.: The ordered bicontinuous double-diamond morphology in diblock copolymer/homopolymer blends. Macromolecules **25**, 422–428 (1992)
117. Winey, K.I., Thomas, E.L., Fetters, L.J.: Swelling of lamellar diblock copolymer by homopolymer: influences of homopolymer concentration and molecular weight. Macromolecules **24**, 6182–6188 (1991)
118. Mayes, A.M., Russell, T.P., Satija, S.K., Majkrzak, C.F.: Homopolymer distributions in ordered block copolymers. Macromolecules **25**, 6523–6531 (1992)
119. Leibler, L., Orland, H., Wheeler, J.C.: Theory of critical micelle concentration for solutions of block copolymers. J. Chem. Phys. **79**, 3550–3557 (1983)
120. Rodríguez-Hernández, J.: Nano/micro and hierarchical structured surfaces in polymer blends. In: Thomas, S., Shanks, R., Chandrasekharakurup, S. (eds.) Nanostructured Polymer Blends, pp. 357–421. William Andrew, Oxford (2014)
121. Ansari, I.A., Clarke, N., Hutchings, L.R., Pillay-Narrainen, A., Terry, A.E., Thompson, R.L., et al.: Aggregation, adsorption, and surface properties of multiply end-functionalized polystyrenes. Langmuir **23**, 4405–4413 (2007)
122. Palacios-Cuesta, M., Liras, M., Labrugère, C., Rodríguez-Hernández, J., García, O.: Functional micropatterned surfaces prepared by simultaneous UV-lithography and surface segregation of fluorinated copolymers. J. Polym. Sci. A Polym. Chem. **50**, 4902–4910 (2012)
123. Yokoyama, H., Miyamae, T., Han, S., Ishizone, T., Tanaka, K., Takahara, A., et al.: Spontaneously formed hydrophilic surfaces by segregation of block copolymers with water-soluble blocks. Macromolecules **38**, 5180–5189 (2005)
124. Zhang, R., Seki, A., Ishizone, T., Yokoyama, H.: Reduced hydrophobic interaction of polystyrene surfaces by spontaneous segregation of block copolymers with oligo (ethylene glycol) methyl ether methacrylate blocks: Force measurements in water using atomic force microscope with hydrophobic probes. Langmuir **24**, 5527–5533 (2008)
125. Shull, K.R., Kramer, E.J., Hadziioannou, G., Tang, W.: Segregation of block copolymers to interfaces between immiscible homopolymers. Macromolecules **23**, 4780–4787 (1990)
126. Shull, K.R., Kramer, E.J., Hadziioannou, G., Tang, W.: Segregation of block copolymers to interfaces between immiscible homopolymers. Macromolecules **23**(22), 4780–4787 (1990)
127. Shull, K.R., Winey, K.I., Thomas, E.L., Kramer, E.J.: Segregation of block copolymer micelles to surfaces and interfaces. Macromolecules **24**, 2748–2751 (1991)
128. Ibarboure, E., Bousquet, A., Toquer, G., Papon, E., Rodriguez-Hernandez, J.: Tunable hierarchical assembly on polymer surfaces: combining microphase and macrophase separation in copolymer/homopolymer blends. Langmuir **24**, 6391–6394 (2008)
129. Bousquet, A., Ibarboure, E., Drummond, C., Labrugere, C., Papon, E., Rodriguez-Hernandez, J.: Design of stimuli-responsive surfaces prepared by surface segregation of polypeptide-b-polystyrene diblock copolymers. Macromolecules **41**, 1053–1056 (2008)
130. Bousquet, A., Ibarboure, E., Teran, F.J., Ruiz, L., Garay, M.T., Laza, J.M., et al.: PH responsive surfaces with nanoscale topography. J. Polym. Sci. A Polym. Chem. **48**, 2982–2990 (2010)
131. Bousquet, A., Ibarboure, E., Papon, E., Labrugère, C., Rodríguez-Hernández, J.: Structured multistimuli-responsive functional polymer surfaces obtained by interfacial diffusion of amphiphilic block copolymers. J. Polym. Sci. A Polym. Chem. **48**, 1952–1961 (2010)
132. Russell, T.P.: Surface-responsive materials. Science **297**, 964–967 (2002)
133. Bousquet, A., Pannier, G., Ibarboure, E., Papon, E., Rodríguez-Hernández, J.: Control of the surface properties in polymer blends. J. Adhes. **83**, 335–349 (2007)

Chapter 6
Template Guided Structuration of Polymer Films

David Coffey and Joseph Wei

6.1 Introduction

As detailed in previous chapters, polymer films can evolve by means of dewetting and phase segregation. Importantly, these features can be induced, inhibited, and guided through patterning of the underlying substrate. Figure 6.1 provides one such example. In the experiment represented in these images, Jeong and coworkers used a templated substrate to guide the phase separation of block copolymers into aligned lines 8 nm in width [1]. They then demonstrated that these structures could be used as templates for inorganic materials and cross-layered to produce grids. Such films present opportunities in optical coatings, waveguides, and photonics crystals. More importantly, this is an example of a fabrication technique that is both potentially cheaper than comparable standard photolithographic techniques and can produce structures that are impossible to generate with standard photolithographic techniques.

To go beyond this preliminary example and understand more generally how such structured films are created, we must delve into several questions: What methods are available for structuring polymer films? What physical processes do these methods employ to guide this structuring? What are the limits to these processes—both in terms of resolution and flexibility? What are the current and future applications of structured polymer films?

D. Coffey (✉)
Department of Chemistry and Physics, Warren Wilson College,
PO Box 9000, Asheville, NC 28815, USA
e-mail: dcoffey@warren-wilson.edu

J. Wei
Lam Research Corporation, 11155 SW Leveton Drive,
Tualatin, OR 97062, USA
e-mail: jhcwei@gmail.com

J. Rodríguez-Hernández, C. Drummond (eds.), *Polymer Surfaces in Motion*,
DOI 10.1007/978-3-319-17431-0_6

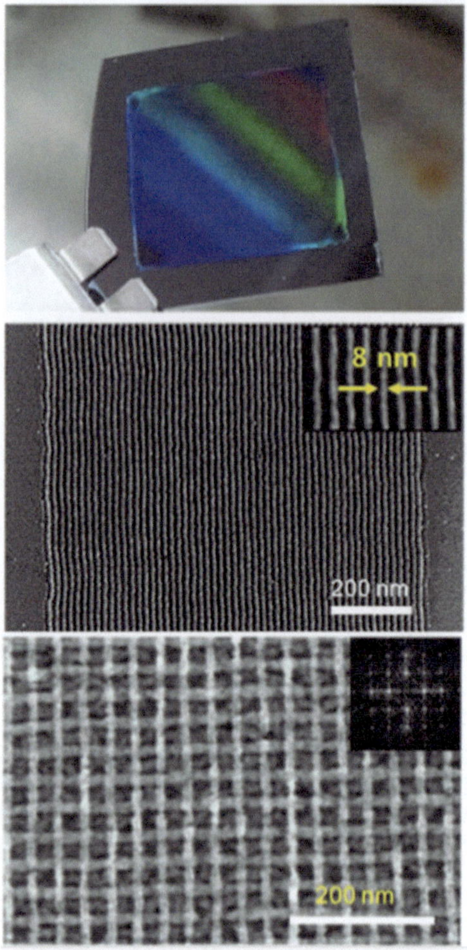

Fig. 6.1 Substrate templating was used to produce this polymer film that possesses aligned, cylindrical domains (*top*). Similar films could be oxidized to produce films of aligned, inorganic nanowires (*middle*) and printed sequentially at right angles to form grids (*bottom*). Reprinted with permission [1]

To address these questions, this chapter is organized to cover the two most commonly used procedures for substrate templating of polymer films—surface chemistry templating and topographic templating. Within these sections, we will discuss the underlying mechanisms, limits, and applications of these techniques.

6.2 Templating: Overview

When attempting to realize specific characteristics in a polymer film, an experimenter will often encounter a dizzying array of experimental choices. For example, when making a polymer film simply through drop casting (coating a substrate with a solution of polymer dissolved in solvent, and letting the solvent dry to leave behind a polymer film), the film characteristics can be impacted by the polymer type, the polymer molecular weight distribution, the purity and defects in the polymer, the amount of solution used, the solvent type, the processing temperature, the substrate cleaning procedure, the substrate size, and the substrate chemistry and roughness. As more sophisticated deposition procedures are used or as combinations of solvents and polymers are used, the number of relevant experimental variables only increases.

As detailed in previous chapters, these experimental variables can be selected to generate films that exhibit spatial features such as dewetting or phase separation. However, while the appearance of these features can be somewhat controlled by changing experimental conditions—for example, switching to a more volatile solvent during drop casting can shrink dewetting features by reducing the time during which the dewetting proceeds—such variations will not induce long range order in these features or place them in specific locations. One might be able to tailor the characteristic size and separations of film features, but the exact size and positions will vary.

The primary means for controlling film features is templating of the underlying substrate. This is usually done by modifying the surface chemistry or topography of the substrate the film will be deposited upon. This can be achieved through photolithography, transfer printing (soft lithography or dip-pen nanolithography), or soaking the substrate in a solute of molecules that self-assemble on the surface [2–5].

Figure 6.2 demonstrates the role the substrate can play. In this example, the same polymer blend solution (polystyrene and polythiophene dissolved in a chlorobenzene solution) was deposited with the same procedure (spin coating) [6]. The only difference between samples was a modification of the surface chemistry

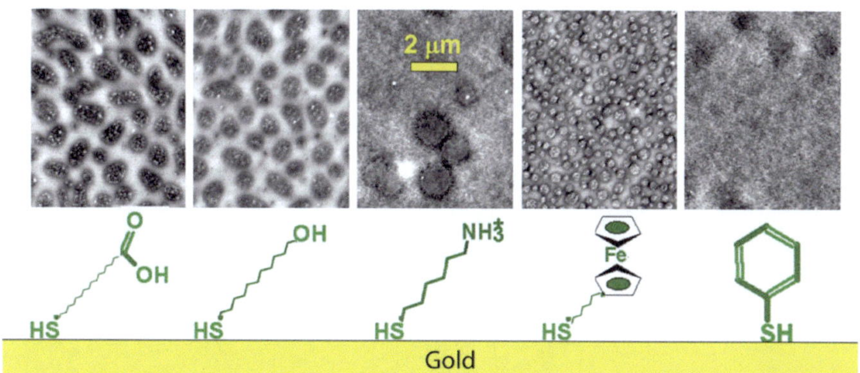

Fig. 6.2 This sequence of images highlights the impact of the underlying substrate on an overlaying polymer film. The films were made from a polymer blend solution where the polymer blend ratio and the processing conditions were kept constant. The images are atomic force microscopy height images and show where the polymers have segregated. In each sample the substrate surface chemistry, originally gold, was coated with a different self-assembled molecule layer to form various surface chemistries. Reprinted with permission [6]

between substrates. The images show atomic force microscope (AFM) height images of a small section of each film. The white/black regions represent higher/lower topography and correspond to local enrichments of one of the two polymers, also known as phase separation or phase segregation. The phase separation of a polymer blend film will depend on the chemistry of the two polymers, their molecule weights, their ratio, the solvent, the deposition conditions, and, as shown in this figure, the substrate chemistry. The five substrates used in this sample were silicon wafers coated with a thin layer of evaporated gold. The chemistry of the exposed gold was modified by dipping the gold in one of five different solution with thiol-functionalized molecules that self-assemble on gold. As can be observed, the polymer film responded uniquely to each surface chemistry.

Figure 6.3 shows how intelligently controlling the substrate chemistry can allow one to spatially guide structures in polymer films. In the late 1990s, Boltau et al. [7] and other researchers pioneered methods whereby the natural phase separation in a polymer blend film can be guided by a substrate with a patterned surface chemistry. The film shown in Fig. 6.3 was produced by casting a blend solution of polystyrene and polyvinylpyridine from a solvent onto a substrate. The AFM height image shows the final polymer film topography (as the solvent evaporated, the polymers phase separated then vitrified). The high/white regions are enriched in one polymer and the low/black regions are enriched in the other polymer. In the rectangular region where the underlying substrate surface was patterned with more and less wettable lines (not shown in this image), the phase separation is ordered. It is worth noting that the template size and surface chemistry had to be matched to the material system and deposition conditions. For example, Boltau would have been

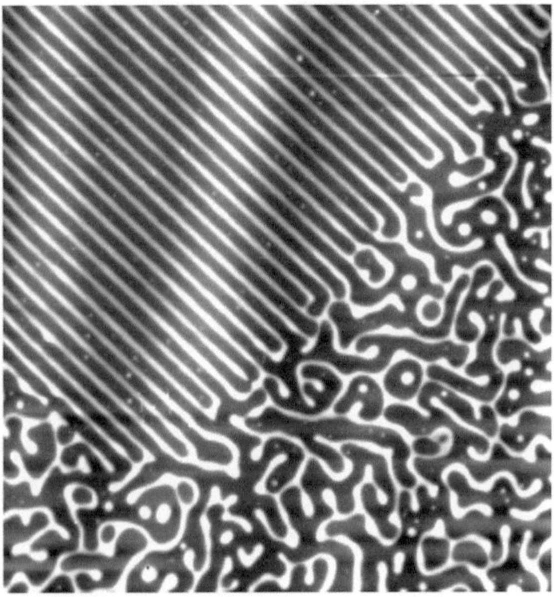

Fig. 6.3 $80 \times 80\ \mu m^2$ AFM height image of a polymer blend film. The phase separation of the film is ordered in a rectangular region by the underlying substrate (*top portion of the figure*), which had a patterned surface chemistry. The unpatterned film area (*bottom portion of the figure*) shows unpatterned phase separation. Reprinted with permission [7]

unable to generate lines ten times smaller or larger with the same polymer system processed in the same way. The following sections of this chapter will discuss this and other patterning rules. For now it is sufficient to recognize that a patterned substrate can induce patterns in a polymer film.

Researchers have templated structures in numerous classes of polymer films using a variety of methods. The next sections of this chapter will summarize two particularly important templating methods—surface chemistry templating and topographic templating—for single polymer films, polymer blends, and block copolymer films.

6.3 Surface Chemistry Templates and Guided Dewetting

Dewetting in polymer films has been detailed in previous chapters, but can be summarized here. A polymer film can be thermodynamically stable, unstable, or metastable. In an unstable film, capillary fluctuations will spontaneously grow in a process called spinodal dewetting. In a metastable film, the film will remain intact except where nucleation events, possibly instigated by a contaminate (heterogeneous nucleation) or by a concentration fluctuation (homogeneous nucleation) trigger localized dewetting. In both the metastable and unstable cases, a uniform film will tend to evolve by drift and diffusion into a mottled film of polymer droplets and dry patches with both feature types tending to grow over time.

The viscosity of the film is also an important variable in the film's evolution. Thermal annealing or solvent annealing will speed the process. Alternatively, lowering the temperature or drying a film will slow or stall a film's evolution. For example, a polymer film held at low temperature can be energetically unstable but, effectively, never evolve to a more favorable state. Unstable and metastable films will kinetically evolve with a speed dependent on the temperature and, if included, solvent concentration.

In all practical fabrication procedures, the evolution of a film structure is a complicated and dynamic process that can only be partly understood through equilibrium cases. In spin coating a polymer film, for example, the evolution of the film will progress through several stages. The polymers that start fully dispersed and weakly interacting will begin to near each other and interact as the solvent evaporates. As the interactions increase, however, so does the viscosity. Further, the solvent evaporation will not be uniform throughout the thickness of the film. Eventually, before thermal equilibrium is fully established, the film will vitrify. Images of dewet films, such as Fig. 6.3, can usefully be considered snapshots in time of an evolving film that has vitrified.

The evolution of a simple film on a homogeneous substrate will be driven by the surface tension of the film with respect to air and the surface tension of the film with respect to the substrate. This interaction can either be stabilizing or destabilizing. If the substrate is chemically heterogeneous, a driving force will arise that will tend

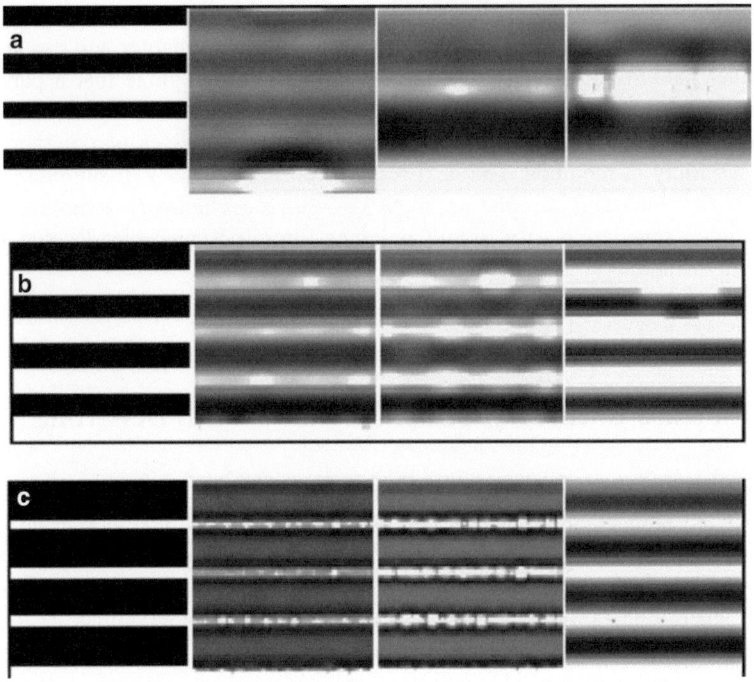

Fig. 6.4 Three simulations of how substrate surface chemistry influence polymer film dewetting. The *left column* indicates the surface chemistry of more wettable regions (*white*) and less wettable regions (*black*). The following columns indicate the evolution of an overlaying polymer film in time as the film is annealed. The difference between the three rows is the spacing of the wettable lines from $0.75\lambda_h$ (**a**, *top row*), $1.00\lambda_h$ (**b**, *middle row*), and $3\lambda_h$ (**c**, *bottom row*). Reprinted with permission [11]

to guide the polymer drift from the areas of the substrate it finds less wettable to the areas of the substrate it finds more wettable.

In the early 2000s, Kargupta and Sharma performed a variety of simulations of polymers dewetting on chemically patterned substrates [8–11]. Figure 6.4 shows one such result of three polymer films dewetting on chemically patterned substrates [11]. The white and dark regions of the first column represent more and less wettable regions, respectively, with the distance between the wettable lines increasing. The next three columns show the time evolution of the films with the color scale representing the thickness of the polymer: the white regions represent thicker accumulations and the black regions represent thinner dewetting regions. From top to bottom (Fig. 6.4a–c), the width of the wettable lines remains constant while the pitch of these lines goes from $0.75\lambda_h$, $1.00\lambda_h$, and $3\lambda_h$, where λ_h is termed the characteristic length scale of instability. λ_h is a simulation determined parameter of the film that, for instance, increases as the film thickness increases. Note that the area represented by each image correspondingly increases with the pitch.

The annealing sequences in Fig. 6.4 show polymer dewetting being guided by the underlying surface chemistry, with the quality of the patterning dependent on the pitch. The $0.75\lambda_h$ pitch works poorly at guiding the dewetting while the larger pitches work well. Kargupta and Sharma concluded that the dewetting follows the underlying pattern if the surface energy contrast is sufficient and if the pattern pitch is greater than λ_h.

It is worth highlighting that neither a chemically or topographically patterned substrate can guarantee a well-patterned film. As exhibited in the top row of this figure, features that are too small relative to the thickness of the film and the characteristics of the polymer (such as λ_h) direct dewetting poorly. Analogously, features much larger than the thickness of the film can similarly be imperfect guides. Important factors that will typically enhance guided dewetting include a template pitch that matches the characteristic feature length scale of the film, a greater contrast in surface energy of the surface chemistry template, a greater height difference or sharpness in a topographic template, thinner films, higher annealing temperatures, and longer annealing times.

6.4 Surface Chemistry Templates and Guided Phase Separation

When one goes from patterning a film composed of a single polymer to a film composed of multiple polymers (or a block copolymer), the number of interactions multiply as do the number of possible final film structures. For a single polymer film, the factors that must be considered include the polymer-substrate interaction, the polymer-atmosphere interaction, and the entropy of the polymer configurations. When adding a second polymer, the additional parameters include the second polymer's interaction with the substrate and atmosphere, the interaction between the polymers, the entropy of the second polymer's configurations, and the entropy of the two polymer's mixing (as can be described by Flory-Huggins Theory) [12]. Such polymer blends can exhibit not only dewetting of one or both polymers but also segregation from each other (sometimes called phase separation and sometimes called phase segregation). The polymer-polymer phase segregation can occur vertically and horizontally within the film relative to the substrate and can occur on several length scales simultaneously.

The surface chemistry of the substrate, even if the substrate is homogeneous, will affect the film phase segregation (see Fig. 6.2). In a two component blend film of two polymers—let us call them polymer A and polymer B—if the surface tension of polymer A and the substrate is less than the surface tension of polymer B and the substrate, polymer A will tend to sink towards the substrate and polymer B will tend to rise from the substrate, producing a vertically phase segregated film. Substrate interactions tend to dominate the atmospheric interactions, though a similar process can lead to vertical phase separation at the atmospheric interface.

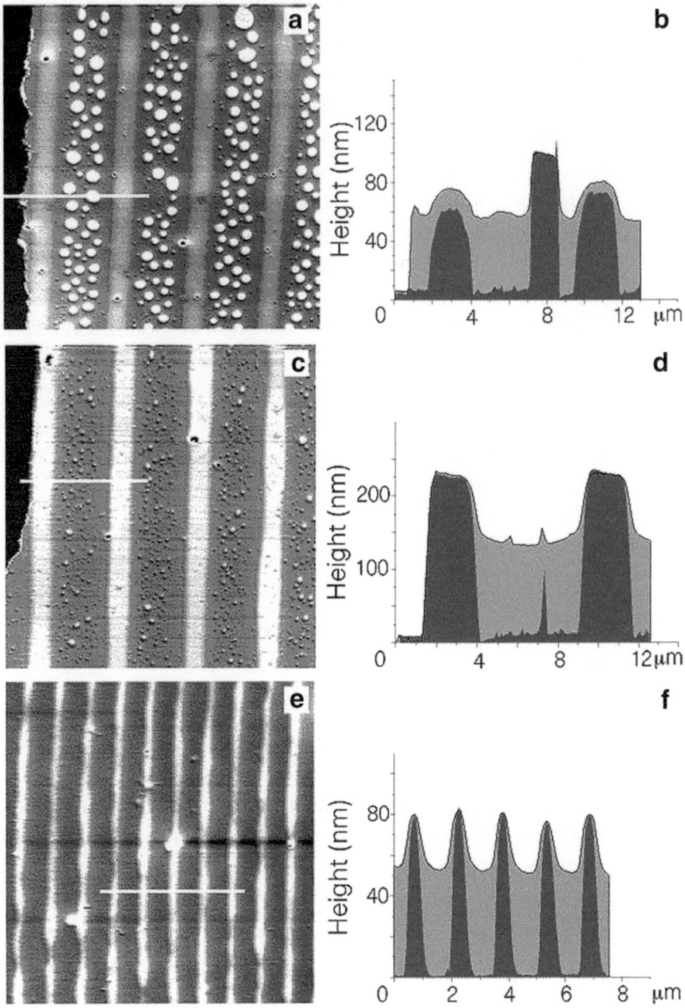

Fig. 6.5 AFM height images of three polymer blend films guided in their phase separation by underlying templates with varying surface chemistry. The traces at the right of each image show the height profile of the two polymers (PS and brPS) along the trace indicated on the AFM image. Reprinted with permission [7]

The key to guiding the phase separation of polymer components through a patterned substrate is that surface chemistry/surface energies will affect the component polymers differently. Figure 6.5, from Boltau et al., shows guided phase segregation in a polystyrene(PS)/partially brominated polystyrene(brPS) film [7]. To produce these samples, solutions of PS/brPS were deposited on chemically templated substrates and allowed to dry. The underlying substrate included stripes of alternating silicon oxide and hydrogen terminated silicon fabricated through photolithography. The line traces in Fig. 6.5 were generated by imaging a trace of the film with atomic force microscopy before and after solvent-selected removal of the PS, showing the final locations of the two polymers. These traces show that before vitrification the brPS preferentially drifted over the silicon oxide while the PS drifted over the hydrogen terminated silicon. It should be noted that while the line traces suggest a complete separation of polymers, in reality, phase separation typically leads only to enrichments. For example, the "PS domains" might be 90 % PS and 10 % brPS mixed on a molecular length scale while the "brPS domains" might be 90 % brPS and 10 % PS [13].

Similar to the necessity of matching the template length scale to the characteristic length scale of instability, λ_h, when patterning dewetting, the template length scale for patterning a polymer blend film must be approximately matched to the length scale of phase separation observed in an unpatterned film and the associated film thickness. These length scales are best found experimentally. The middle and bottom figures by Boltau show that imperfect templating (Fig. 6.5a, b) can be improved by either increasing the film thickness (Fig. 6.5c, d) or reducing the template line pitch (Fig. 6.5e, f).

Template substrates can guide the evolution of polymer blend film phase separation in a variety of ways. Figure 6.6 schemes two such potential mechanisms [14]. In case 1, the left-hand figures, the predominant surface chemistry (the substrate

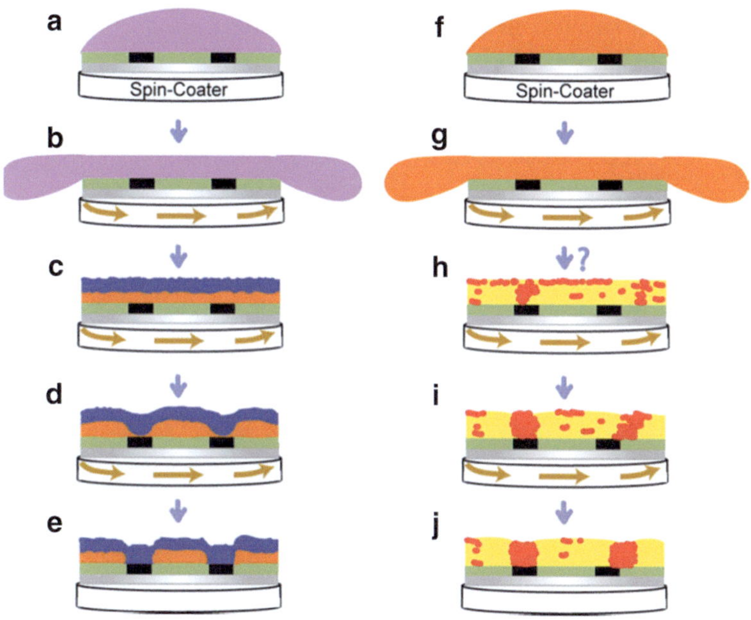

Fig. 6.6 Schemes of two possible mechanisms whereby surface chemistry templates can guide the phase separation in polymer blend films. In both cases, the substrate is predominately one surface chemistry—*light green areas*—with a second surface chemistry covering a minority of the substrate—*black areas*. In (**a**)–(**e**), the film vertically phase separates as induced by the majority surface chemistry, and then the minority surface chemistry induces dewetting of the lower polymer. In (**f**)–(**j**), the minority surface chemistry nucleates phase separation at precise locations before spontaneous nucleation can occur [6, 14]

indicated in light green) drives vertical phase separation of a polymer blend solution (b–c), and the minority surface chemistry (the substrate indicated in black) drives nucleated dewetting of the lower but not the upper polymer (d–e). In case 2, the right-hand figures, the predominant surface chemistry does not drive significant vertical or horizontal phase separation. However, the minority surface chemistry increases the rate of nucleation of phase separation, creating enrichments of one polymer (indicated in red). While these cases represent templated, nucleated dewetting (case 1) and templated, nucleated phase separation (case 2), analogous cases might be schemed for templated, spinodal dewetting; templated, spinodal phase separation; or combinations thereof [15].

Similar strategies can be used to pattern the phase separation of block copolymers (BCP). As described in previous chapters, BCP films can exhibit a wonderful diversity of phase separated patterns. Similar to standard polymer blend films, the phase separation that is exhibited is affected by the underlying substrate chemistry and can be guided by patterned templates. Figure 6.7 shows such an example for a PS-b-PMMA block copolymer film [16]. The underlying substrate on the left side of the image is unpatterned while the underlying substrate on the right side of the image is patterned with photolithographically defined self-assembled monolayers of varying chemistry. While the techniques of the templating method are similar to what has been described above, the advantages of BCP phase separation, smaller and more regular features, are also evident.

Fig. 6.7 SEM image of a diblock copolymer film following a template (underlying the film on the *right side of the image*) composed of lines of alternating surface chemistry. Reprinted with permission [16, 17]

The force that initiates dewetting or phase separation arises from the gradient in hydrophobicity/surface energy on the template substrate. Experimentally, this requires choosing templating materials that can be considered high surface energy/low surface energy, wettable/less wettable, or hydrophobic/hydrophilic to the particular polymers and solvents deposited. Determining the magnitude of the surface tension difference induced by the substrate surface energy difference is difficult to achieve quantitatively but can be understood qualitatively. The higher the surface energy of the substrate, the more easily it will be wet by polymers and solvents alike. Generally, surfaces composed of covalently bonded materials or metals will have a higher surface energy than a surface material held together by van der Walls bonds.

Figure 6.8 exemplifies the importance of the magnitude of surface energy difference of a template as well as the template length scale [18]. In this example, surface chemistry templates composed of polymer brushes were used to guide block copolymers phase separation, with both the template chemistry and the template length scales (pitch) being varied. Both the template polymer brushes and the deposited block copolymers (BCP) were constructed from polystyrene and PMMA units. What was varied chemically was the percentage of the template brushes composed of PS monomers. As the brush composition was varied from 50:50 PMMA/PS (left images, where the BCP PMMA and PS blocks would experience less wetting competition on the brush regions) to 0:100 PMMA/PS (right images, where the PS blocks would wet the brush regions more readily than the PMMA blocks). As expected, as the wettability contrast is increased (going from the left to right images), the template is increasingly successful at guiding the phase separation.

It is worth repeating that the film features produced by such surface templates are not directly written such as with ink-jet printing and are only guided with a success rate that is often statistical in nature. For a particular polymer film combination and thickness, there are limitations to the size, spacing, and type of features. Figures 6.9, 6.10, and 6.11 demonstrate some of these fidelity issues [6].

Figure 6.9 demonstrates what can happen when the template features are too fine to guide the phase separation of a particular polymer film [20]. A series of template polymer blend films are shown as imaged by atomic force microscopy. The substrate surface chemistry was modified through dip-pen nanolithography to include small, circular, hydrophilic regions. The predominant surface chemistry of the substrate produces a seemingly homogenous film, light gray areas, that in actuality are areas where the polymers are separated vertically like the layers of a sandwich (such as schemed in Fig. 6.6c). The dark gray circles are induced by the hydrophilic surface chemistry dots through nucleated dewetting/phase separation (such as schemed in Fig. 6.6d). In this particular series, the templates with pitch lengths greater than 1 μm guide the phase separation successfully whereas the templates with pitch sizes smaller than 1 μm generate features that overlap. This overlapping of features occurs, even though underlying template features do not overlap, because the characteristic feature size of these polymer films (the size of a nucleated dewetting/phase segregation) is around 1 μm and much smaller features cannot be realized.

Figure 6.10 highlights analogous patterning issues in films that exhibit widespread natural lateral phase separation. These images again display AFM height images of a series of polymer blend films (polystyrene and Poly[2-methoxy-5-(2-ethylhexyloxy)-1,4-phenylenevinylene]) cast from solutions onto substrates with templated surface chemistry. In this case, the template guidance is fair when the length scale of the template pitch is close to the natural pitch (Fig. 6.10a) but is increasingly poor when the pitch length and template feature size deviate from the nature length and size (Fig. 6.10b, c). In this last case, only the edges of the template circles generate a significant effect (where the wettability gradient would be largest).

Figure 6.11a presents a combinatorial array of template films that highlight the limitations of template patterning: the probability of initiating/guiding

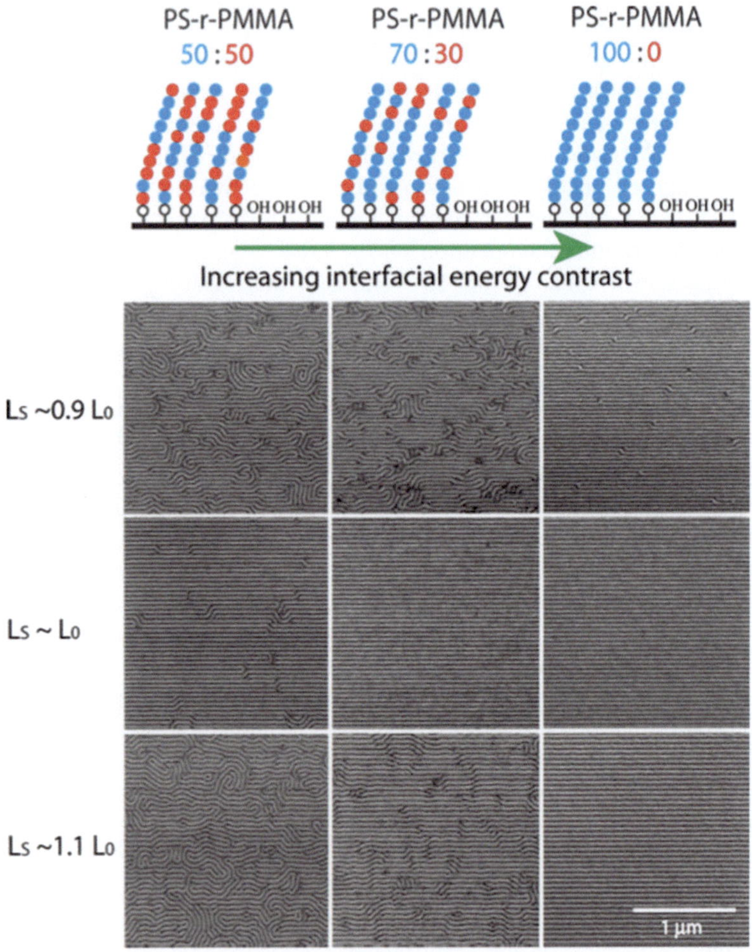

Fig. 6.8 The SEM images show the response of BCP films to templates of varying chemistry and pitch. The schemes above show the template chemistry. Reprinted with permission [18, 19]

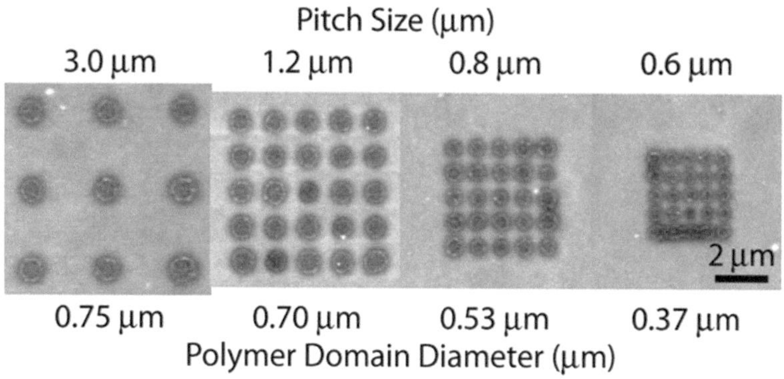

Fig. 6.9 AFM height images of polymer blend films guided by chemical templates of different pitches. The smallest pitches lead to poor feature differentiation. Reprinted with permission [6, 20]

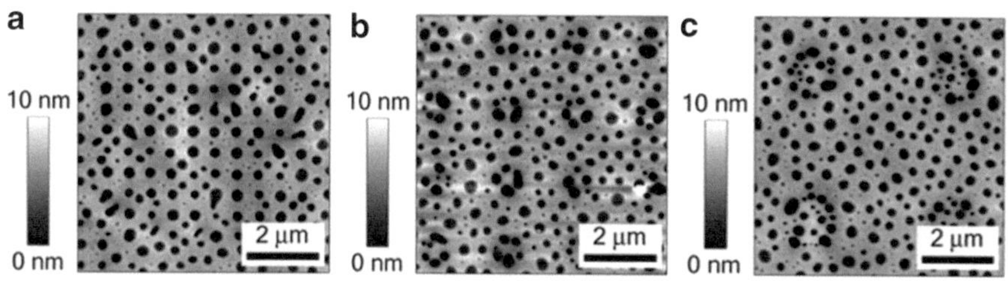

Fig. 6.10 AFM height images of polymer blend films guided by chemical templates of different pitches. In this case, the natural phase separation is guided only when the template pitch is similar to the natural feature pitch. Reprinted with permission [6]

phase separation. Thirty-two templated polymer blend films are shown, as imaged by atomic force microscopy. In each case the underlying substrate surface chemistry was patterned with dots of increased hydrophilicity through dip-pen nanolithography [20]. In the regions without the hydrophilic dots, a laterally homogenous film with vertical phase separation is produced. Hydrophilic dot features of sufficient size nucleate localized phase segregation. While the dot pitch was kept constant, the dot diameter was varied as well as the ratio of the two polymers (Polythiophene and PS).

The adjoining Fig. 6.11b plots the probability of successful nucleation as a function of template feature size as compiled from Fig. 6.11a and similar data sets. Several observations can be made from this array of data. First, a minimum template feature size of 50–200 nm is necessary to have a chance of nucleating phase separation, and the probability of successful nucleation is statistical in nature. Second, the probability of successful nucleation and the size of the induced features depend on the polymer blend ratio. While this data set was only acquired for one type of polymer blend film that follows a nucleated phase separation schemed in Fig. 6.6a–e, the size effects and the statistical nature

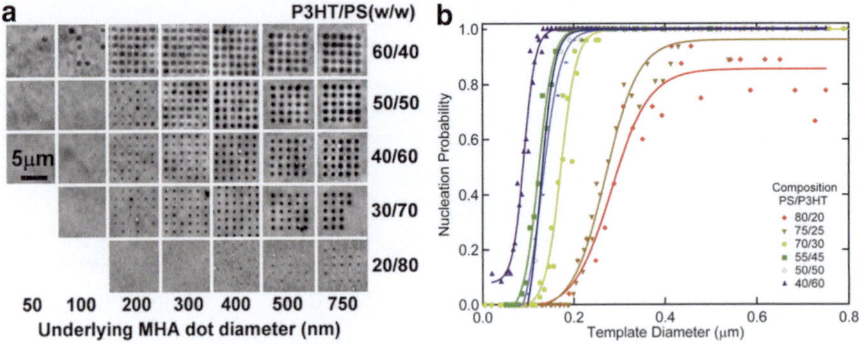

Fig. 6.11 (**a**) AFM height images of P3HT/PS polymer blend films of varying blend ratio guided by chemical templates composed of features with varying diameter. In each sample 25 template dots were written (except the rightmost 30/70 template that was missing several template dots). The probability of the template feature nucleating phase separation increases with increasing dot size and P3HT concentration. (**b**) The probability of nucleating phase separation is plotted as a function of template feature size. Each data point represents the probability as observed from a grid of 25 template dots. The sigmoidal line traces are not fits and are only included to guide the eye. Reprinted with permission [6, 20]

of templating polymer films structures extend more generally. Careful selection of the film components, processing, and templating is necessary to achieve a desired structure and not all structures are feasible.

6.5 Topographic Templates

Substrates that are chemically uniform but possess varying topography can also guide the dewetting or phase separation of polymer films [21–24]. Substrate topography will initiate and/or guide thickness variations in the film. The thinner areas of the film and the regions near the edges of the substrate will exhibit higher rates of nucleated dewetting and capillary fluctuation. Further, after the dewetting begins, the substrate topography can guide the evolution of the dewetting.

Figure 6.12 provides an example of both such guidance mechanisms [22]. In this example, a PMMA polymer film was cast on a PDMS elastomeric substrate covered in square microwells. The resulting film was solvent annealed and the film optically imaged over time. Figure 6.12a shows two nucleated holes that have dewetted in the film. The upper hole sits between wells (likely generated because the film would have been thinner here and more likely to dewet) and the lower hole surrounds a well (likely generated because dewetting is more likely at feature edges). Over time these dewetted features grow to encompass numerous wells (Fig. 6.12b–d), leaving behind an array of droplets sitting in the microwells. Thus, in this example, the substrate topography influenced the location of the initial dewetting and the growth of the dewetting, as described in Chap. 2.

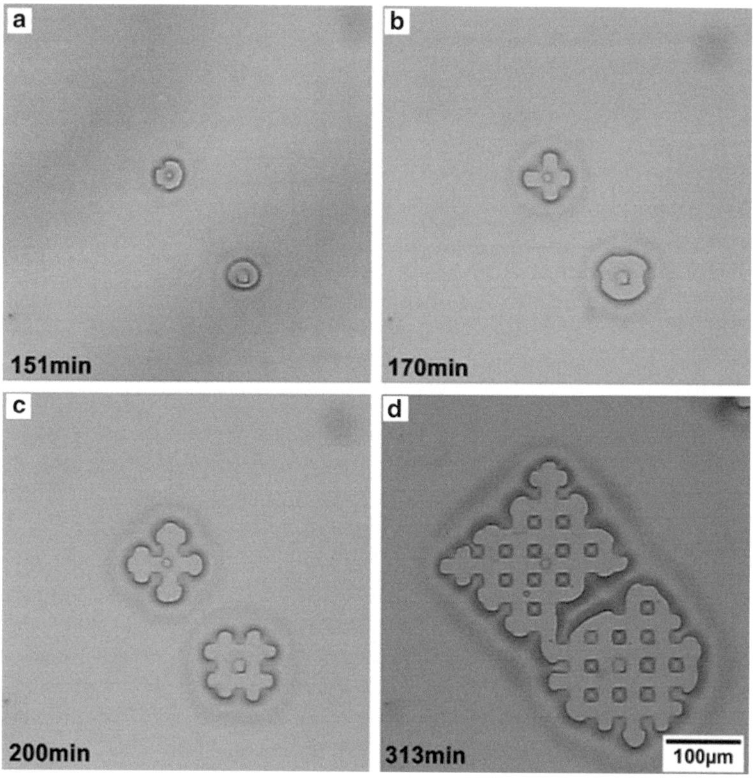

Fig. 6.12 Template induced dewetting: optical microscopy images of a polymer film dewetting over time on a topographic template consisting of square wells. Reprinted with permission [22]

Variation in the ratio of template height to film thickness will often change the type and density of features. Figure 6.13 demonstrates an edifying example where substrate trenches of varying depth were used to guide film dewetting [23]. The deepest substrate trenches (leftmost figure) induced dewetting in the corresponding thinnest regions of the film, leaving lines of droplets within the trenches. Films deposited on substrates with thinner channels (thinner than the film thickness) exhibited progressively less ordered features. The shortest channels (rightmost figure) did not guide the film, producing randomly distributed droplets.

In addition to simulations on chemical templating by Sharma and Kargupta discussed earlier, they also ran a variety of simulations on topographic templates [10, 11]. They discuss how the evolution of an unstable polymer film on a topographically patterned substrate will depend on numerous variables, but is particularly dependent on (a) the ratio of the starting film thickness to the height of the substrate topographic features and (b) the ratio between the substrate pattern pitch and the natural spinal dewetting length scale. As one varies these ratios, the resulting films will express different regimes. Figure 6.13 provides a nice example

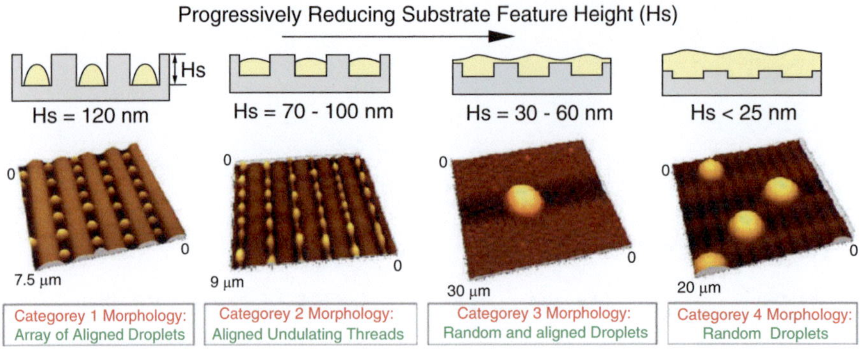

Fig. 6.13 Scheme and AFM images of four classes of topographic template induced dewetting. The topographic features were channels of varying depth. Reprinted with permission [23]

of the first ratio dependence: the films go from guided droplets, to guided lines to unguided droplet as the ratio of channel height to the film thickness decreases. With regard to the second ratio, Sharma and Kargupta state that successful patterning requires a template pitch no less that 0.8 times the natural spinodal dewetting length scale. These ratio rules are analogous to the requirement for successful chemistry-templated patterning. The topographic feature height plays a somewhat analogous role to the surface chemistry contrast. The reader is encouraged to read the series of papers Sharma and Kargupta wrote in the early 2000s to familiarize themselves with the theory guiding these simulations and the variety of possible film patterning regimes [8–11].

Just as topographic features can be used to control dewetting in single component polymer films, such templates also can be used to control polymer blend and block copolymer phase segregation. A difficulty with BCPs, however, is generating a template with sufficiently small feature sizes to match the natural phase separation length scale of a BCP film [25].

Figure 6.14 shows a topographic patterning strategy for BCP films unavailable with standard blends. As opposed to a fine template guiding every feature in the film, a larger scale template guides mesoscale ordering [24]. In this study, the researchers deposited PS-b-PEP block copolymer films on substrates with topographic channels 95 nm wide and 600 nm deep and observed the ordering as the film was annealed. Figure 6.14e–i shows the evolution of the film over the course of 33 h. Figure 6.14e is consistent with the principle previously discussed that a film cast on a substrate with a template length scale much larger than the natural length scale of phase separation will not result in complete guidance. Indeed, only the polymer domains nearest the channels edges are aligned to the template. However, as the individual domains in a BCP film do not grow during annealing, a longer range ordering can be achieved. Over the course of the annealing period, the aligned regions propagate progressively deeper into the channel until the film's phase separation is entirely ordered in the channel.

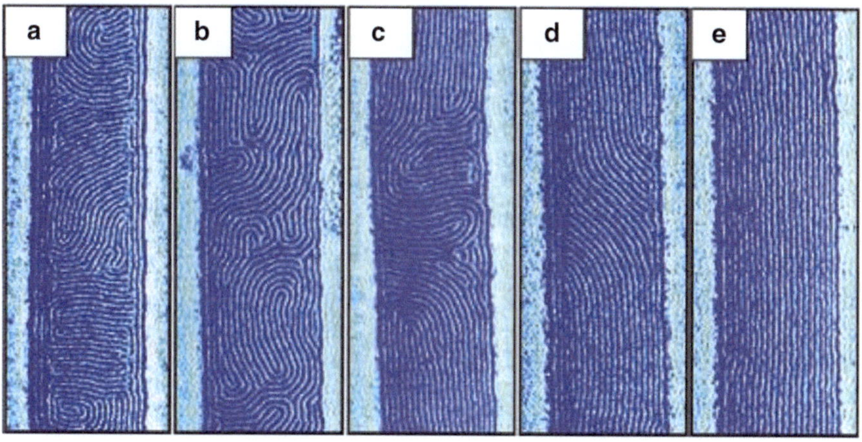

Fig. 6.14 SEM images of a diblock copolymer film confined in a topographic channel of width 600 nm. Figures (**a**)–(**e**) show the film evolution during 33 h of annealing. Reprinted with permission [24]

6.6 The Promise and Limitations of Patterning Polymer Films

The examples given throughout this chapter have shown that tailored variations in substrate surface chemistry or substrate topographic height can guide structures in polymer films. Unlike other polymer structuring methods, template structuring only gives order to features that naturally occur in the film. For example, block copolymer naturally phase separate but they do so in a way that lacks larger order. While templates can provide this long range order, a template will not radically change the length scale of the individual domains.

So what are the resolution limitations to the template structuring of polymer films? In one sense, the resolution of polymer film template structuring has been improving, but much of this improvement has arisen from improvements in template resolutions. Many of the original templates employed photolithography or microcontact printing (with stamps generated through photolithography), and the template features were about a micron or larger. Researchers have since used e-beam lithography and dip-pen nanolithography with their improved resolution, and achieved structuring at the resolutions of these lithography techniques [3, 15, 24]. However, the mechanisms harnessed are not substantially different with these finer templates. The resolution of these techniques is as much tied to the polymers used as the templates used. In general, higher resolution templates, thinner films, and films with smaller characteristic feature length scales are all necessary to produce finer structuring.

Another potential limitation of template structuring of polymer films is associated with generating arbitrary feature shapes. In the examples given above, polymer

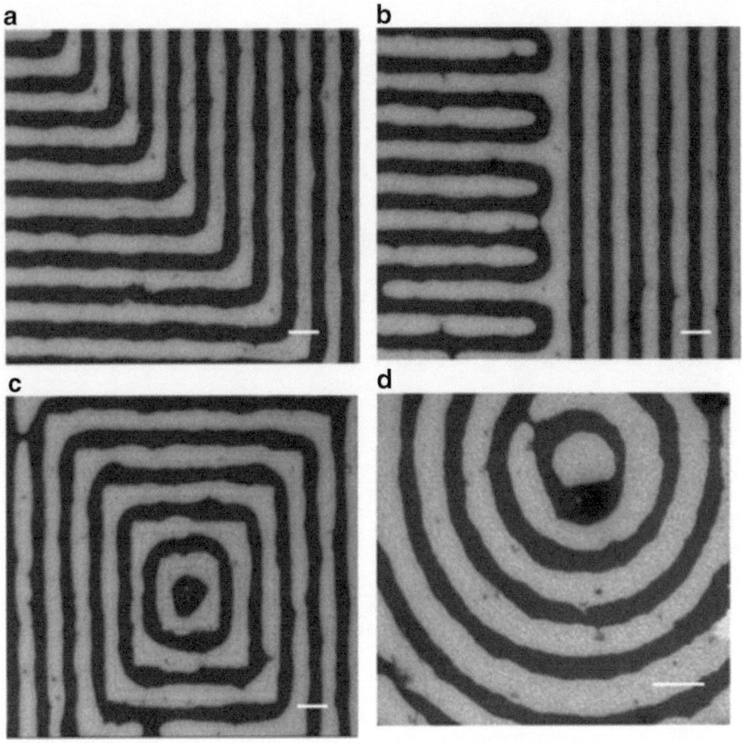

Fig. 6.15 SEM images of polymer blend film patterned by a substrate chemical template. Scale bar 1 μm. Reprinted with permission [26]

film structures have either included aligned lines or dots. However, experiments that are more recent have shown that geometries that are more complicated are possible. Nucleated features, such as in Fig. 6.9, allow one to generate essentially any pattern, though with a pixelated resolution limited to the individual feature size. Space filling phase separation such as shown in Fig. 6.7 cannot be formed into any arbitrary pattern, but groups have demonstrated the creation of nonregular features such as corners, circles, squares, and other simple, space filling patterns (see Fig. 6.15) [26]. In these cases, the corners of these shapes cannot be sharper than the natural feature size, and larger patterns will always be filled with the smaller scale phase separation.

A variety of additional techniques are being developed that combine substrate templating and other methods [27, 28]. In Fig. 6.1, for instance, the originally structured polymer film was combined with oxidation and printing to fabricate a layered mesh. As another example, many researchers have included nanoparticles within their patterned polymer film in an effort to structure, not the polymer, but the nanoparticles. Further, complex structures also can be achieved in combination with selective etching, electric fields, or molding.

In summary, the methods born in recent decades to pattern polymer films by means of an underlying templated substrate have grown in sophistication and flexibility. These techniques harness the natural tendencies of polymer films to dewet and phase separate, and through a simple means guide these natural processes. In all cases, the template and the desired structure must be matched to the properties of the polymer film and the processing conditions, but the final films can exhibit a variety of designs over many length scales.

References

1. Jeong, J.W., Park, W.I., Do, L.-M., Park, J.-H., Kim, T.-H., Chae, G., Jung, Y.S.: Nanotransfer printing with sub-10 nm resolution realized using directed self-assembly. Adv. Mater. **24**, 3526–3531 (2012)
2. Germack, D.S., Checco, A., Ocko, B.M.: Directed assembly of P3HT:PCBM blend films using a chemical template with sub-300 nm features. ACS Nano **7**, 1990–1999 (2013)
3. Ginger, D.S., Zhang, H., Mirkin, C.A.: The evolution of dip-pen nanolithography. Angew. Chem. Int. Ed. **43**, 30–45 (2004)
4. Gates, B.D., Xu, Q.B., Stewart, M., Ryan, D., Willson, C.G., Whitesides, G.M.: New approaches to nanofabrication: molding, printing, and other techniques. Chem. Rev. **105**, 1171–1196 (2005)
5. del Campo, A., Arzt, E.: Fabrication approaches for generating complex micro- and nanopatterns on polymeric surfaces. Chem. Rev. **108**, 911–945 (2008)
6. Coffey, D.C.: Characterizing the local optoelectronic performance of organic solar cells with scanning-probe microscopy. Ph.D. dissertation, University of Washington, Seattle, WA (2007)
7. Boltau, M., Walheim, S., Mlynek, J., Krausch, G., Steiner, U.: Surface-induced structure formation of polymer blends on patterned substrates. Nature **391**, 877–879 (1998)
8. Kargupta, K., Sharma, A.: Mesopatterning of thin liquid films by templating on chemically patterned complex substrates. Langmuir **19**, 5153–5163 (2003)
9. Kargupta, K., Sharma, A.: Morphological self-organization by dewetting in thin films on chemically patterned substrates. J. Chem. Phys. **116**, 3042–3051 (2002)
10. Kargupta, K., Sharma, A.: Creation of ordered patterns by dewetting of thin films on homogeneous and heterogeneous substrates. J. Colloid Interface Sci. **245**, 99–115 (2002)
11. Kargupta, K., Sharma, A.: Dewetting of thin films on periodic physically and chemically patterned surfaces. Langmuir **18**, 1893–1903 (2002)
12. Gennes, P.G.: Scaling Concepts in Polymer Physics. Cornell University Press, New York (1979)
13. McNeill, C.R., Watts, B., Thomsen, L., Belcher, W.J., Greenham, N.C., Dastoor, P.C.: Nanoscale quantitative chemical mapping of conjugated polymer blends. Nano Lett. **6**, 1202–1206 (2006)
14. Wei, J.H., Coffey, D.C., Ginger, D.S.: Nucleating pattern formation in spin-coated polymer blend films with nanoscale surface templates. J. Phys. Chem. B **110**, 24324–24330 (2006)
15. Park, L.Y., Munro, A.M., Ginger, D.S.: Controlling film morphology in conjugated polymer: fullerene blends with surface patterning. J. Am. Chem. Soc. **130**, 15916–15926 (2008)
16. Kim, S.O., Solak, H.H., Stoykovich, M.P., Ferrier, N.J., de Pablo, J.J., Nealey, P.F.: Epitaxial self-assembly of block copolymers on lithographically defined nanopatterned substrates. Nature **424**, 411–414 (2003)
17. Stoykovich, M.P., Nealey, P.F.: Block copolymers and conventional lithography. Mater. Today **9**, 20–29 (2006)

18. Edwards, E.W., Montague, M.F., Solak, H.H., Hawker, C.J., Nealey, P.F.: Precise control over molecular dimensions of block-copolymer domains using the interfacial energy of chemically nanopatterned substrates. Adv. Mater. **16**, 1315–1319 (2004)
19. Cheng, J.Y., Ross, C.A., Smith, H.I., Thomas, E.L.: Templated self-assembly of block copolymers: top-down helps bottom-up. Adv. Mater. **18**, 2505–2521 (2006)
20. Coffey, D.C., Ginger, D.S.: Patterning phase separation in polymer films with dip-pen nanolithography. J. Am. Chem. Soc. **127**, 4564–4565 (2005)
21. Higgins, A.M., Jones, R.: Anisotropic spinodal dewetting as a route to self-assembly of patterned surfaces. Nature **404**, 476–478 (2000)
22. Xing, R., Luo, C., Wang, Z., Han, Y.: Dewetting of polymethyl methacrylate on the patterned elastomer substrate by solvent vapor treatment. Polymer **48**, 3574–3583 (2007)
23. Roy, S., Mukherjee, R.: Ordered to isotropic morphology transition in pattern-directed dewetting of polymer thin films on substrates with different feature heights. ACS Appl. Mater. Interfaces **4**, 5375–5385 (2012)
24. Sundrani, D., Darling, S.B., Sibener, S.J.: Guiding polymers to perfection: macroscopic alignment of nanoscale domains. Nano Lett. **4**, 273–276 (2004)
25. Fasolka, M.J., Harris, D.J., Mayes, A.M., Yoon, M., Mochrie, S.: Phys. Rev. Lett. **79**, 3018–3021 (1997)
26. Wei, M., Fang, L., Lee, J., Somu, S., Xiong, X., Barry, C., Busnaina, A., Mead, J.: Directed assembly of polymer blends using nanopatterned templates. Adv. Mater. **21**, 794–798 (2009)
27. Verma, R., Sharma, A., Kargupta, K., Bhaumik, J.: Electric field induced instability and pattern formation in thin liquid films. Langmuir **21**, 3710–3721 (2005)
28. Misner, M.J., Skaff, H., Emrick, T., Russell, T.P.: Directed deposition of nanoparticles using diblock copolymer templates. Adv. Mater. **15**, 221–224 (2003)

Chapter 7
Electrohydrodynamic Lithography of Functional Soft Materials for Advanced Applications

Pola Goldberg Oppenheimer

7.1 Introduction

A polymer is a chemical compound or mixture of compounds formed by polymerisation and consisting essentially of repeating structural units. The name is derived from the Greek, *πολυ*, polu, "many"; and *μέρος*, meros, "part". DNA, proteins and starches in foods, the wheels on our skateboards, the tyres on our bikes and cars are all polymers. In fact, we are constantly surrounded by polymers wherever we may be. Thin polymer films on top of rigid or soft substrates have been investigated thoroughly during the last few years. Generally, instabilities in thin films are undesirable in nature and technology. Defect-free smooth films in the form of insulating layers or photoresists, e.g. coatings, lubricants or protective layers are essential for most important applications in the microelectronics industry. Liquid films are stable, unless an external destabilising force is applied at the surface, because of the minimisation of overall surface energy. The precise control of instabilities in films can however be utilised to produce novel structures by exploiting the destabilisation of an initially homogeneous layer for spontaneous structure formation process, and further decrease its length-scale to technologically interesting feature sizes. The capacity to construct nanostructures is essential for the development of functional devices that incorporate nanoscale features. The thin film instabilities can be generated either by intermolecular interactions, such as *van* der Waals forces (vdW), or by externally applied forces, e.g. electric fields or temperature gradients. Patterned thin polymer films have been implemented in applications ranging from molecular electronics [1–3] and optical devices [4–6], through bio-sensors [7, 8] to etching prevention [9].

P.G. Oppenheimer (✉)
School of Chemical Engineering, University of Birmingham, Edgbaston, Birmingham B15 2TT, UK
e-mail: p.goldbergoppenheimer@bham.ac.uk

J. Rodríguez-Hernández, C. Drummond (eds.), *Polymer Surfaces in Motion*, DOI 10.1007/978-3-319-17431-0_7

7.2 Surface Patterning via Electrohydrodynamic Instabilities

Intense interest in polymer patterning in the past decade originated from the diversity of the existing polymers, the relatively low cost, the convenient mechanical properties, and the compatibility of polymers with most patterning techniques. There are different bottom-up and top-down lithographic methods, such as photolithography, soft lithography imprinting, self-assembly of block copolymers, and instability-induced patterning. In an instability-driven approach to polymer patterning, application of a high (10^8 V/m) electric field across thin films at temperatures above the glass transition temperature of the polymer generates variations in charge density, destabilises the film and creates lateral, topographic features [10–13]. The emerging ordered patterns are a direct result of the interaction of a vertically applied electric field with a flow of fluid, i.e. it is named the electrohydrodynamic (EHD) patterning process. The physical principles that describe the deformation of a liquid surface induced by an electric field were first noted in 1897 [14], and were followed by the initial theories on the EHD instabilities [15, 16]. Ever since these ground-breaking studies, extensive investigations have been carried out on these issues [17–20].

Consider placing a liquefied insulator in a plate capacitor geometry and subsequently applying an electric field perpendicular to the film surface (Fig. 7.1). This causes an amplification of low amplitude capillary waves and EHD instabilities with a characteristic wavelength λ develop. First, the non-linear equation that governs the pattern formation process is formulated, followed by a simplified description in terms of a linear stability analysis leading to the derivation of a dispersion relation. At temperatures above the glass transition temperature (T_g), polymers can be described as incompressible viscous fluids. Given the geometry of a supported thin polymer film on a rigid substrate we can derive the equation of motion for this film.

For the typical experimental set-up schematically shown in Fig. 7.1, the natural choice is rectangular Cartesian coordinates. The y-coordinate of the film's interface is given by $h = h(x.t)$, where x is the lateral coordinate, t is the time and h is the film thickness.

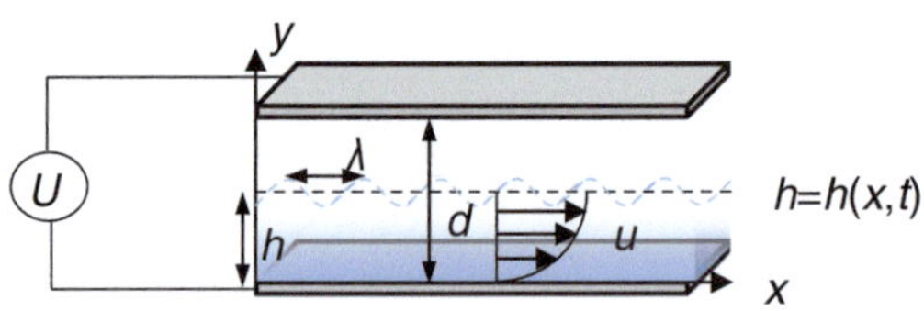

Fig. 7.1 Schematic geometry of a thin supported polymer film of thickness h. Temperature fluctuations induce surface waves with wavelength λ. The Poiseuille-type flow is indicated by a parabolic flow profile with velocity u

The *Navier–Stokes equation of motion* for an incompressible Newtonian fluid is:

$$\rho\left(\frac{\partial u}{\partial t}\right)+u\left(\frac{\partial u}{\partial x}\right)+v\left(\frac{\partial u}{\partial y}\right)+w\left(\frac{\partial u}{\partial z}\right)=-\frac{\partial p}{\partial x}+\eta\left(\frac{\partial^2 u}{\partial x^2}+\frac{\partial^2 u}{\partial y^2}+\frac{\partial^2 u}{\partial z^2}\right)+\rho g_x \tag{7.1}$$

The required *equation of continuity* is:

$$\frac{\partial u}{\partial x}+\frac{\partial v}{\partial y}+\frac{\partial w}{\partial z}=0 \tag{7.2}$$

For flow in thin films, the N-S equation can be simplified in the following way:

1. High viscosities of the considered fluids result in a low flow velocity: the convective term in Navier–Stokes equation can be neglected $\rightarrow$ $u\left(\frac{\partial u}{\partial x}\right)+v\left(\frac{\partial u}{\partial y}\right)+w\left(\frac{\partial u}{\partial z}\right)=0$
2. Slow dynamics: quasi-steady state is assumed for the system $\rightarrow \rho\left(\frac{\partial u}{\partial t}\right)=0$
3. Thin-film geometry: gravity is negligible $\rightarrow \rho g_x=0$
4. Flow in the film is in the x direction $\rightarrow$ v $= w = 0 \rightarrow \frac{\partial u}{\partial x}=0$ and the velocity dependent only on the y-coordinate.

Taking into account all the above considerations and substitution of the continuity equation leaves us with:

$$0=-\frac{\partial p}{\partial x}+\eta\left(\frac{\partial^2 u}{\partial y^2}\right)\quad\Rightarrow\quad\frac{1}{\eta}\left(\frac{\partial p}{\partial x}\right)=\frac{\partial^2 u}{\partial y^2} \tag{7.3}$$

and

$$0=\frac{\partial p}{\partial y} \tag{7.4}$$

meaning that the pressure p is uniform across the film's depth. Integration of this differential equation leads to a parabolic velocity profile:

$$u=\frac{1}{2\eta}\left(\frac{\partial p}{\partial x}\right)y^2+c_1 y+c_2 \tag{7.5}$$

The *boundary conditions* are that the velocity is equal to zero at the solid interface and that there are no stresses at the interface:

1. $y=0$; $u=0$ (non-slip boundary condition)
2. $y=h \quad \sigma_{xy}=\eta\frac{\partial u}{\partial y}=0$

Therefore, the constants that follow from the boundary conditions are:

$$c_1 = -2\frac{\partial p}{\partial x}y\big|_{y=h} = -2\frac{\partial p}{\partial x}h, \quad c_2 = 0$$

Substitution of which in Eq. (7.5) gives:

$$u = \frac{1}{2\eta}\left(\frac{\partial p}{\partial x}\right)y(y-2h) \tag{7.6}$$

In the lubrication assumption, the lateral Poiseuille volume flow rate induced in the direction of decreasing pressure is obtained:

$$j_x = \int_0^h u dy = -\frac{1}{3\eta}\frac{\partial p}{\partial x}h^3 \tag{7.7}$$

The overall pressure is uniform all over the depth of the film. However, the pressure does depend on the film thickness, and its distribution at the film surface can be written as:

$$p = p_0 - \gamma\frac{\partial^2 h}{\partial x^2} + p_{\mathrm{el}}(h) + p_{\mathrm{dis}}(h) \tag{7.8}$$

where p_0 is the ambient air pressure, γ is the surface tension and the second term represents the Laplace pressure. For an applied voltage, U, the third term is the destabilising *electrostatic pressure* exerted on the interface by the interaction of the electric field, E_{p}, with the polarisation charges at the interface [21, 22] and given by:

$$p_{\mathrm{el}} = -\varepsilon_0\varepsilon_{\mathrm{p}}(\varepsilon_{\mathrm{p}} - 1)E_{\mathrm{p}}^2 = -\varepsilon_0\varepsilon_{\mathrm{p}}(\varepsilon_{\mathrm{p}} - 1)\frac{U^2}{\left[\varepsilon_{\mathrm{p}}d - (\varepsilon_{\mathrm{p}} - 1)h\right]^2} \tag{7.9}$$

where ε_0 is the dielectric permittivity of the vacuum, ε_{p} is the dielectric constant of the polymer and d is the capacitor plate spacing. The initial stage of structure formation is a sinusoidal undulation (Eq. 7.10) with an initial film thickness h_o and low-amplitude undulation A

$$h(x,t) = h_0 + A\mathrm{e}^{(iqx+t/\tau)} \tag{7.10}$$

The last term of the Eq. (7.8) is the disjoining pressure which is a result of the intermolecular forces exerted on the thin film [23] which plays a role in the near surface regions only. Due to a strong electric field generated in a capacitor device, only the electrostatic pressure and the Laplace term need to be taken into consideration.

A mass balance enforced by the continuity equation governs the equation for the air–polymer interface profile:

$$\frac{\partial h}{\partial t} + \frac{\partial}{\partial x}\left(\frac{h^3}{3\eta}\frac{\partial p}{\partial x}\right) = 0 \tag{7.11}$$

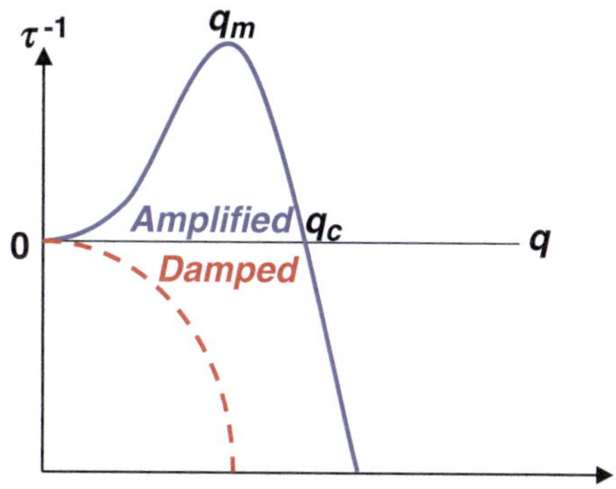

Fig. 7.2 Graphic representation of the dispersion relation (Eq. 7.12). While in the absence of or for a positive electrostatic pressure, all modes are damped ($\tau < 0$), the dispersion relation yields a dominant mode q_m with corresponding growth rate time $\tau_m{}^{-1}$, for a negative p_{el}

A linear stability analysis [21, 24–26] leads to the establishment of the *dispersion relation* for the system relating the time constant τ with the wave vector q of a sinusoidal perturbation of the film:

$$\frac{1}{\tau} = -\frac{h_0^3}{3\eta}\left[\gamma q^4 + \frac{\partial p_{el}}{\partial h} q^2\right] \tag{7.12}$$

The predictions are schematically shown in Fig. 7.2 While for $\frac{\partial p_{el}}{\partial h} \geq 0$, $\tau < 0$ for all q - fluctuations are damped, meaning that all growth rates are negative and the film is *stable*, for $\frac{\partial p_{el}}{\partial h} < 0$ all modes with $\tau > 0$ and $q < q_c = \sqrt{-\frac{1}{\gamma}\frac{\partial p_{el}}{\partial h}}$ are amplified and the film is *unstable*.

With time, the fastest mode of growth given by the maximum of Eq. (7.12):

$$q_m = \sqrt{-\frac{1}{2\gamma}\frac{\partial p_{el}}{\partial h}} \tag{7.13}$$

will dominate, governing the most unstable wavelength of the system:

$$\begin{aligned} \lambda_m &= 2\pi\sqrt{\frac{2\gamma}{-{}^{\partial p}/_{\partial h}}} = 2\pi\sqrt{\frac{\gamma U}{\varepsilon_0\varepsilon_p(\varepsilon_p - 1)^2}}E_p{}^{-3/2} \\ &= 2\pi\sqrt{\frac{\gamma\left[\varepsilon_p d - (\varepsilon_p - 1)h\right]^3}{\varepsilon_0\varepsilon_p(\varepsilon_p - 1)^2 U^2}} \end{aligned} \tag{7.14}$$

The associated maximal growth scales with the forth power of the dominant growing wave vector, proportional to the surface tension and depending inversely on the viscosity for a given λ:

$$\frac{1}{\tau_m} = \frac{\gamma h_0^3}{3\eta} q_m^4 \tag{7.15}$$

As seen from Eq. (7.14), the theory suggests that a reduced feature size can be achieved by a thinner film, h, smaller plate spacing, d, lower surface tension and higher electric field.

The introduction of the reduced variables allows the superimposition of different experimental data sets. Defining characteristic variables of the instability:

$$\lambda_0 = 2\pi\left[\varepsilon_0\varepsilon_p(\varepsilon_p - 1)U^2\right]/\gamma \tag{7.16}$$

$$E_0 = 2\pi U/\lambda_o \tag{7.17}$$

$$\tau_0 = 3\eta/h_0^3 q_0^4 \tag{7.18}$$

Equation (7.14) can be written in simplified form:

$$\frac{\lambda}{\lambda_0} = \left(\frac{E_p}{E_0}\right)^{-3/2} \tag{7.19}$$

And Eq. (7.15) along with Eq. (7.18) provides the dependence of the time constant on the electric field:

$$\frac{\tau}{\tau_0} = \left(\frac{E_p}{E_0}\right)^{-6} \tag{7.20}$$

In which λ_0 is the characteristic wavelength, E_0 is the corresponding characteristic electric field and τ_0 is a characteristic time constant of the instability. As can be seen from Eqs. (7.16)–(7.20), the most unstable mode dependence on the electric field is $\lambda \propto E^{-3/2}$, characteristic time constant for the instability to set in has a very strong dependence on the electric field $\tau \propto E^{-6}$ and scales with the electrode spacing, d, to the sixth power ($\tau \propto d^6$). It also scales linearly with the film viscosity $\tau \propto \eta$. For a given film thickness, the characteristic lateral structure size exhibits a power-law dependence as a function of increasing electric fields, corresponding to a decrease in the electrode spacing.

To compare the destabilisation of materials with differing viscosities, it is useful to introduce an additional set of scaling relations. The inverse of the time constant defines a rate for the lateral displacement of material over a distance λ. Multiplied by η this defines a characteristic shear stress:

$$\sigma = \frac{\eta}{\tau} \tag{7.21}$$

Defining:

$$\sigma_0 = \frac{\eta}{\tau_0} \tag{7.22}$$

introduces the further scaling equation:

$$\frac{\sigma}{\sigma_0} = \left(\frac{E_{\mathrm{P}}}{E_0}\right)^6 \tag{7.23}$$

7.2.1 EHD Patterning Under a Homogeneous Electric Field

The physical basis of the film instability can be understood by considering the competition of forces acting at the polymer–air interface. A polymer when heated above its glass transition temperature begins to flow as a viscous liquid. Subsequently, applying a voltage to the capacitor device (Fig. 7.3a) results in the development of an undulatory instability with a characteristic wavelength (Fig. 7.3b). The liquefied polymer experiences surface tension and gravity, which stabilise an initially homogeneous film, in contrast to the electrostatic pressure at the interface which destabilises the film. In the thin-film geometry, gravity is usually negligible and can therefore be disregarded.

For high electric fields, the destabilising electrostatic force overpowers the stabilising effects of surface tension acting at the film–air interface. With time, this force imbalance causes an amplification of film undulations induced by the electric field (Fig. 7.3b), until they span the capacitor gap, thus leading to rearrangement to the energetically favoured configuration of the liquid. It is characterised by monodispersed column diameters and a local hexagonal symmetry, which arises from the repulsion of wave maxima and minima, caused by the differing polarisation charges at these locations (Fig. 7.3c). The inter-column spacing is determined by the field strength and the polymer surface tension, and the column diameter is a function of the initial film thickness. Since the vertical liquid conformation (Fig. 7.3c) has lower electrostatic free energy than layered configurations (Fig. 7.3b), the electrostatic force drives the liquefied polymer to grow upward focusing the instability in the direction of the highest field.

Spacers of different height cause a gradual variation of the plate spacing, estabishing a wedge geometry, in which the plate spacing varies by a few micrometres over a lateral distance of a few centimetres (Fig. 7.4). With a varying plate spacing in a lateral direction, the electric field varies as well. Since the characteristic time constant for the instability to set in has a very strong dependence on the electric field, $\tau \propto E^{-6}$ and scales with the electrode spacing, d, to the sixth power $\tau \propto d^6$, larger electrode spacing corresponds to a lower electric field. Therefore, for a given annealing time, different stages of instability can be observed on a single sample (Fig. 7.4, left). While in the wedge geometry, the spontaneous instability propagates towards decreasing confinement, the density and ordering of the columns increases (Fig. 7.4).

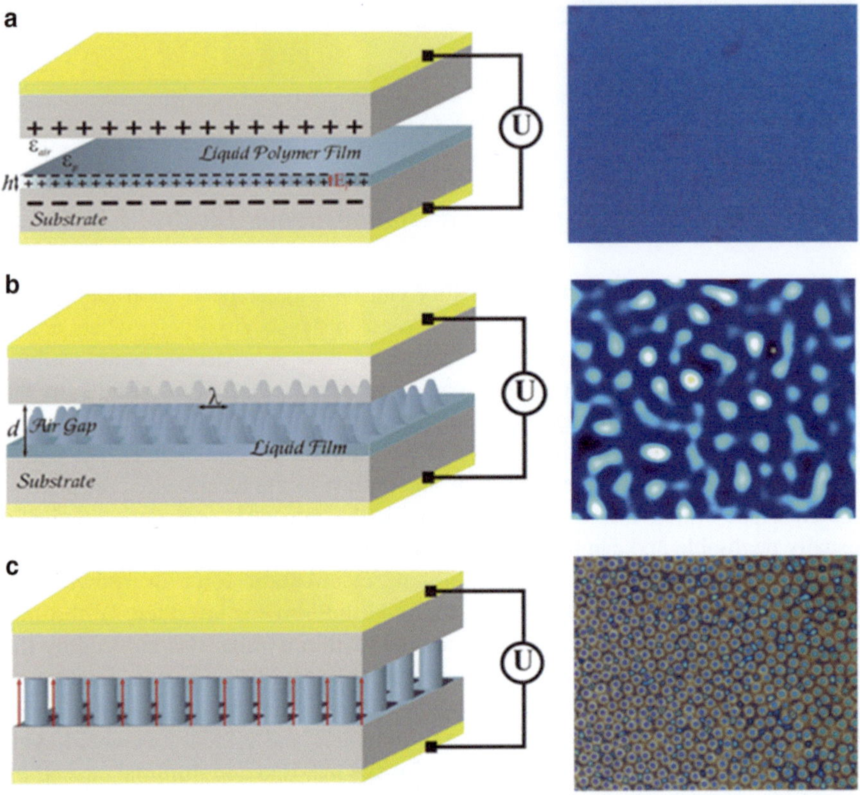

Fig. 7.3 Representative experimental set-up (*left*) with an evolving EHD instability of a polymer film under a laterally homogeneous electric field and the corresponding optical micrographs (OM) of the EHD pattern formation (*right*). (**a**) Applying a voltage through the capacitor assembly, results in the amplification of a surface undulation with a characteristic wavelength λ, which leads to the formation of hexagonally ordered columns (**c**). The origin of the destabilising electrostatic pressure is schematically shown in (**b**): the electric field causes the energetically unfavourable build-up of displacement charges at the dielectric interface (**a**), and in (**c**) the alignment of the dielectric interface parallel to the electric field lines lowers the electrostatic energy

7.2.2 Electrohydrodynamic Lithography

Substitution of the planar electrode by a topographically structured top plate (Fig. 7.5a) results in an inhomogeneous electric field, which focuses the instability towards the increased electric field, i.e. underneath the protruding structures, with an increased associated growth rate (recalling that $\tau \propto E^{-6} \propto d^6$, and therefore the polymer is destabilised first at the locations with the smallest d). Since $p_{\mathrm{el}} \propto 1/d^2$ (Eq. 7.9) the electrostatic pressure is much stronger for smaller electrode distances, pattern replication of topographically structured upper electrode proceeds at a shorter time scale than the regular EHD structure formation.

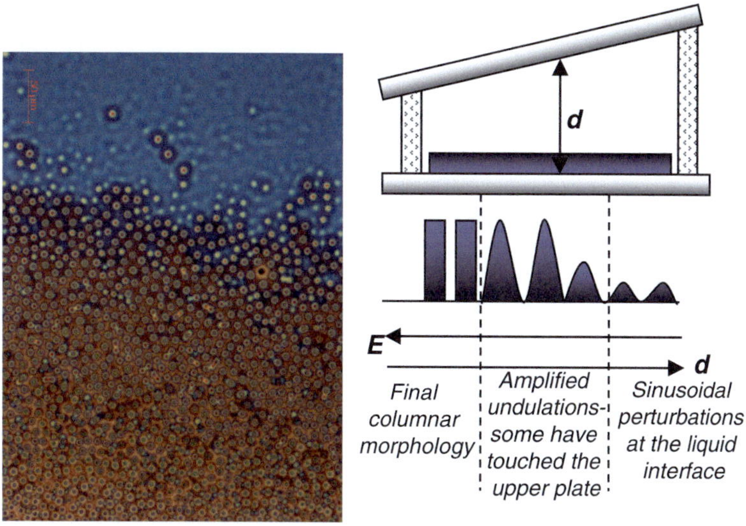

Fig. 7.4 *Left*: OM micrograph in which different stages of the instability are all present simultaneously. While on the top of the image undulations coexist with columns, the decreasing plate spacing towards the bottom increases the density of the columns, and thereby the inter-column interactions and ordering. *Right*: Largely exaggerated schematic representation of wedge geometry with the variation of the gap between the electrodes which cause the non-uniform strength of the electric field in the experiment and allows visualising the progress of the instabilities as they propagate from low to high field. Locally the configuration is nearly parallel

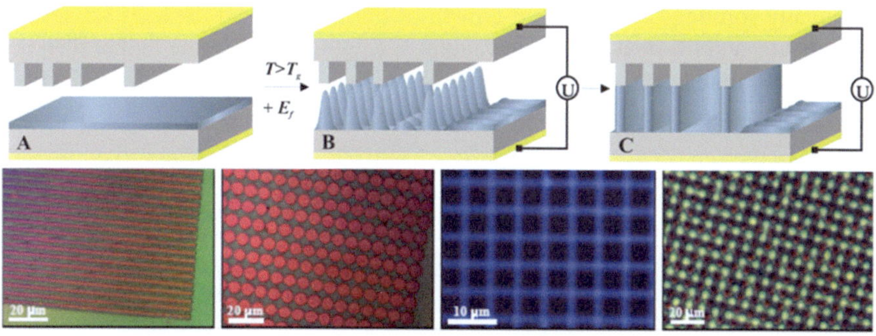

Fig. 7.5 *Top*: Schematic of experimental device during the EHD lithography process under a laterally heterogeneous electric field. (**a**) A structured upper plate creates a heterogeneous force field focusing the instability towards the protruding structures (**b**). In (**c**) a positive replica of the master pattern is transferred into the polymer. In unstructured regions, the film remains stable on a much longer time scale. *Bottom*: High fidelity EHD lithography showing the robustness of pattern replication with respect to the parameters of the electrode

The instability is guided towards the protruding patterns of the top mask, and the liquid polymer is drawn towards these protrusions. Consequently, the final structures are a positive replication of the master patterned electrode (Fig. 7.5). Pattern replication succeeds only if the length scale of the master pattern matches the instability wavelength. Using this electrostatic lithography process, varying structures can be replicated, ranging from lines, through circles and squares to letters and numbers.

7.3 Formation of Additional Instability Patterns

7.3.1 Under a Featureless Mask: Nucleation and Viscous Fingering Instabilities

When placing a featureless mask at a certain distance above the polystyrene film in a presumably laterally homogeneous electric field, additional intriguing patterns can be observed (Fig. 7.6). The image shows the nucleation effect at the edge of the film, which implies heterogeneity in the electric field, caused either by the edge of the upper electrode or by a dust particle. A strong lateral gradient in the electric field is induced from the presence of the edge, leading to the nucleation of the EHD instability along the edge [22]. However, instead of nucleated columns with a characteristic wavelength, λ, an alternating sequence of columns and wave maxima were observed: only every other maximum touches the upper electrode to form a pillar. This phenomenon was also previously observed by Schäffer et al. and was described as a competition between nucleation, depletion and the non-linearities in the system. A brief indication of the nucleation pathway is an initial amplification of a continuous rim parallel to the edge, which in turn breaks-up into waves eventually forming columns. The growth of the first column depletes its nearest neighbours, and the next nearest maximum is amplified first, being higher than the directly adjoining one. This chain effect propagates along the rim, and given enough time, all waves eventually will develop into columns.

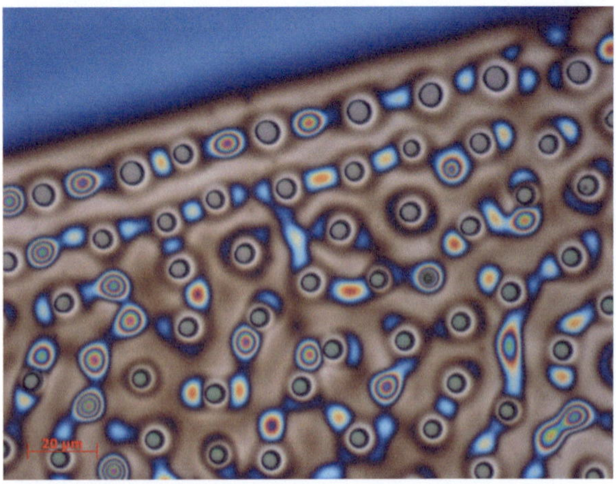

Fig. 7.6 Optical micrograph of the nucleation of alternating columns and wave maxima induced by a straight edge

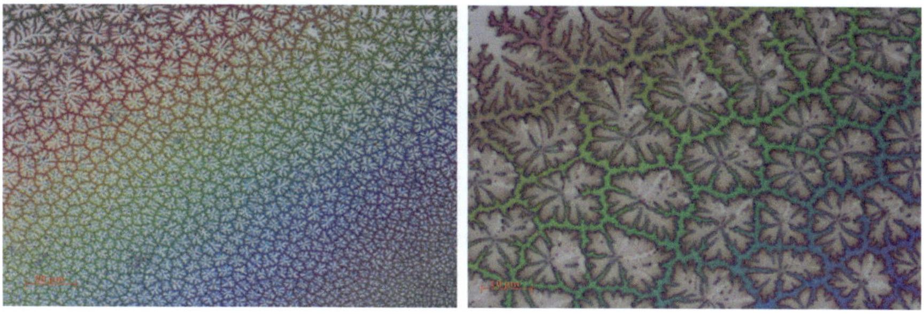

Fig. 7.7 Optical images of viscous fingers of trapped air appearing under the flat mask

When a fluid pushes a more viscous fluid into a very narrow gap between parallel plates, the interface between the two fluids develops an instability leading to the formation of finger-like patterns called viscous fingers (Fig. 7.7) [27]. Viscous fingering instability belonging to fractal geometry is a result of air trapped underneath the flat mask, which is the less viscous fluid driving the polymers melt. The physical source of this instability lies in the geometry of the moving interface [28].

7.3.2 Periodicity Mismatch Under Laterally Varying Electric Field

Imperfect replication of the template can occur however, in part of the lithography processes. These are explained based on the three replication cases which have been previously theoretically established [21] and experimentally observed [29]. In the presence of the laterally varying confinement (from d' to d), the liquid morphologies are organised according to the ratio of the plate spacing and the initial amount of polymer in the capacitor gap, i.e. the filling ratio, $f (f = h_0/d)$. While the pattern selection during the early stage of the process is a sinusoidal surface undulation, for all samples, filling ratio establishes the late stage of pattern formation. The final morphology of the replicated pattern is determined by the partial coalescence of the initial pattern (Fig. 7.8). Since the shape of the generated pattern depends on the ratio of the intrinsic film undulation wavelength, λ_i, and the lateral periodicity of the master electrode, L_p, three replication mechanisms can be verified:

1. $\lambda_i \ll L_p$: *Periodicity mismatch-small wavelength regime*. The initial structure formation is followed by lateral coarsening of polymer material λ_i, leading to a partial positive replication.
2. $\lambda_i \approx L_p$: *Periodicity match-similar wavelength regime*. Perfect positive replica of the templates is obtained.
3. $\lambda_i \gg L_p$: *Periodicity mismatch-large wavelength regime*. The film pattern develops large number of defects and every protrusion of the electrode cannot produce a liquid column.

Fig. 7.8 A heterogeneous electric field is formed under the structured upper electrode, focusing the instability towards the protruding structures. (**a**–**c**) Optical microscopy images show rather imperfect replications of different structure of the electrodes under the periodicity mismatch conditions (e.g. in (**a**) $d = 500$ nm, $h/d' = 0.37$, $h = 83$ nm)

To obtain a faithful replication of the master topography, three criterions have to be fulfilled: (1) The filling ratio h/d' must be comparable to the surface area ratio, S (fraction of the template surface that protrudes towards the polymer film, $S = w/l_o$) of the topography, $h/d' \approx S$. (2) Periodicity match, $\lambda_i \approx L_p$ and, (3) The ratio of the interstitial spacing and the wavelength is smaller or equals to one.

7.4 Rapid EHD Patterning

While the origin of the underlying electrohydrodynamic (EHD) surface instabilities has long been established [14–16], its exploitation for high-throughput lithography is novel, and provides a way to replicate sub-micron scale patterns in a simple and robust fashion. One of the limiting aspects has, however, been the slow pattern-formation dynamics. Typical polymeric resists require prolonged-time span for the pattern formation to complete [30]. This is caused by the interplay of the factors that most resists consist of high molecular-weight (M_w) polymers which have a high melt viscosity in their accessible temperature window and the ability of accurate control over patterning onset and cessation. While too long annealing times result in coalescence of small-scale structures, which degrades pattern fidelity, too short an exposure to the electric field gives rise to an incomplete pattern replication. Since the electric field strength is limited by the dielectric strength of the materials in the narrow capacitor gap, the obtainable strain rates are generally low. Reducing the M_w of the resists typically leads to the reduction of the polymer's glass transition temperature, T_g, prerequisiting the solidification above room-temperature.

A number of low-viscosity materials have been studied and exploited for fast EHD patterning, aiming to resolve the above intrinsic problems [31]. Different from the previously used polymers, the chosen materials had both a thermally accessible T_g as well as low viscosities upon softening, which resulted in a considerable reduction of the completion of the EHL to a few seconds only rather than hours, thus increasing the technological appeal of this technique. Low-viscosity epoxy materials were also studied by Dickey et al., who reported

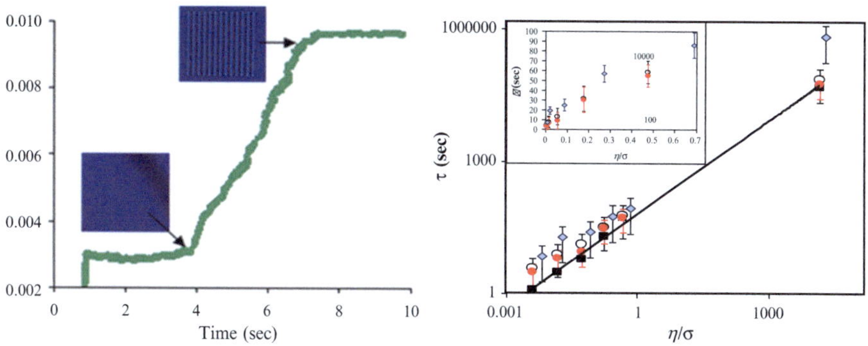

Fig. 7.9 (**a**) Overall-circuit current across capacitor devices versus time during the EHD patterning process. Ethyl Cellulose with $\eta = 0.084$ Pa s in a laterally varying electric field caused by a topographic grating ($h \approx 100$ nm, $d \approx 250$ nm (average), $U = 40$ V, and $T = 135$ °C). The *insets* show OMs of the sample at the various stages. (**b**). Variation of τ with viscosity in terms of Eq. (7.21). The data obtained by current monitoring (*full red circles*: $\tau = t_i$) agree within their error margins and are well described by the prediction of Eq. (7.21) (*line*). The data obtained from the termination times (*diamonds*, $\tau = t_f$) follow the same trend, but are offset towards longer times

similarly low patterning times [32]. A rapid EHD lithography, however, requires precise termination of the pattern formation process. While Dickey and co-workers achieve this by cross-linking of the resists, Goldberg Oppenheimer et al. demonstrated a method for the in-situ control of the EHL process during this short time span by monitoring the current across the patterning device [31]. Since the current is passed through the leaky dielectric once it spans the capacitor gap, the overall current is a good measure for electrohydrodynamic patterning completion.

Nevertheless, the acquired speed of the EHD instability requires an enhanced monitoring of the patterning process. Thus, minute but finite current that flows through the capacitor during pattern formation has been exploited. The current trace exhibits a sigmoidal shape that defines the characteristic onset and completion times of the EHD process (Fig. 7.9a). While t_i corresponds closely to the characteristic time of the instability τ, t_f indicates the completion of the pattern process.

To prevent a deterioration of the pattern by coarsening, it is essential that the sample is solidified at $t \approx t_f$. Because of the rapid patterning, quick sample quenching is essential. In situ optical investigation with a transparent electrode was thus exploited to correlate the structure formation process with the output current. Figure 7.9b shows the variation of the film destabilisation time with the viscosity of the polymer. To compare samples with different experimental parameters, the data is plotted as a function of η/σ, where all other experimental parameters are absorbed in the shear stress σ. While the optically detected onsets of the instabilities and of the current increase τ_i are well described by the prediction of Eq. (7.22), the completion of patterning τ_f is offset towards higher values compared to the prediction of τ. To be noted that, despite the significantly more rapid patterning time, the EHD process is similar to that of high viscosity polymers and is therefore well described by a linear stability analysis.

7.5 EHD Patterning of Conductive Polymers

Since the total potential difference generated by the dipole layers at the interface is suppressed across the liquid layer, the driving force of the pattern formation in the case of a *leaky dielectric* polymer subjected to the EHL patterning lies in the electric field in the air gap, E^1 [25]. A sub-ambient pressure within the film balances the electrostatic force due to the field in the air gap on the polymer–air interface, placing the film in tension, and therefore generating the origin of the EHL instability.

For EHL patterning of conductive polymers however, free charges in the film (which substantially modify the electric field distribution in the film-air double layer) have to be taken into consideration. The following discussion is based on the formalism proposed by Pease and Russell [25] for charge-driven EHL patterning of leaky dielectric films.

p scales with the square of U

$$p = -\frac{1}{2}\frac{\varepsilon^{g}\varepsilon_0 U^2}{(d - h_0)} \tag{7.24}$$

with the gap fluid ε^{g} (which is an air gap in our case, i.e. $\varepsilon^{g} = \varepsilon_1$). The conductivity of the leaky dielectric suppresses the electric field in the film and the field in the gap drives the EHL pattern formation. This yields the dimensionless conductivity, Σ representing the ratio of a time scale for free charge conduction to the process time scale

$$\sum = \frac{\sigma\eta\gamma d^3}{(\varepsilon_1)^3\varepsilon_0^3 U^4} \tag{7.25}$$

With the limit of $\Sigma \gg 1$ (in our study Σ is on the order of 10^7) the most dominant (characteristic) wavelength

$$\lambda = \frac{(2\sqrt{2})\pi}{U}\sqrt{\frac{\gamma(d - h_0)^3}{\varepsilon_0}} \tag{7.26}$$

is given by a force balance between the destabilising electrostatic pressure due to the field in the gap acting on the polymer–air interface with respect to the interfacial height, h_0 and the γ acting to minimise the surface area, thus suppress the height variations. The characteristic time constant for the instability is given by

$$\tau = \frac{12\gamma\eta(d - h_0)^6}{U^4 h_0^3 \varepsilon_0^2} \tag{7.27}$$

Despite the shorter destabilisation time and faster growing modes of a fully conducting polymer compared to a perfect dielectric, the patterning process is reminiscent of the well-studied case of generic polymers [33–35], thus confirming the same underlying physical mechanism.

To assess the electrolyte-gated field effect transistor (FET) performance, EHL-generated structure arrays were fabricated with gate length of 700 nm and pitch of 500 nm. Liquid-ion gate FET geometry was constructed using a potassium chloride (KCl) solution and a tungsten needle as a contacting electrode (Fig. 7.10a). The electrical drain current (I_{DS})-drain voltage (V_{DS}) characteristics of the device as a function of different gate voltages (V_G) are shown in Fig. 7.10b. Inset ii of Fig. 7.10b, presents the transfer characteristics at a constant drain-source voltage.

The positive gate voltages decrease the current and the values of I_{DS} increase upon raising the positive V_G at a negative V_{DS} indicating that the device exhibits *p*-type FET behaviour with the holes being the major charge carriers. Along with promising electrical transistor characteristics, the use of low-cost EHL lithographic technique and simple gate definition process steps could make such devices suitable candidates for next generation technology nodes.

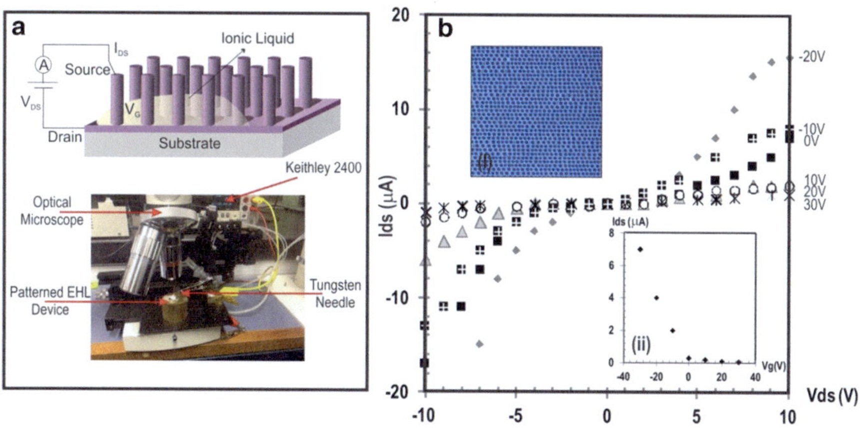

Fig. 7.10 FET performance of EHL-polypyrrole (PPY) architectures. (**a**) Schematic representation (*top*) and an overview image (*bottom*) of the configuration of a liquid-ion gate vertical FET using the EHL-generated structures on top of Si-SiO_2 substrate. A probe tip comprised of a tungsten needle was employed under a 1,000× magnification of an optical microscope to establish good connection between the source and the drain. (**b**). Drain current versus drain voltage characteristics of PPY electrolyte-gated transistor based on EHL fabricated pillars shown in a top-view optical image (*inset* **b**,i) and gate voltage performance (inset **b**, ii) of the PPy-FET described in (**a**)

7.6 EHD-Generated Architectures in Thin Polymer Films for Applications in Surface Enhanced Raman Scattering (SERS) and Solar Cell Devices

7.6.1 Hierarchical EHD Architectures as SERS Platforms

Using electrohydrodynamic instabilities, controlled self-organised hierarchical architectures can also be fabricated. Hierarchical electrohydrodynamic (HEHD) patterning uses a sequential instability in multilayer thin films caused by an applied electric field to guide the layered material to form designed structures in a one step process. Parameters can be altered to change the dimensions and the spacing of the structures during the HEHD patterning which are then covered in a plasmon-active-metal (in this case gold) to enhance the electromagnetic field for SERS applications. Using HEHD multi-scale three-dimensional structured arrays SERS substrates can be realised [35].

The structure formation in a multilayer thin-film assembly is controlled by carefully designing the layer sequence. In a polymer($P1$)–polymer($P2$)–polymer($P3$)–air trilayer (Fig. 7.11), an uncontrolled dewetting of the multilayer is prevented allowing sequential destabilisation of the assembly layer by layer. The trilayer is assembled according to the Tg of the individual layers, starting from the highest for the bottom thin layer and the lowest for the top film, allowing each layer to be liquefied individually while preserving the underlying film in a stable state. The resulting hierarchical structures are an outcome of sequential electrostatic destabilisation. Since, the wetting properties of the three materials come also into play once the composite columns have formed, organising the resists according to the increasing surface tension stabilises the final coaxial architectures.

The various HEHD patterns show structure dependent SERS. For an optimised aspect ratio and geometry they give a 100-fold enhancement in SERS signals. Considering that each of the individual structures gives such high enhancements the HEHD patterned surfaces are a promising platform for multiplexed SERS detection. The demonstrated hierarchical structures open up a range of further possible implementations including for instance advanced photo-catalysis and bio-analytics.

7.6.2 Combined Bottom-Up *and* Top-Down *Approach Based on EHD Formation of Surface Patterns in Thin Films*

As depicted in Chap. 3, the ability to pattern functional polymers at different length scales is important for a range of applications and developments including the fabrication of micro plastic electronics, the production of optical components and bio-medicinal related research. EHD patterning can also be exploited for the fabrication of hierarchical functional patterns using electric field induced

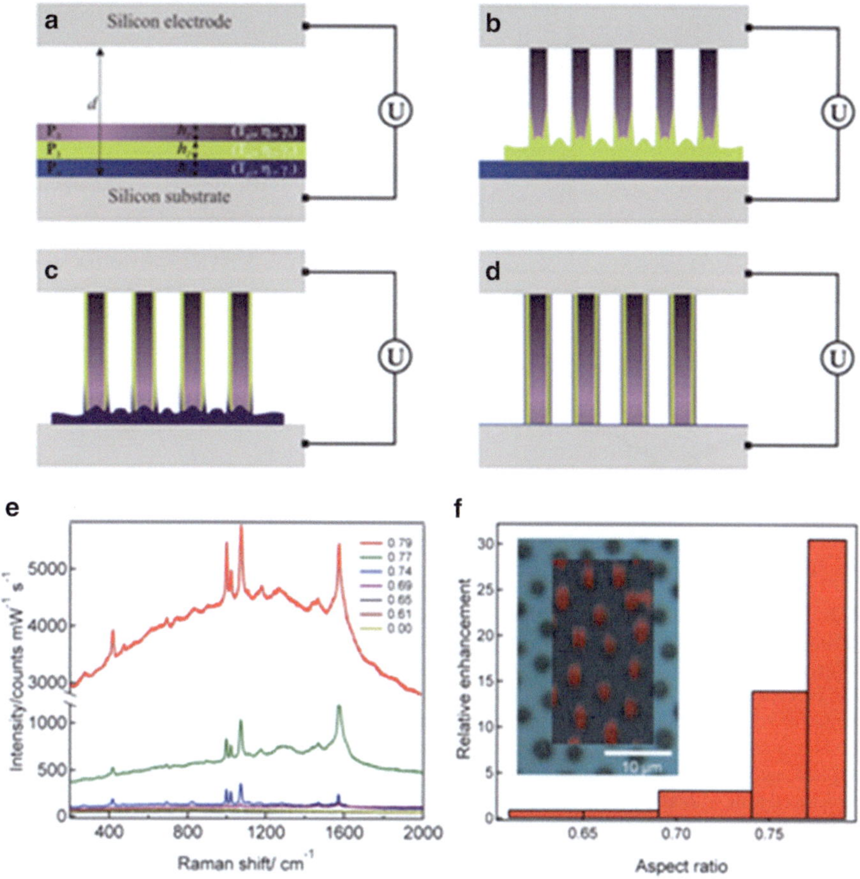

Fig. 7.11 Mechanism of the HEHD patterning process. (**a**) The initially assembled polymer trilayer is destabilised by applying a voltage. (**b**) The primary instability yields pillars which span the capacitor gap. The secondary instability arises from the deformation of the intermediate layer, P_2 occurring at the P_2–P_3 contact line during the primary columns formation process. The amplification of the secondary instability guides the middle layer upwards forming a surrounding shell outside the primary pillar. (**c**) Subsequently, a third instability takes place at the bottom film and the liquified P_1 polymer is drawn upwards by the EHD torque while forming an additional mantle around the bilayer column. (**d**) The final configuration is comprised of a primary P_3 pillar, coated by additional concentric P_2 cylinder and the outer P_1 mantle. (**e**) SERS spectra of benzenethiol recorded from HEHD-generated substrates with pillars of various aspect ratios. (**f**) High aspect ratio structures show high SERS signals. The *inset* shows a SERS map over the optical image of a patterned area. The SERS signal for the 1,070 cm^{-1} peak is in register with the column locations

instabilities to pattern a fully functionalised electron-conducting block copolymer [36]. This method combines the *bottom-up* self-assembly of block copolymer films (with internal structures on a scale of tens of nanometres) with the induced self-organisation of the intermolecular structure of the constituent crystalline blocks

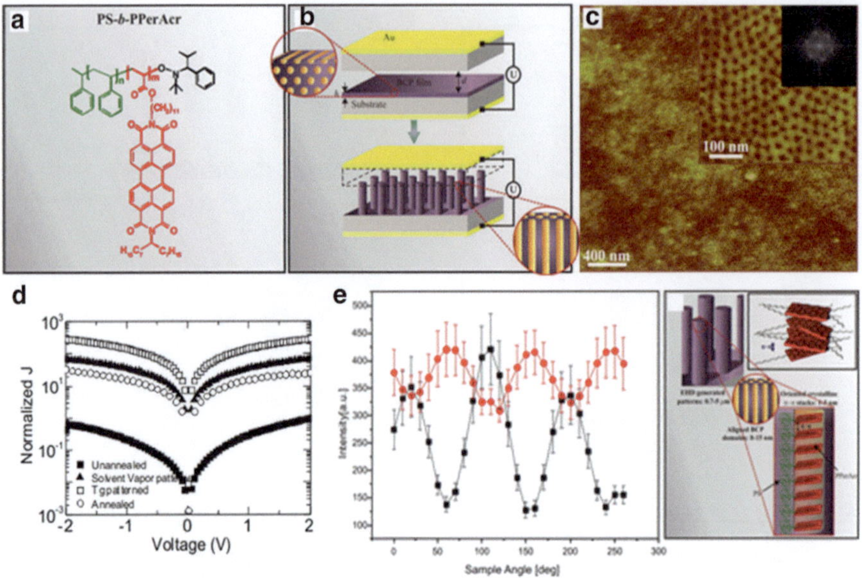

Fig. 7.12 (**a**) Molecular structure of the Polystyrene-block-poly(perylene bismide acrylate) (PS-b-PPerAcr) BCP comprised of an inert block of polystyrene and a crystalline side chain PPerAcr. (**b**) EHD patterning process with the zoomed-in schematics showing internal structure of the BCP film before (lying cylinder) and after patterning (vertical cylinders) process. (**c**) Representative AFM height image of the improved packing of cylinders with an average pore diameter of 14 ± 3 nm, formed inside PS-b PPerAcr matrix when swollen in chloroform vapour with 80 V applied across the capacitor electrodes for 5 h. The *inset* in (**c**) displays a FFT pattern of the structured film with a sixfold symmetry representing hexagonal packing of the cylindrical nanodomains. (**d**). J–V characteristics (DC) of ITO/BCP/ITO sandwiches non-annealed (*black squares*), EHD patterned using thermal annealing (*white squares*) and patterned in a solvent-vapour atmosphere (*black circles*) show improved charge transport properties in the patterned BCP films. (**e**) *Left*: Optical anisotropy analysis of internally oriented PS-b-PPerAcr patterns. Intensity dependence on the sample angle in a cross (*black squares*) and parallel (*red circles*) polarisation configuration. *Right*: Schematic image of the structural hierarchy

(in the range of 2–3 nm, Fig. 7.12e) inside the EHD driven microstructures, to produce functional patterns with a structural hierarchy. Such generation of structures which span several length scales is predominantly based on the interplay of molecular self-assembly and the structural control exerted by electrostatic forces. The polarisation of dielectric interfaces is the driving torque for the electrically induced copolymer domain orientation and the EHD guided instability. Block copolymer (BCP) microstructures also couple to the dielectric contrast between the different domains. The destabilising electrostatic pressure resulting from the electric field causes the energetically unfavourable build-up of displacement charges at the dielectric interface causing the alignment of the interface along the electric field lines, lowering the electrostatic energy. Amplification of the interfacial fluctuations caused by the interchange of the destabilising p_{el} and the restoring γ gives rise to the pattern formation.

The micrometre-sized patterns contain an additional sub-nanostructure arising from the microphase-separated BCP morphology aligned parallel to the electric vector. Oriented nano-domains constitute the BCP crystalline stacks on the sub-10 nm scale with the π–π interactions oriented perpendicular to the E_f direction. These electro-conductive nano-domains within the pillars act as half-wave plate polarisers. The resulting morphology represents the optimal orientation of the copolymers nanostructure with respect to the substrate, exhibiting enhanced charge carrier percolation between the two electrodes (Fig. 7.12d). The charge transport across the entire device renders these aligned nano-morphologies technologically appealing for organic photovoltaic cells.

7.7 Concluding Remarks

In summary, the many interesting aspects of instabilities in thin liquid films are illustrated which are both of scientific and technological importance. While thin film instabilities are typically undesired in industrial and technological applications, they lead to novel surface morphologies and are applicable in alternative (soft-) lithographic techniques. In this context, EHD patterning exploits instabilities in softened thin films stemming from externally applied electrostatic forces to hydrodynamically generate topographic architectures with a controllable length scale in the technologically interesting sub-micrometre and sub-100 nm range. An externally applied field generates forces orders of magnitude larger than the electromagnetic fluctuations (long-ranged vdW forces) and the thermal vibrations (which are both comparatively week), thus dominating film destabilisation. In such a limit, the EHD instability of a thin film is clearly attributed to the applied electric field and can be exploited as a lithographic technique. The advantages of this patterning technique are also noteworthy. The facile EHD patterning technique provides the ability for a straightforward design, manufacture and control of highly reliable and robust submicron patterns at low cost. Its versatility and the applications appeal are emphasised by the possibility of patterning a wide range of materials, single and multilayer film assemblies yielding varieties of hierarchical and anisotropic substructures. Therefore, the inclusion of the studied lithographic route into the advanced miniaturised devices manufacture processes can be envisioned.

Finally, EHD lithography is a versatile patterning tool on the sub-micrometre and nanometre length scales of functional materials. The aesthetic beauty of the generated structures, the applied technological potential and the underlying theories based on basic principles render this technique appealing, intriguing and set to become a pathway for advanced devices and applications.

References

1. Burn, P.L., Kraft, A., Baigent, D.R., Bradley, D.D.C., Brown, A.R., Friend, R.H., Gymer, R. W., Holmes, A.B., Jackson, R.W.: J. Am. Chem. Soc. **115**, 10117 (1993)
2. Dai, L.M., Griesser, H.J., Hong, X.Y., Mau, A.W.H., Spurling, T.H., Yang, Y.Y., White, J.W.: Macromolecules **29**, 282 (1996)
3. Nishizawa, M., Shibuya, M., Sawaguchi, T., Matsue, T., Uchida, I.: J Phys Chem **95**, 9042 (1991)
4. Burn, P.L., Holmes, A.B., Kraft, A., Bradley, D.D.C., Brown, A.R., Friend, R.H., Gymer, R.W.: Nature **356**, 47 (1992)
5. Healey, B.G., Foran, S.E., Walt, D.R.: Science **269**, 1078 (1995)
6. Xia, Y.N., Kim, E., Zhao, X.M., Rogers, J.A., Prentiss, M., Whitesides, G.M.: Science **273**, 347 (1996)
7. Liu, Y.L., Zhao, M.Q., Bergbreiter, D.E., Crooks, R.M.: J. Am. Chem. Soc. **119**, 8720 (1997)
8. Wells, M., Crooks, R.M.: J. Am. Chem. Soc. **118**, 3988 (1996)
9. Xia, Y.N., Mrksich, M., Kim, E., Whitesides, G.M.: J. Am. Chem. Soc. **117**, 9576 (1995)
10. Schaffer, E., Harkema, S., Roerdink, M., Blossey, R., Steiner, U.: Macromolecules **36**(5), 1645–1645 (2003)
11. Schaffer, E., Harkema, S., Roerdink, M., Blossey, R., Steiner, U.: Adv. Mater. **15**(6), 514 (2003)
12. Schaffer, E., Thurn-Albrecht, T., Russell, T.P., Steiner, U.: Nature **403**(6772), 874 (2000)
13. Schaffer, E., Thurn-Albrecht, T., Russell, T.P., Steiner, U.: Europhys Lett **53**(4), 518 (2001)
14. Swan, J.W.: Proc R Soc **62**, 38 (1897)
15. Frenkel, J.: Phys. Z. Sowjetunion **8**, 675 (1935)
16. Tonks, L.: Phys. Rev. **48**, 562 (1935)
17. Melcher, J.R.: Phys. Fluids **4**, 1348 (1961)
18. Melcher, J.R.: MIT Press, Cambridge, MA (1963)
19. Reynolds, M.: Phys. Fluids **8**, 161 (1965)
20. Taylor, G.I., McEwan, A.D.: J Fluid Mech **22**, 1 (1965)
21. Schaffer, E.: PhD thesis (2001)
22. Voicu, N.E., Harkema, S., Steiner, U.: Adv. Funct. Mater. **16**, 926 (2006)
23. Teletzke, G.F., Davis, H.T., Scriven, L.E.: Rev Phys Appl **23**, 989 (1988)
24. Wyart, F.B., Martin, P., Redon, C.: Langmuir **9**, 3682 (1993)
25. Pease, L.F., Russel, W.B.: J Chem Phys **118**, 3790 (2003)
26. Pease, L.F., Russel, W.B.: Langmuir **20**, 795 (2004)
27. Lindner, A., Coussot, P., Bonn, D.: Phys. Rev. Lett. **85**, 314 (2000)
28. Bensimon, D., Kadanoff, L.P., Liang, S.D., Shraiman, B.I., Tang, C.: Rev. Mod. Phys. **58**, 977 (1986)
29. Harkema, S.: PhD thesis (2006)
30. Leach, K.A., Lin, Z.Q., Russell, T.P.: Macromolecules **38**(11), 4868 (2005)
31. Goldberg Oppenheimer, P., Steiner, U.: Small **6**, 1248 (2010)
32. Dickey, M.D., Collister, E., Raines, A., Tsiartas, P., Holcombe, T., Sreenivasan, S.V., Bonnecaze, R.T., Willson, C.G.: Chem. Mater. **18**(8), 2043 (2006)
33. Harkema, S., Steiner, U.: Adv. Funct. Mater. **15**(12), 2016 (2005)
34. Morariu, M.D., et al.: Nat. Mater. **2**(1), 48 (2003)
35. Goldberg Oppenheimer, P., Mahajan, S., Steiner, U.: Adv. Mater. **24**(23), OP175 (2012)
36. Goldberg Oppenheimer, P., Kabra, D., Vignolini, S., Huttner, S., Sommer, M., Neumann, K., Thelakkat, M., Steiner, U.: Chem. Mater. **25**(7), 1063 (2013)

Chapter 8
Elastic Instability and Surface Wrinkling

Pascal Damman

8.1 Introduction

The stability of constrained slender structures is a very old topic, that has started with Galileo, who questioned the stability of a beam supporting a heavy load and solved by Leonhard Euler with the Elastica [1]. With the emergence of new architectural design, this stability problem rapidly becomes a major subject for mechanical engineers. Surprisingly, physicists and engineers have rediscovered the specific problem of thin sheets instability at the end of twentieth century. This re-discovery was triggered by two seminal papers reporting experiments about the formation of regular wrinkles in constrained systems. Tanaka et al. showed that the outer surface of a gel confined in a petri dish swollen by solvent vapors becomes unstable and exhibits periodic structures (Fig. 8.1a) [2]. Bowden et al. have beautifully explained the formation of tiny wrinkles during the thermal evaporation of a metal thin film on a soft foundation (Fig. 8.1b) [3]. They performed a linear stability analysis including the bending of the rigid metal layer and the elastic deformation of the soft foundation that clearly demonstrates the instability of the outer surface.

From countless examples, it is now crystal clear that the homogeneous (often flat) states of constrained thin sheets are very often unstable. Since the seminal works of Tanaka and Bowden, the final states were broadened in a zoo of complex morphologies made of wrinkles, creases, crumples, folds, and blisters, several morphologies sometimes co-existing in a single experiment. Transition from one morphology to another can be observed depending on various experimental conditions. For instance, by decreasing adhesion, you switch from regular wrinkles to blister. Increasing compression could also reveal nonlinear regimes with increasing complexity. As fluids, thin elastic sheets appear to be very promising systems that

P. Damman (✉)
Laboratoire Interfaces & Fluides Complexes, Université de Mons,
20 Place du Parc, B-7000 Mons, Belgium
e-mail: pascal.damman@umons.ac.be

J. Rodríguez-Hernández, C. Drummond (eds.), *Polymer Surfaces in Motion*,
DOI 10.1007/978-3-319-17431-0_8

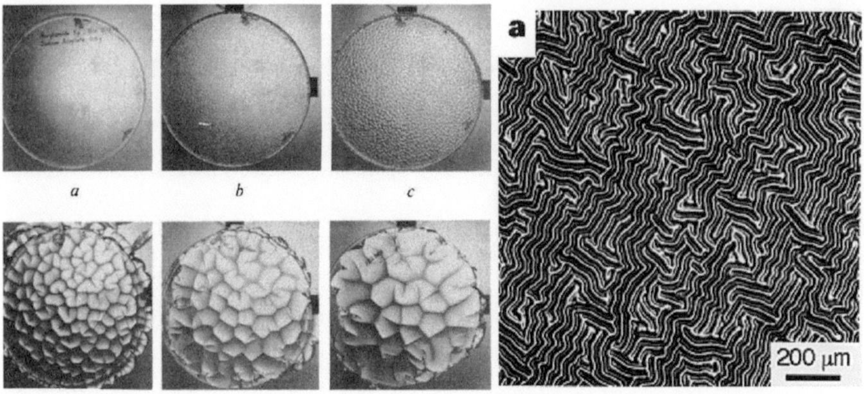

Fig. 8.1 (*Left*) ionized acrylamide gel formed in a petri dish allowed to swell in water, (**a**)–(**c**) show the evolution of the morphology with time [2]. (*Right*) optical micrographs showing representative patterns of wrinkles that formed when a nanometric gold layer is evaporated onto warm (110 °C) polydimethylsiloxane. The *pattern* appears when the sample is cooled to room temperature [3]

bear similarity with classical problems of linear and nonlinear pattern formation such as period-doubling bifurcations. Their study opens new prospects to understand the emergence of complexity, breaking of symmetry and singularities. Even the simple buckling of a sheet shows intrinsic behaviors that raise fundamental questions such as, why do films become folded upon confinement whereas a thick slab of an identical material favors creases? Why does paper sheets crumple into singularities whereas rubber sheets would smoothly wrinkle? Answering these questions (and many others) is important, first for our natural curiosity, and to understand the emergence of complex shapes and patterns in Nature.

Understanding thin sheets behavior is also extremely important for many technological applications, specially for those involving micro-patterning of surfaces. The design of new materials combining extreme mechanics with optical, electronic, or chemical properties is very often achieved with specific coatings on thin sheets. In this case, the failure of the coating, or even the thin sheet itself, should be avoided. The opposite is also true! These complex features can be very interesting for some applications, essentially in micro- and nano-technology. Indeed, understanding how complex patterns emerge spontaneously under featureless forces may inspire efficient methods for tailoring a desired surface pattern to achieve the required property (e.g., reversible superhydrophobicity, flexible electronics).

8.2 Wrinkling in Constrained Free-Sheets

Usually, wrinkles are associated with multilayers, including materials with very contrasted elastic properties. The archetype of these systems that will be discussed in the next section is the rigid/inextensible thin sheet glued on an elastomer or a

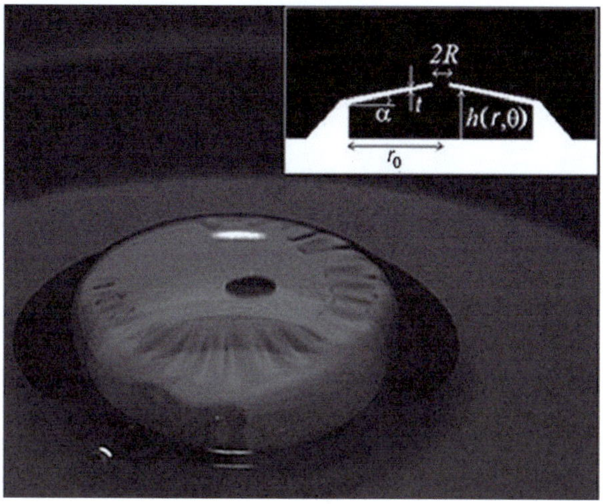

Fig. 8.2 Image of a collapsing viscous bubble. The bubble loses its axisymmetric shape, small amplitude ripples grow. The *inset* displays a schematic side view of the essentially conical deflating bubble at the onset of the instability [5]

fluid foundation and subjected to a uniaxial constraint. It is however possible to generate wrinkles in free sheets provided you are using the proper geometrical constraints. The very typical example of this geometrically induced wrinkling is the collapse of a viscous bubble (Fig. 8.2). During the collapse, the shape of the bubble drastically changes from a purely spherical to a flat shape, due to surface tension at the edge of the hole created in the bubble. With this sudden change of morphology, a hoop stress builds that induces the wrinkling of the fluid thin layer (high viscosity slows the relaxation of the wrinkles allowing their visualization) [4, 5].

Geometrical constraints can also be applied on rectangular sheets. Indeed, thin sheets compressed at one edge to follow a periodic sinusoidal profile and free at the other, i.e. a curtain-like morphology (Fig. 8.3), develop a self-similar hierarchy of folds [6, 7]. We will now demonstrate a universal method based on a scaling approach to find the morphology of constrained thin sheets.

As shown in Fig. 8.3, sheets made of various materials constrained at one edge develop a regular hierarchical pattern of folds that follow simple power laws. These patterns consist of a hierarchy of successive generations of folds whose wavelength gradually increases along x. For the sake of clarity, the structure of the deformed sheet is described by a periodic function $z(x, y) = A(x) \sin q(x)y$, with z the amplitude of out-of-plane deflections, y being parallel to the edge ($q = 2\pi/\lambda$). The hierarchical patterns are characterized by the evolution of the average wavelength, $\lambda(x)$. For rigid sheets, the amplitude of the folds is determined by the compression ratio and the inextensibility of the sheet, $A \sim \lambda\sqrt{\delta}$. Since, inextensibility ensures that

$$W_0 = \int_0^W ds \cos\,\theta \simeq \int_0^W ds\left(1 - \frac{1}{2}\dot{z}^2\right) \simeq W - \frac{1}{2}W\left(\frac{A}{\lambda}\right)^2$$

and $\delta = (W - W_0)/W$.

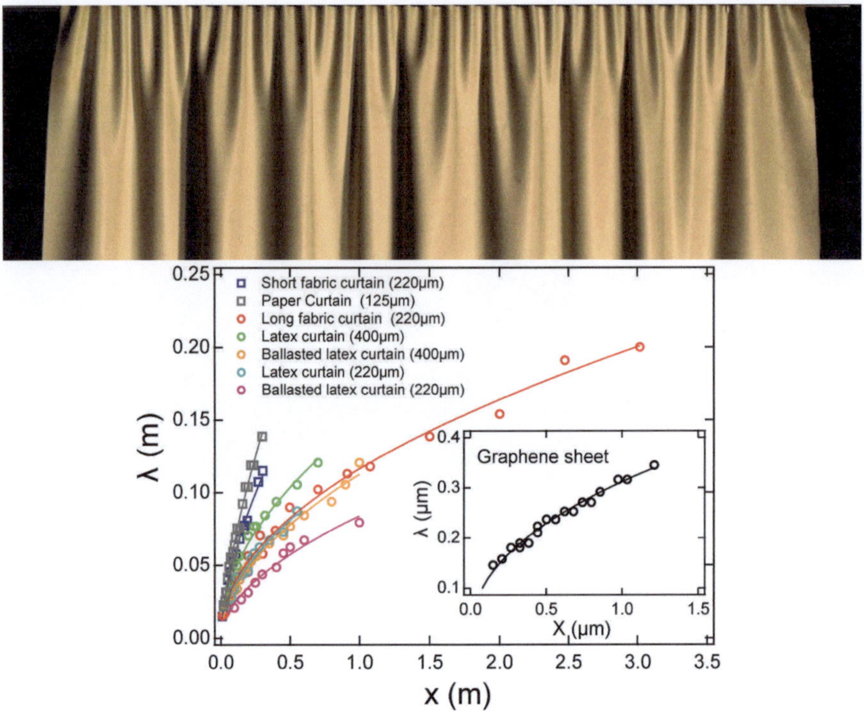

Fig. 8.3 *Top* of a rubber curtain constrained at one edge with an imposed sinusoidal deformation $z(0, y) = A(0)\ \sin(q(0)y)$ and power laws describing the evolution of the wavelength with the distance from the constrained edge [6]

The bending energy density related to a fold is thus given by

$$u_b \sim Eh^3\ \kappa^2 \sim Eh^3\delta/\lambda^2.$$

Since u_b is proportional to $1/\lambda^2$, the membrane adopts the largest possible wavelength compatible with the imposed constraints. The minimization of bending energy is thus the "driving force" toward larger and larger folds and is the source of the observed hierarchy. Figure 8.3 shows that curtains made of various materials with contrasted properties can be sorted in two classes with different exponents $\sim 2/3$ for "light" sheets and $\sim 1/2$ for "heavy" sheets (we will see later the meaning of light and heavy). What does determine these exponents, is it related to pure geometry or to deformation energy of the sheet? To increase the wavelength, adjacent folds should merge. Looking carefully a curtain, you would probably observe the merging of two, three, and very rarely fourfolds, some folds remaining almost unaltered. We will however make the assumption, first proposed by mathematicians Jin and Sternberg, that the observed morphology can be described by successive period-doubling transitions constituting the building-blocks of the global pattern [8]. The hierarchy is obtained by stitching these

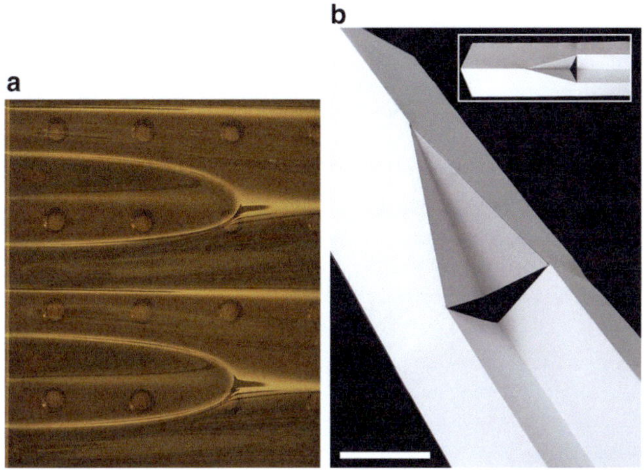

Fig. 8.4 (**a**) Wrinklon morphology. (**b**) Origami model of the $\lambda - 2\lambda$ transition [6]

building-blocks. The key feature of a single block element, here after named wrinklon, is its length, L, i.e., the sheet length required to accommodate the $\lambda - 2\lambda$ transition. If the energy is also involved (and not only geometry), this length should be determined by material properties, E, h the constraint/compression ratio δ and the wavelength. The power law describing self-similar patterns can be built from this length according to $d\lambda/dx \sim \lambda/L$. To estimate the length of the wrinklon, we use a scaling approach where energetic penalty must be compensates by energetic gain. Close inspection of the wrinklon morphology reveals the occurrence of a curved ridge at the tip of the merging folds (Fig. 8.4). Such curved ridge is characterized by a non-vanishing Gauss curvature. From the *Theorema Egregium* [9], the sheet around these curved ridge should concentrate on stretching energy (i.e., the surface is no more isometric of a flat surface). The energetic penalty involved into a single $\lambda - 2\lambda$ transition should be related to local stretching, the energetic gain being related to the decrease in curvature.

The stretching energy can be estimated from the slope of the sheet which determines the strain ε induced by the change of amplitude $A - 2A$ (related to the change of wavelength since inextensibility ensures that $A \sim \sqrt{\delta\lambda}$). The strain given by $\varepsilon \sim A^2/L^2 \sim \delta\lambda^2/L^2$ yields the stretching energy $U_s \sim Eh\, L\, \lambda\, \varepsilon^2 \sim Eh\, \delta^2\lambda^5 L^{-3}$. The wrinklon length results then from a balance of this stretching energetic penalty and bending energy, $U_b \sim Eh^3\, L\, \lambda\, \kappa^2$ which yields $L(\lambda) \sim h^{-1/2}\delta^{1/4}\lambda^{3/2}$. The scaling for the wavelength describing the whole hierarchical pattern of folds is obtained by the integration of equation $d\lambda/dx \sim \lambda/L$,

$$\frac{\lambda(x)\delta^{1/6}}{h} \sim \left(\frac{x}{h}\right)^{2/3}.$$

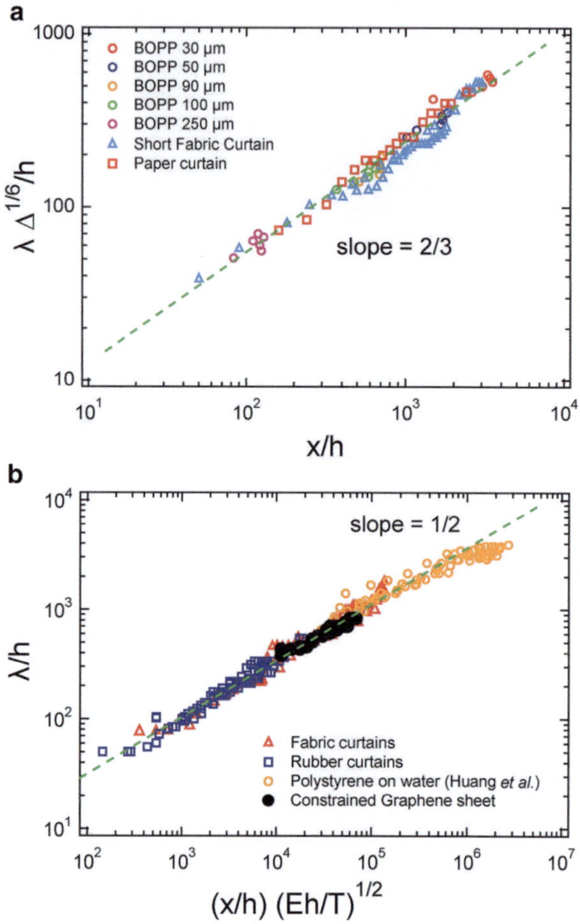

Fig. 8.5 Master curves, normalized wavelength vs. normalized distance from edge for the "light" (**a**) and "heavy" (**b**) curtains [6]

The class of patterns related to "light" curtains, 2/3 exponent, is in close agreement with this scaling model. In addition to yielding the proper exponent, this relation enables the comparison of the data obtained from seemingly disparate systems, over a wide range of lengthscales. Figure 8.5 provides a remarkable collapse of the data measured with paper, fabric, and various plastic sheets. Interestingly, the elastic modulus of the material used to build the sheet does not appear in this relation. This was expected since both stretching and bending energies linearly depend on the elastic modulus.

As shown in Fig. 8.3, "heavy" curtains, made of nanometric films of polystyrene on water [7], rubber sheets, and constrained graphene do not follow the 2/3 scaling. Instead, their dynamics obey $\lambda \propto \sqrt{x}$). The main difference between both families is related to the lack or the occurrence of a significant tensile force, T. For all "heavy" curtains, an additional tensile force is acting on the sheet. For graphene sheets, this tension is related to the longitudinal tensile strain induced by thermal manipulations of the compression device [10]. For rubber curtains, the tension is determined by gravity ($T \sim \rho_c ghH$, where ρ_c and H are the density and height of the curtain). For compressed nanometric polystyrene films on water, a tensile force is exerted by the surface tension of water at the free edges of the polymer film [7].

The tension per unit width imposes an additional stretching energy $U_t \sim T\alpha^2 L\lambda \sim T \delta \lambda^3 L^{-1}$ where α is the slope of the sheet within the wrinklon ($\alpha \sim A^2/L^2$). This energy becomes dominant when $U_t > U_s$, that is when $T > Eh^2\delta/A$. Neglecting the stretching term, the total energy of the distorted membrane becomes $U_{\mathrm{tot}} = U_t + U_b$. The wrinklon length which minimizes U_{tot} (balancing tension and bending energies) becomes $L(\lambda) \sim \frac{\lambda^2}{h}\sqrt{\frac{T}{Eh}}$. As expected, the tensile force increases the length of wrinklons for a given wavelength and can thus be used to tune the energetic penalty associated with the $\lambda - 2\lambda$ transitions. Considering the equation $d\lambda/dx \sim \lambda/L$, we obtain the scaling for the wavelength along a heavy sheet

$$\frac{\lambda(x)}{h} \sim \left(\frac{Eh}{T}\right)^{1/4} \left(\frac{x}{h}\right)^{1/2}. \tag{8.1}$$

This scaling is in excellent agreement with the power laws observed for heavy curtains and graphene bilayers (Fig. 8.3). The data of various macroscopic curtains, graphene bilayers, and nanometric polystyrene thin films indeed collapse onto a single master curve without any fitting parameters (see Fig. 8.5) which highlight the universality of our description. Our formalism is thus validated for seven orders of magnitude in thickness from graphene sheets to rubber and fabric curtains.

There is obviously various method to geometrically constrain a free sheet. Consider, for instance, a thin film of PS of nanometric thickness deposited on a water droplet (Fig. 8.6). The edge of the circular sheet is decorated with a set

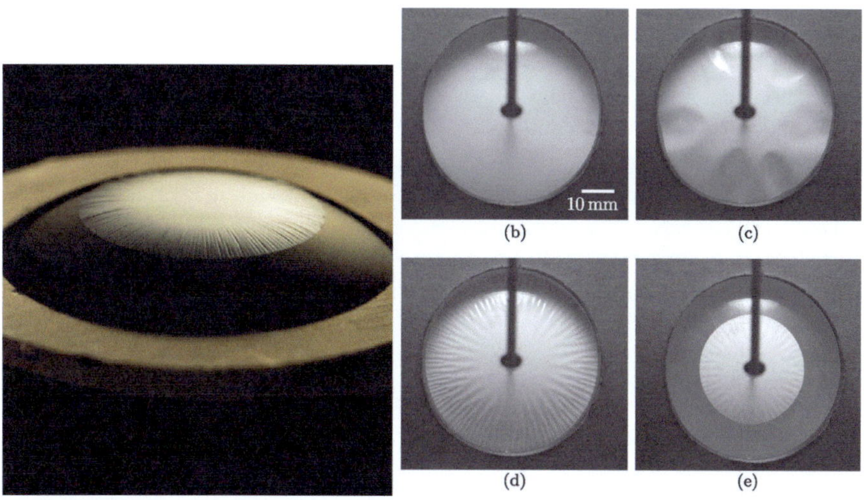

Fig. 8.6 (*Left*) image of a ultrathin sheet of polystyrene deposited on a water droplet [11]. (*Right*) sheets confined between two spheres [12]

wrinkles [11]. These wrinkles should grow to accommodate the stress arising when a flat sheet is forced to adopt a spherical shape. In fact, the change of shape induces a change in Gaussian curvature and thus a stretching. Since the surface tension exerted by water at the edge is not strong enough to overcome this stretching force, the thin sheet slightly retracts and wrinkles to avoid compression along the edge.

8.3 Compression of Thin Sheets on a Soft Foundation

In the previous section, we showed how to generate wrinkles by playing with geometrical constraints applied on a free thin sheet. This method is very powerful to generate hierarchical wrinkles, as shown in Figs. 8.3 and 8.6. In fact constrained free-sheets always tend to develop cascades of wrinkles with an increasing wavelength whatsoever the way geometrical constraints are applied. This natural tendency is directly related to the minimization of bending energy through the curvature ($U_b \sim \kappa^2 \sim 1/\lambda^4$). Surprisingly, the growth of a pattern made of regular parallel wrinkles with a constant wavelength is far from obvious! The solution is to balance the bending energy with another energy that decreases when curvature increases. This can be achieved by gluing the thin sheet on a soft substrate, able to follow the shape of the constrained sheet. This strategy was successfully applied to fluid [13] and to elastomers [3, 14]. As shown in Fig. 8.7, very regular parallel wrinkles are observed when a rigid sheet deposited on a soft substrate is compressed uniaxially. The wavelength is determined by the elastic properties of the membrane and the foundation (Fig. 8.7). The exact shape of the sheet can be obtained either via variational methods through Euler–Lagrange equation [15–17] or from a linear stability analysis of the homogeneous flat state [2, 3, 13, 15].

A linear stability analysis for systems consisting in a rigid layer, E_m on top of an elastomer, E_f with $E_m \gg E_f$ can be easily achieved. When uniaxially compressed by a factor $(\Delta = L - L_0)/L_0)$, energy can be stored in two ways, (1) as pure compression, the sheet remaining flat, U_c, or (2) as bending of the sheet plus elastic deformation of the soft foundation, U_b, U_f in the wrinkled state. The comparison of these energy yields $\Delta U = U_b + U_f - U_c$. The profile of the wrinkled state is periodic $(q = 2\pi/\lambda)$ and can always be described through a Fourier expansion, $y(x) = \sum A_m \cos mqx$. For a linear analysis, we only consider single mode, the mean curvature of a sinusoidal profile is $< y'' > \sim A^2 q^2$. We also assume inextensible boundary condition for the rigid sheet which implies that $\Delta = A^2 q^2$. The difference in energy can be written as

$$\Delta U = \Delta\left(E_m h^3 q^2 + E_f \frac{1}{q} - E_m h \Delta \right).$$

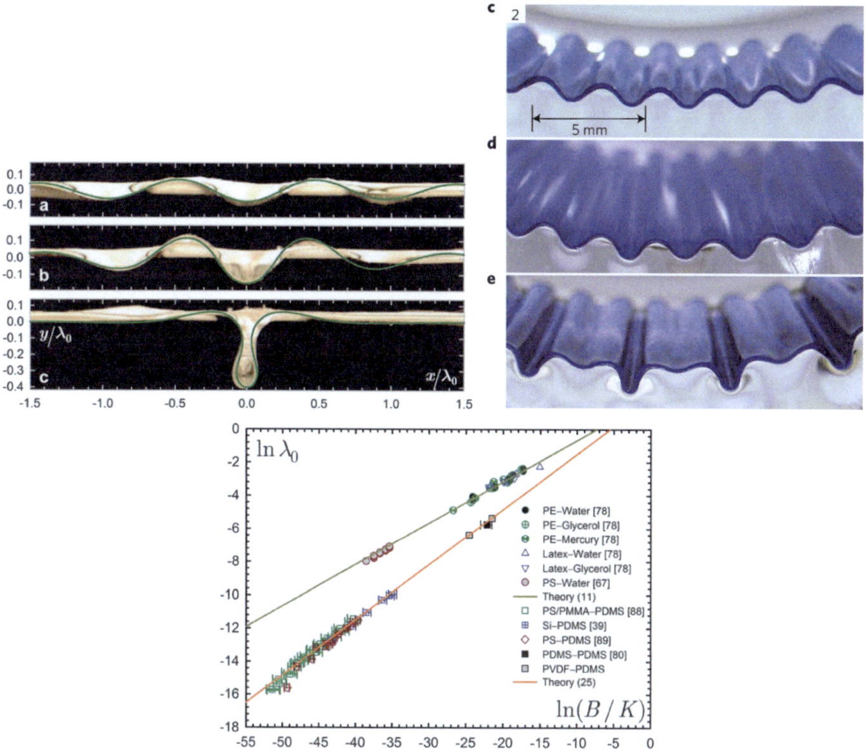

Fig. 8.7 (*Top*) images of wrinkled state of a rigid sheet compressed on water for increasing compression ratios [13]. Uniform wrinkled state and period-doubling morphology observed for compressed sheet deposited on elastomer [14]. (*Bottom*) evolution of the wrinkles wavelength for fluid (*circles*) and elastomer (*square*) foundations [16]

The flat state becomes unstable if $\Delta U < 0$. For each wavenumber q we have thus a critical compression, $\Delta_c = h^2q^2 + (E_f/E_m h)q^{-1}$. Close to the threshold, the observed wrinkles period should correspond to the minimum of Δ_c which yields the following wavelength and critical strain

$$q = \left(\frac{E_f}{E_m h^3}\right)^{1/3} \qquad \lambda = h\left(\frac{E_m}{E_f}\right)^{1/3} \tag{8.2}$$

$$\Delta_c^* = 2\left(\frac{E_f}{E_m}\right)^{2/3}. \tag{8.3}$$

The predicted evolution of the wavelength with the material properties is in very good agreement with observed data for a large variety of experimental systems (Fig. 8.7). A similar method can be used to study the stability of compressed sheets on fluid substrates provided you replace the energy of substrate deformation by the

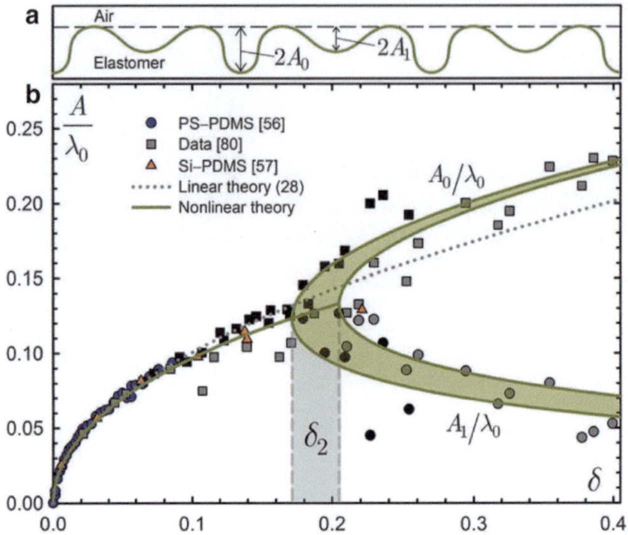

Fig. 8.8 (**a**) Definitions of the wrinkles amplitudes A0 and A1. (**b**) Comparison between experimental and theoretical evolutions of A0 and A1 as a function of the compression ratio for polystyrene and PDMS [16]

relation $U_f \sim \rho g A^2$, involving the hydrostatic pressure of the fluid [13, 18]. The wavelength for fluid substrate becomes $\lambda \sim (E_m h^3/\rho g)^{1/4}$ in very good agreement with experimental observations (Fig. 8.7).

Obviously such a linear analysis is only valid very close to the threshold, i.e. when A is infinitely small. This model suggests that above (but close to) the threshold, the wrinkles amplitude should scale as $\sqrt{\Delta}$ with the compression ratio. Figure 8.8 shows that this law remains however valid for unexpectedly large compression ratios, $\Delta \leq 0.2$, well above the threshold for this peculiar system, close to 0.02. This observation highlights the robustness of the single wavelength wrinkled morphology, there is no emergence of super-harmonic modes (nq, with n integer) as expected for classical pattern formation mechanisms [19].

Close to the threshold for wrinkling, there is no way to discriminate wrinkles on fluid or solid/elastomer substrates. Increasing the compression ratio changes the rules, a drastic modification of morphology is indeed observed with different responses for fluid and elastomers. For fluid substrates, the extra length due to compression concentrates in a single fold, while compressed sheets on elastomers always stay periodic. This difference should reflect the emergence of different nonlinear terms that breaks the symmetry of the wrinkled state. In a similar manner that buckling breaks the symmetry of the homogeneous flat state. For fluids, the nonlinearity essentially arises from the bending energy, through the curvature, $\kappa = \ddot{y}/\sqrt{1-\dot{y}^2}$. This nonlinearity generates a quartic term in the expression of energy that will break the longitudinal symmetry keeping the transversal symmetry of the deformed sheet, i.e., folds up and down are energetically equivalent. As shown by Diamant and Witten, the resulting nonlinear Euler–Lagrange equation describing these systems is reminiscent of the sine-Gordon family equations explaining the growth of solitons [17].

For elastomers, a bifurcation due to the nonlinearities is observed with the emergence of a sub-harmonic mode, $q/2$ (Fig. 8.8). This bifurcation clearly corresponds to a period-doubling sequence similar to what is observed for Faraday instability of shaken fluid layers, except that this bifurcation refers to time and not space [20]. For the Faraday instability, the period-doubling is usually described with a nonlinear parametric oscillator model. Interestingly, the Euler–Lagrange equation describing the minimization of energy for the compressed sheet on elastomer bears also some resemblance with a parametric oscillator [14].

It should be noticed however that the period-doubling behavior is only observed for elastomer substrates without pre-strain. When the compression is achieved through a pre-stretching of the elastomer slab prior deposition of the thin sheet, the nonlinear elasticity of the PDMS rubber comes into play and other morphology characterized by periodic cusps is observed [21].

8.4 Wrinkling Coupled to Diffusion, Swelling, Thermal Constraints, . . .

Wrinkles are however not limited to a set of parallel periodic folds, 2D wrinkled patterns can also be produced. Obviously, there is no way to generate 2D morphology from simple uniaxial compression of the thin sheets. We will now consider radial compression, symmetric or not. The simplest method to produce 2D patterns is to apply compression in the rigid sheet along two orthogonal directions. Different morphologies, with periodic nipples, squares, or labyrinthine can be achieved. However, several experimental studies highlight that the resulting morphology strongly depends on the history of deformations, applying two subsequent uniaxial compressions is clearly not the same as applying simultaneously compression along two directions (Fig. 8.9).

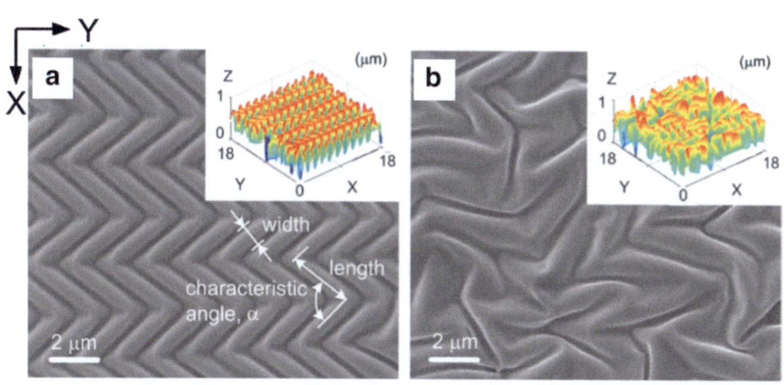

Fig. 8.9 SEM and AFM images of surface patterns when compressing an oxidized PDMS film either (**a**) sequentially or (**b**) simultaneously [22]

Interestingly, 2D compression stress can be applied through unconventional methods involving, swelling, or surface tension. 2D stress can also be applied by subsequent inflating/deflating of a spherical bump in a microfluidic device (Fig. 8.10) [23]. Swelling can be used in two different ways, either by a direct compression of a confined slab as shown by Tanaka et al. [2], or indirectly, the solvent changing the rheological properties of the foundation [24, 25]. This last method was used to induce wrinkling in bilayers of titanium thin film deposited on polystyrene (PS). Wrinkling was induced by immersing the multilayers in toluene vapors. Toluene is a good solvent of PS, it can swell the polymer layer located below the Ti membrane by diffusing through tiny defects, either resulting from the deposition process or obtained by AFM indentation.

As shown in Fig. 8.11, immersion of a polystyrene film capped with a thin titanium layer in toluene vapors leads to the formation of wrinkle domains in the

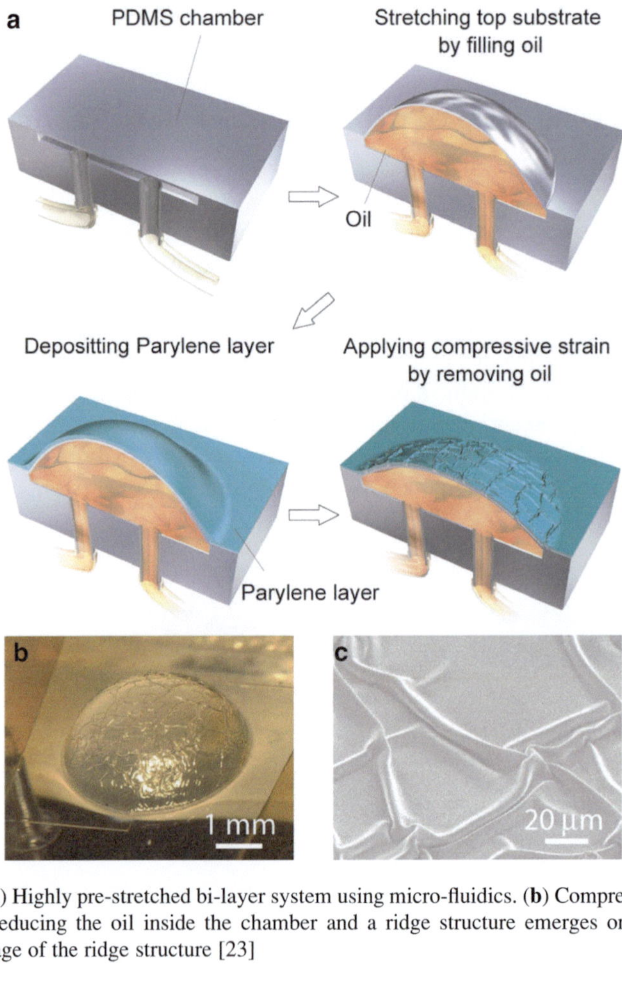

Fig. 8.10 (**a**) Highly pre-stretched bi-layer system using micro-fluidics. (**b**) Compressive strain is applied by reducing the oil inside the chamber and a ridge structure emerges on the surface. (**c**) SEM image of the ridge structure [23]

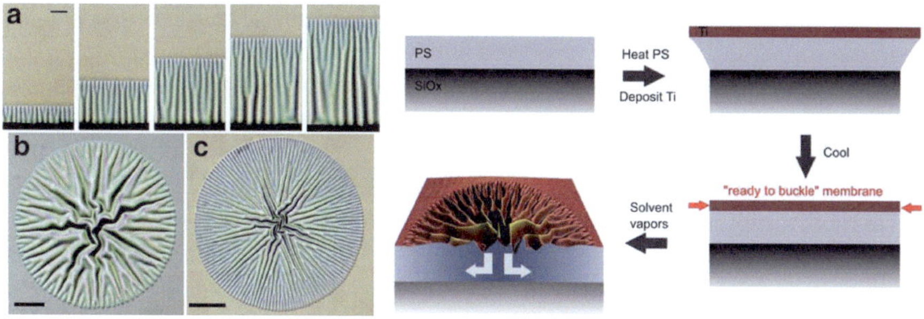

Fig. 8.11 (*Left*) parallel and radial wrinkles morphologies observed when solvent diffuses in the PS layer from an edge or a point-like defect, respectively. (*Right*) schematic representation of the wrinkle mechanism induced by solvent diffusion. First, the thermal deposition process generates compression in the upper membrane. Subsequently, solvent diffusion triggers the transition from unbuckled to buckled state

metallic membrane with various morphologies. Parallel wrinkles, herringbones, splaying-fan like morphologies, circles, and circles decorated with tiny dots can be observed when the solvent diffuses inside the polymer layer. These metastable patterns emerge spontaneously depending on the geometry of the diffusion front and the layer thicknesses.

For the diffusion assisted wrinkling process, the compressive stress at the origin of the patterns is related to the thermal deposition of the metal film on the polymer substrate. Due to the very high elastic modulus of glassy PS the metal surface while in compression remains flat. The wrinkling instability is then induced by immersing the samples in solvent vapors of the polymer, since swelling induces a drastic decrease of the elastic modulus (Fig. 8.11). Strikingly, the resulting wrinkling patterns do not show the usual labyrinthine morphology [3] (see Fig. 8.1) but are clearly determined by the geometry of the diffusion process (Fig. 8.11). Parallel wrinkles are observed when the solvent diffuses from the edge yielding a linear front. Radial organization of wrinkles arises from point-like diffusion starting at tiny holes randomly distributed in the thin metal layer (i.e., defects resulting from the deposition process of the metallic film). The wrinkles thus always grow preferentially perpendicular to the wavefront. The relevance of molecular diffusion in the observed phenomena is obviously supported by the dynamics of the process that follows the classical Fickian diffusive behavior, distance $\propto t^{1/2}$.

As a consequence, multilayers made up of low elastic modulus elastomers, such as polydimethylsiloxane (PDMS, $E_p \sim 10^6$ Pa), buckle with very small critical stresses. It was thus not surprising that metal surfaces deposited by thermal evaporation wrinkle during the deposition process [3]. The PDMS layer, thermally expanded during metal deposition, induces a compressive stress in the rigid membrane when cooled to ambient temperature. In contrast, replacing the elastomer with a high modulus glassy polymer, such as PS, $E_p \sim 10^9$ Pa, increases the critical stress by one order of magnitude. It could become so large that the thin metal surfaces, while stressed, remain perfectly flat after the thermal deposition. These stressed but flat titanium surfaces can thus be considered as ready to buckle membranes.

The Ti/PS/SiO_x multilayers are however highly sensitive to chemical stimulation thanks to the polymer layer. Indeed, the diffusion of a good solvent in PS leads to a drop of the glass transition temperature together with a drastic decrease of the elastic modulus (e.g., adding 15 % w/w of toluene is enough to obtain elastomeric PS at room temperature). To some extent, we could consider that solvent diffusion is equivalent to a local increase of temperature for the PS layer. Since the formation of wrinkles is fully determined by the critical stress, strongly dependent on polymer elastic modulus, this diffusion process should trigger a transition from unwrinkled to wrinkled state (Fig. 8.11).

The relation between the wavelength and materials properties can be obtained from a scaling energetic approaches, based on a balance of the membrane bending energy with the penalty associated with the deformation of the foundation [26]. For micrometric polymer film thicknesses, the wrinkle wavelength follows the allometric relation $\lambda \propto (hH_p)^{1/2}$ or $\lambda^* \propto H_p^{*1/2}$ using variables without dimensions, $\lambda^* = \lambda/h$ and $H_p^* = H_p/h$.

Interestingly, we observe a drastic deviation from the expected behavior for very thin polymer films (nanometric thickness), Fig. 8.12. Instead of a continuous decrease of λ^* with H_p^*, a "V shape" curve with a slope reversal is observed. Since the deviation appears for polymer films thinner than 50 nm, we add van der Waals (VDW) interactions between the silicon substrate and the titanium membrane in our model. These interactions cannot be neglected when both surfaces are separated by a distance smaller than 100 nm. The VDW energy (per unit surface area) between two surfaces separated by a distance z is given by $P(z) = A_H/12\pi z^2$ [27]. The Hamaker constant, A_H, corresponding to the interactions of two surfaces, 1 and 2, through a medium 3, can be computed from the dispersive component of the individual surface tensions, γ_i^D, by using the relation $A_{132} \simeq (\sqrt{\gamma_1^D} - \sqrt{\gamma_3^D})$

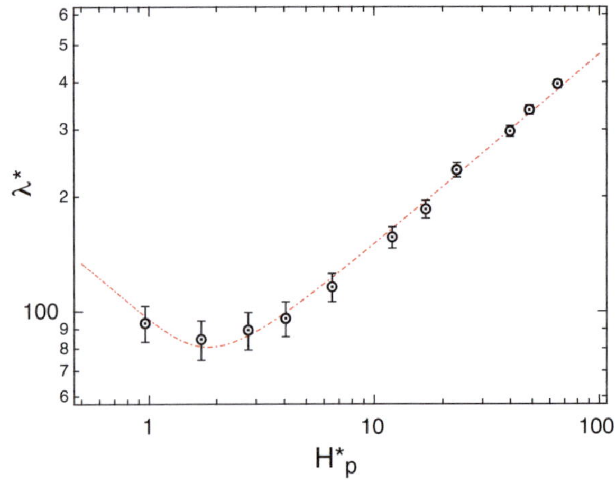

Fig. 8.12 Evolution of the normalized wavelength, λ^*, with the normalized foundation thickness H_p^*, for trilayers Ti/PS/SiO_x immersed in toluene vapors. The *line* corresponds to the solution of Eq. (8.5)

$(\sqrt{\gamma_2^D} - \sqrt{\gamma_3^D})$ [28]. For two materials of high surface energy separated by a polymer medium, $A_H \simeq -4.10^{-19}$ J. For a wrinkled surface with a slowly varying thickness $z(x)$, the corresponding energy can be estimated from a Taylor expansion limited at the second order of $P[z(x)]$ that yields

$$\int_0^\lambda \frac{1}{2}\left(\frac{d^2P}{dz^2}\right)_{z=z_0} [z(x) - z_0]^2 \, dx,$$

where z_0 is the average film thickness.

For a sinusoidally wrinkled polymer layer with $z(x) = H_p + A \sin 2\pi x/\lambda$ and $z_0 = H_p$, the total energy of the system per unit surface area can be written as:

$$U_T \simeq \frac{Eh^3A^2}{\lambda^3} + \frac{E_p\lambda^3A^2}{H_p^3} - \frac{|A_H|A^2\lambda}{H_p^4}, \tag{8.4}$$

where the first term is the bending energy of the upper Ti membrane, the second term is the elastic energy stored in the polymer layer, and the third term is the contribution of the VDW energy. For such very high molecular weight polymer ($M_w = 1.4 \times 10^6$ Da), we can consider a purely elastic foundation. Indeed, the time scale of the experiment ($\simeq 100$ s) remains very small with respect to the disentanglement time τ_d that determines the transition between elastic and viscous behaviors. Minimizing the total energy with respect to λ^*, considering λ^* and H_p^* variables, yields the relation

$$-\frac{E}{\lambda^{*4}} + \frac{E_p\lambda^{*2}}{H_p^{*3}} - \frac{|A_H|}{h^3H_p^{*4}} = 0. \tag{8.5}$$

The solution of this equation is plotted with the experimental data in Fig. 8.12, showing the good agreement between the model including VDW energy and the experimental results.

Wrinkling from solvent diffusion was also used for other systems [29]. For instance, combining UV exposure and solvent diffusion, a zoo of morphology: flowers, concentrated rings, and labyrinthine patterns can be obtained (Fig. 8.13). It should be noted however that there is no theoretical model to explain the transition between all these rather exotic patterns.

Combining wrinkling and diffusion provides thus an interesting method to produce complex patterns with tunable dimensions. However this physical method by itself is not suitable as a patterning technique due to the randomness of the wrinkle nucleation events. Indeed, the random distribution of the wrinkled domains is related to the uncontrolled localization of defects in the metal membrane. To solve this problem, we use thicker titanium layers and an AFM tip (Fig. 8.14) to make small holes in the metal layer with a specific geometry. As shown in

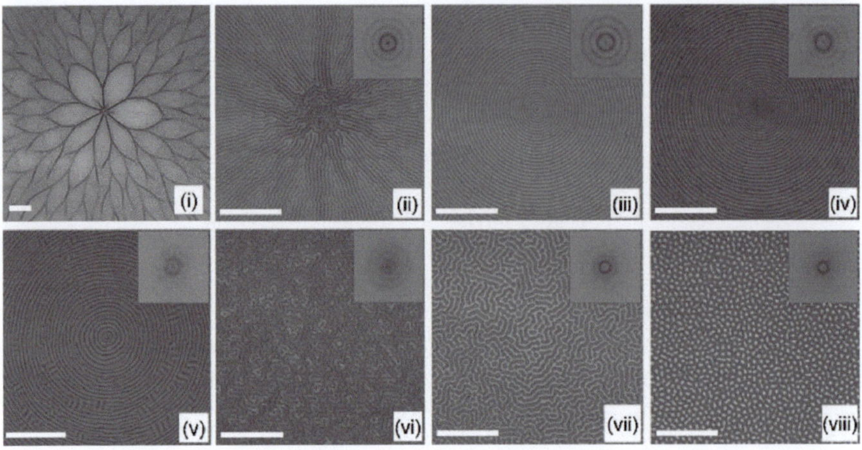

Fig. 8.13 Variety of morphological patterns via swelling-induced surface wrinkling. Optical microscopy images of various wrinkling patterns produced on 500 nm thick PS films with different UVO exposure times from 2–40 min [29]

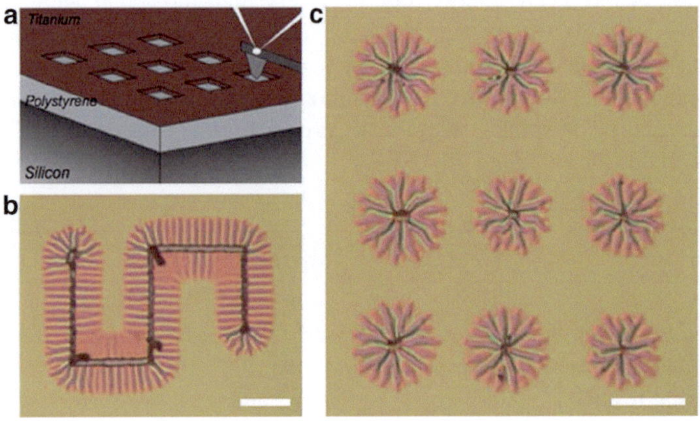

Fig. 8.14 Optical images of wrinkled Si/PS/Ti multilayers, previously patterned with an AFM tip and then exposed to toluene vapor for 2 min. (**a**) scheme of the patterning method, observed patterns for squares 2 μm (**b**) and lines 2 μm wide (**c**). Scale bars correspond to 10 mm [25]

Fig. 8.14c, this nano-indentation process prior to solvent exposure provides a fine control of the spatial layout of the wrinkled domains. Furthermore, since the wrinkle orientation is determined by the diffusion front, it becomes possible to generate tailor-made wrinkled patterns by tuning the geometry of the carving. Wrinkles initiated at the corners and tips of the engraved area (Fig. 8.14b) exhibit radial orientation while parallel wrinkles develop from the linear parts.

8.5 Conclusions

Playing with materials properties and the way the compression stress is applied, various topography with parallel wrinkles (with a cascade of wavelength, or a superposition of several modes, or the growth of a soliton-like solution), radial wrinkles, chaotic labyrinthine, and many others patterns can be generated. For application purposes, it is usually more convenient to generate in situ the rigid thin sheet on the elastomer by using UV irradiation, for instance. Oxydation of the polymer surface creates a crust that wrinkles when a compression stress is applied. As shown by Eq. (8.2), the thin sheet thickness and thus irradiation time determines the wavelength that could range from hundred of nanometers to centimeters. These surfaces were used for various applications including the design of superhydrophobic surfaces, new non-permanent adhesive [30], flexible electronic devices [31]. Wrinkled surfaces exhibiting sub-micrometric features were also used for optical devices such as anti-reflective coatings, optical cavity [32, 33].

Acknowledgements I would like to acknowledge the researchers who actively participate to the various research described here, H. Vandeparre, F. Brau, A. Abbas, A. Boudaoud, C. Gay, P. Reis, B. Roman, and J. Bico.

References

1. Lieven, R.: The Elastica: a mathematical history. Technical Report No. UCB/EECS-2008-103 (2008)
2. Tanaka, T., Sun, S.T., Hirokawa, Y., Katayama, S., Kucera, J., Hirose, Y., Amiya, T.: Mechanical instability of gels at the phase transition. Nature **325**, 796–798 (1987)
3. Bowden, N., Brittain, S., Evans, A.G., Hutchinson, J.W., Whitesides, G.M.: Spontaneous formation of ordered structures in thin films of metals supported on an elastomeric polymer. Nature **393**, 146–149 (1998)
4. Debrégeas, G., de Gennes, P.-G., Brochard-Wyart, F.: The life and death of bare viscous bubbles. Science **279**, 1704 (1998)
5. da Silveira, R., Chaïeb, S., Mahadevan, L.: Rippling instability of a collapsing bubble. Science **287**, 1468–1471 (2000)
6. Vandeparre, H., et al.: Wrinkling hierarchy in constrained thin sheets from suspended graphene to curtains. Phys. Rev. Lett. **106**, 224301 (2011)
7. Huang, J., et al.: Smooth cascade of wrinkles at the edge of a floating elastic film. Phys. Rev. Lett. **105**, 038302 (2010)
8. Jin, W., Sternberg, P.: Energy estimates for the von Karman model of thin film blistering. J. Math. Phys. **42**, 192 (2001)
9. Audoly, B., Pomeau, Y.: Geometry and Elasticity. Oxford University Press, Oxford (2010)
10. Bao, W., et al.: Controlled ripple texturing of suspended graphene and ultrathin graphite membranes. Nat. Nanotech. **4**, 562–566 (2009)
11. King, H., Schroll, R.D., Davidovitch, B., Menon, N.: Elastic sheet on a liquid drop reveals wrinkling and crumpling as distinct symmetry-breaking instabilities. PNAS **109**, 9716–9720 (2012)
12. Hure, J., Roman, B., Bico, J.: Stamping and wrinkling of elastic plates. Phys. Rev. Lett. **109**, 054302 (2012)

13. Pocivavsek, L., et al.: Stress and fold localization in thin elastic membranes. Science **320**, 912–916 (2008)
14. Brau, F., et al.: Multiple-length-scale elastic instability mimics parametric resonance of nonlinear oscillators. Nat. Phys. **7**, 56–60 (2011)
15. Cerda, E., Mahadevan, L.: Phys. Rev. Lett. **90**, 074302 (2003)
16. Brau, F., Damman, P., Diamant, H., Witten, T.A.: Wrinkle to fold transition: influence of the substrate response. Soft Matter **9**, 8177–8186 (2013)
17. Diamant, H., Witten, T.A.: Phys. Rev. Lett. **107**, 164302 (2011)
18. Milner, S.T., Joanny, J.F., Pincus, P.: Europhys. Lett. **9**, 495 (1989)
19. Manneville, P.: Instability, Chaos and Turbulence. Imperial College Press, London (2004)
20. Douady, S.: Experimental study of the Faraday instability. J. Fluid Mech. **221**, 383–409 (1990)
21. Cao, Y., Hutchinson, J.W.: Wrinkling phenomena in neo-Hookean film/substrate systems. J. Appl. Mech. **79**, 031019:1-9 (2012)
22. Lin, P.-C., Yang, S.: Appl. Phys. Lett. **90**, 241903 (2007)
23. Takei, A., Jin, L., Hutchinson, J.W.: Ridge localizations and networks in thin films compressed by the incremental release of a large equi-biaxial pre-stretch in the substrate. Adv. Mater. **26**, 4061–4067 (2014)
24. Vandeparre, H., Damman, P.: Wrinkling of stimuloresponsive surfaces: mechanical instability coupled to diffusion. Phys. Rev. Lett. **101**, 124301 (2008)
25. Vandeparre, H., et al.: Confined wrinkling: impact on pattern morphology and periodicity. Soft Matter **7**, 6878 (2011)
26. Vandeparre, H., et al.: Slippery or sticky boundary conditions: control of wrinkling in metal-capped thin polymer films by selective adhesion to substrates. Phys. Rev. Lett. **99**, 188302 (2007)
27. de Gennes, P.G., Brochard-Wyart, F., Quéré, D.: Capillarity and Wetting Phenomena: Drops, Bubbles, Pearls, Waves. Springer, New York (2003)
28. Léopoldès, J., Damman, P.: From a two-dimensional chemical pattern to a three-dimensional topology through selective inversion of a liquid-liquid bilayer. Nat. Mat. **5**, 957–961 (2006)
29. Chung, J.Y., Nolte, A.J., Stafford, C.M.: Diffusion-controlled, self-organized growth of symmetric wrinkling patterns. Adv. Mater. **21**, 1–5 (2009)
30. Chan, E.P., Smith, E.J., Hayward, R.C., Crosby, A.J.: Surface wrinkles for smart adhesion. Adv. Mater. **20**, 711–716 (2008)
31. Rogers, J.A., Someya, T., Huang, Y.: Materials and mechanics for stretchable electronics. Science **327**, 1603–1607 (2010)
32. Koo, W.H., et al.: Light extraction from organic light-emitting diodes enhanced by spontaneously formed buckles. Nat. Photon. **4**, 222–226 (2010)
33. Kolaric, B., et al.: In situ tuning the optical properties of a cavity by wrinkling. Appl. Phys. Lett. **96**, 043119 (2010)

Chapter 9
Reaction-Diffusion Dynamics Induced Surface Instabilities

Murat Guvendiren

9.1 Introduction

Surface instabilities on polymer surfaces are usually regarded as defects, and appearance of these instabilities is a major problem in many applications. In fact, it is very common to come across phrases like "crease-resistant finishing" or "wrinkle-proof coating". The surface instabilities generally arise either during fabrication of the polymer system, such as due to the reaction dynamics, or post-fabrication in response to external stimuli, such as solvent, reactive agent, pH, temperature, and ion concentration. There is a growing interest to harness these "undesired" instabilities to create user-defined surface patterns. Polymer systems that display spontaneous formation of surface instabilities, or patterns, with controlled properties (such as occurrence, size, order, and complexity) are potentially useful for a wide range of applications, including sensors [1], microfluidic devices [2, 3], responsive coatings [4], smart adhesives [5–7], microlens arrays [8–10], cell culture substrates [11, 12], and even property measurement tools [13]. In this chapter, reaction and/or diffusion dynamics induced surface instabilities are summarized, with particular emphasis on transient and persistent creasing instabilities due to inhomogeneous diffusion of a solvent, wrinkling to creasing transition in response to solvent composition, and self-oscillating instabilities due to reaction dynamics.

M. Guvendiren (✉)
New Jersey Center for Biomaterials, Rutgers – The State University of New Jersey, 145 Bevier Rd., Piscataway, NJ 08854, USA
e-mail: mguvendiren@gmail.com

J. Rodríguez-Hernández, C. Drummond (eds.), *Polymer Surfaces in Motion*,
DOI 10.1007/978-3-319-17431-0_9

9.2 Transient to Persistent Surface Instabilities

Gel is a three-dimensional polymer network with the ability to hold a solvent. Hydrogel is formed when the building block of the network is a water-soluble polymer, and the surrounding solvent is water. These polymer networks can reversibly swell and de-swell (shrink) in response to temperature and solvent composition, which is associated with a three-dimensional volume expansion or contraction, respectively. When a polymer network swells (or de-swells), the volume transition can be as large as a 1,000-fold. This volume transition occurs gradually. It is not a homogenous process but rather occurs discontinuously, creating dynamic heterogeneities within the network, which leads the display of transient patterns [14–18]. Almost three decades ago, Tanaka et al. reported the dynamic pattern formation on uniform ionized acrylamide gels during extensive swelling in water [19]. Initially the pattern was extremely fine (like a frosted glass), grew in size with time by coalescing with each other, and gradually disappeared when the pattern size reached to the size of the gel (Fig. 9.1a). They observed that the pattern started to form as the free surface of

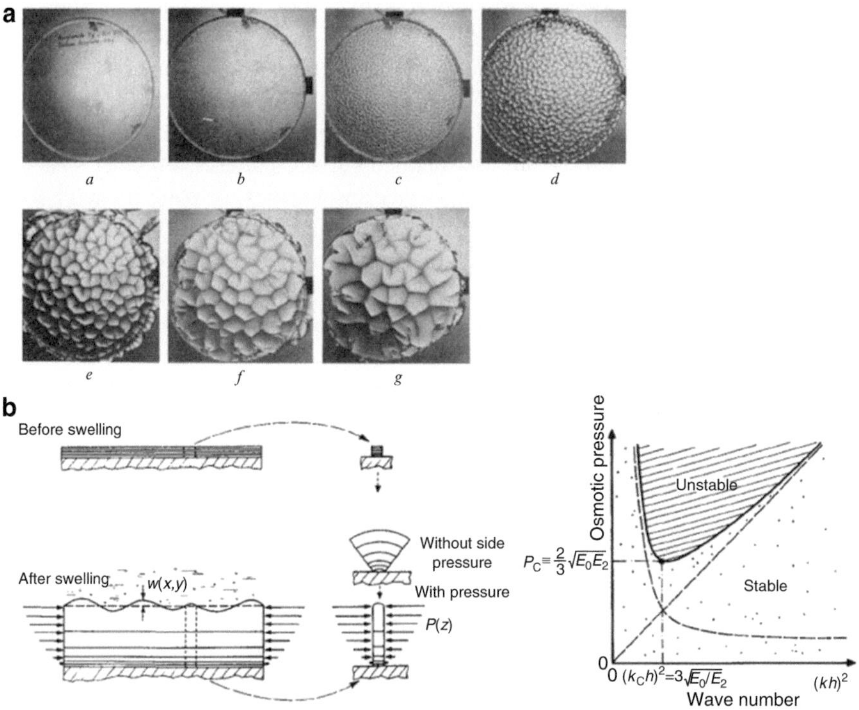

Fig. 9.1 (**a**) Evolution of swelling-induced pattern formation on acrylamide hydrogel formed in a petri dish. (**b**) The schematic of the gel before and after swelling (*left*), and the condition for instability plotted for two independent parameters (*right*). (Reprinted by permission from Macmillan Publishers Ltd: Nature [19], 1987)

the gel began touching each other, forming localized sharp folds. These types of instabilities are generally referred to as creasing instabilities. This phenomenon can be attributed to the collective diffusion of the polymer network into the solvent, governing the kinetics of the swelling [18, 19].

In this approach, the gel is considered as a combination of thin layers, where the swelling process is initiated from the very thin surface layer of the polymer network. This puts each layer under a mechanical constraint as the inner surface of the layer is fixed to the unswollen core whereas the outer surface of the layer is free to swell. The state of the layer is determined by the osmotic pressure in such a way that stretching of the network occurs at lower pressures (gel displays flat surface) whereas the outer layer is forced to buckle (instabilities appear on the surface) when the pressure is high. When the hydrogel reaches equilibrium swelling state, the osmotic pressure becomes independent of the gel depth, resulting in disappearance of the creasing pattern. This phenomenon occurs always in every gel during swelling, whether it is observable or not depends on the kinetics of the process. However, the transient pattern becomes persistent when the bottom surface of the gel is constraint to a substrate, as long as the gel is maintained in the swollen state [17, 19–22]. The creasing pattern disappears when the swelling of the gel is lowered sufficiently, and recovered when the swelling is induced again [20]. This process is reversible, and in most of the cases the location of the individual crease pattern is similar to the initial location. For instance, when the poly(acrylamide) gel was reversibly swollen in the water by changing the solvent composition (i.e., salt concentration), the location of the creases was highly conserved [20]. However, clear changes in crease morphology were reported in some of the regions on the hydrogel, due to the local inhomogeneities in the hydrogel.

For a constraint gel, the osmotic pressure can be expressed as a function of the film modulus (E), the wave vector of the pattern (k), and the thickness of the gel (h) [19]:

$$P = \left(E_2^{1/2}\frac{kh}{3} - E_0^{1/2}\frac{1}{kh}\right)^2 + \frac{2}{3}(E_2E_0)$$

The critical pressure for buckling (P_c) with a critical wavelength (λ_c) is equal to

$$P_c = \frac{2}{3}(E_2E_0)^{1/2}$$

$$k_C = \frac{2\pi}{\lambda_C} = \left(\frac{E_o}{E_2}\right)^{1/4}\frac{\sqrt{3}}{h}$$

In these equations, the critical pressure is roughly proportional to the Young's modulus of a uniform gel ($P_c = E$) and independent of the film thickness (h), whereas the characteristic wavelength (λ_c) of the pattern is proportional to the film thickness and dependent on the modulus (Fig. 9.1b) [19].

To better understand the pattern formation dynamics, the morphological and kinetic evolution of the surface pattern during swelling was studied extensively

[14, 15, 21–23]. For a thin film gel attached to a rigid substrate, the characteristic wavelength of the surface pattern can be assumed to be proportional to the deformation along the *z* axis (normal to the substrate), *u*(*z*,*t*) and can be explained by the linear diffusion theory, or Fickian-type kinetics [21, 23, 24]:

$$\partial u/\partial t = D\partial^2 u/\partial z^2.$$

In this case, the solution for the above equations is equal to

$$u(t)/u(\infty) \approx \lambda(t)/\lambda(\infty)$$

which is equal to

$$\lambda(t) = \lambda_\infty\left(1 - (8/\pi^2)\sum_{n=0}^{\infty}(2n+1)^{-2}\mathrm{e}^{(-t/\tau_n)}\right),$$

where

$\tau_n = \tau/(2n+1)^2$, and $\tau = (\alpha_e h_0)^2/\left[(\pi/2)^2 D\right]$.

Note that α_e is the equilibrium linear expansion and D is the effective diffusion coefficient. This equation can be further simplified for a short and a long period of swelling time, respectively.

$$\lambda(t) = (1 - 1/\alpha)(Dt)^{1/2}$$

$$\lambda(t) = \lambda_\infty\left(1 - (8/\pi^2)\mathrm{e}^{(-t/\tau)}\right)$$

These equations are particularly useful to determine equilibrium wavelength (λ_e) of the pattern and time required for instabilities to reach the equilibrium. For instance, for the poly(2-hydroxyethyl methacrylate) (PHEMA) hydrogels crosslinked with ethylene glycol dimethacrylate (EGDMA), the evolution of the characteristic wavelength of the pattern during swelling in water was found to be determined by the crosslinker concentration and the initial film thickness. The equilibrium wavelength decreased with increasing concentrations of EGDMA and decreasing initial film thickness [25].

When the hydrogel film is covalently attached to a much stiffer substrate, the confinement restricts osmotic swelling at the film/substrate interface during the swelling process. Thus, the film preferentially swells in the direction normal to the surface, affecting the structural and mechanical properties of the hydrogels, such as permeability and solvent uptake [19, 26–29]. Due to this anisotropic swelling, hydrogels experience an in-plane equibiaxial compressive stress, which is related to the equilibrium swelling of the unconstrained hydrogel films in the stress-free state [20, 30, 31]. Therefore, the condition for instability formation, previously explained above with respect to osmotic pressure, can be explained by a model system with respect to mechanical compression (Fig. 9.2)

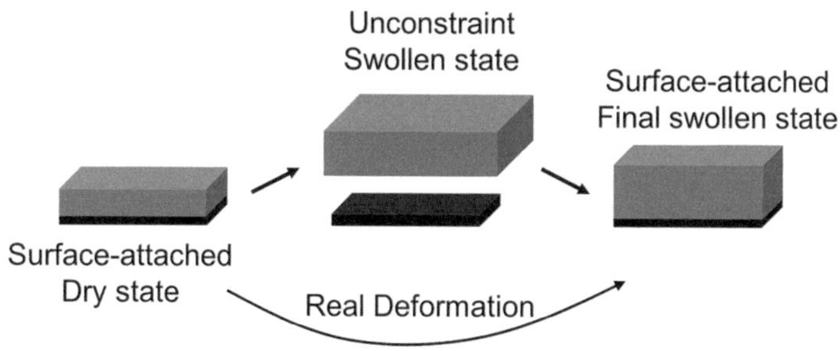

Fig. 9.2 Schematic of the swelling of a surface attached polymer network

[32]. This approach consists of a sequence of the following hypothetical events: (1) the constrained gel is detached from the substrate to a free state, (2) the free gel is swollen to equilibrium, (3) biaxial compression (*x* and *y* axis, perpendicular to the substrate) is applied to the swollen gel to fit the gel back onto the substrate. The gel becomes unstable to form surface instabilities at a critical value of the compressive strain, ε_c, which does not depend on the modulus of the material or the characteristic length scale of the instability [20, 32, 33]. The instability only occurs when the extension ratios parallel to the free surface satisfy $\alpha_x^2\alpha_y = 0.296$ [20, 32, 33]. Note that, $\alpha = l/l_0$, where l_0 and l denote the initial size and the final size, respectively. When the equibiaxial compression ($\alpha_x = \alpha_y = \alpha$) is considered, this leads to $\varepsilon_c = 1 - \alpha = 0.334$. Here, we would like to note that it is also possible to obtain persistent surface instabilities by creating similar heterogeneities on the constraint polymer films via buckling or wrinkling. Although the driving force for both the instabilities is the same, the two modes of instability occur at a very different critical strain, such that wrinkling requires much lower strain [34, 35]. Although wrinkling is the main focus of the previous chapter, it is briefly discussed below.

9.3 Wrinkle to Crease Transition

The majority of work done on surface instabilities involves bilayer films composed of a hard skin [36], e.g., metal [37], polymers [13, 38], and silicate [8, 39–41], on a softer elastomer sheet, such as poly(dimethyl siloxane) (PDMS), which can then be expanded thermally, mechanically, or osmotically [36]. Buckling occurs due to the compressive stress generated as a result of the modulus difference between the skin and bulk. This buckling led to the formation of various wrinkle patterns, including one-dimensional (1D) ripples, two-dimensional (2D) labyrinth and herringbone structures, and their variations [42, 43]. The thickness and the uniformity of the physical properties of each layer can be controlled to tune the morphology, order, and complexity of the wrinkle patterns [36]. The length scales in wrinkling instabilities are typically on the order of submicrons to tens of microns, which is due to the competition between the hard skin favoring the short wavelength wrinkles and the soft elastomer favoring the longer wavelength wrinkles. However, the majority of these studies have focused on pattern formation above the critical stress, and few have studied the onset of wrinkling, which starts with formation of the highly ordered hexagonal patterns [44–46], and much above critical stress, which leads wrinkle to crease transition [34, 47, 48]. In this section the focus is on the wrinkle to crease transition.

Although much work has focused on manipulation of swelling-induced elastic instabilities in elastomeric films to form complex patterns [45, 49–52], there is a growing interest in controlling pattern formation and ordering in much softer systems, such as gels [19, 20, 44, 53, 54]. For elastomer systems with a thin rigid skin, wrinkling instabilities are formed at relatively small compressive strains [37, 42, 43, 55–57], which can often be approximated by the classical plate theory [36, 37, 40, 42]. For hydrogels, on the other hand, the swelling-induced creasing instability arises due to relatively larger compressive strains. Therefore, in addition to the film modulus and the strain level, the crosslinking density and its homogeneity within the film, and film thickness could play important roles in controlling the pattern morphology and the characteristic size [19, 20, 53]. Therefore, polymer systems displaying heterogeneities in their properties (such as crosslinking density) with depth are particularly intriguing as they can provide a much wider range of surface instabilities when exposed to a stimulus, such as solvent or temperature [25, 34, 44, 58, 59].

One of the ways to create a depth-wise gradient is to create a polymer system with two-reactive components: one being reactive to the polymer whereas the other one targeting the first component. In this approach, one component has to be homogenously distributed within the polymer, and the other component has to be initially outside the polymer but diffuses over time during the reaction of the polymer with the first component. For instance, photopolymerization reactions in the presence of oxygen can be used to create depth-wise crosslinking (modulus) gradient for surface attached gels [31, 44, 60]. Here, oxygen acts as a scavenger and inhibits radical polymerization, which is more pronounced on the top surface of the film (Fig. 9.3). In this regard, a wide range of patterns were reported by exposing a photocurable formulation poly(2-hydroxyethyl methacrylate) (PHEMA) to light while open to air and then swelling in water (Fig. 9.4) [44]. The oxygen inhibition of the radical polymerization at the surface created a gradient of the crosslinking with depth, which was confirmed by measuring the double bond conversion at the surface, surface mechanics, and molecule diffusion into the network. The modulus gradient, and hence the osmotic pressure, is controlled by the crosslinker concentration, and the characteristic size of the pattern is determined by the initial film thickness. The pattern was reported to be stable in both swollen and dry states. The formation of stable patterns was explained by the glassy nature of the PHEMA, such that the top surface of the polymer film became rubbery and easily deform when swollen, whereas the relaxation time of the polymer network was much longer than the water diffusion rate during drying, locking (freezing) the pattern in dry state. A similar approach was later reported using reactive silane infusion in toluene [61]. In this case, PHEMA films were exposed to methyltrichlorosilane (MTCS)/toluene solution in a reactor at 90 °C for 2 h. The condensation reactions of

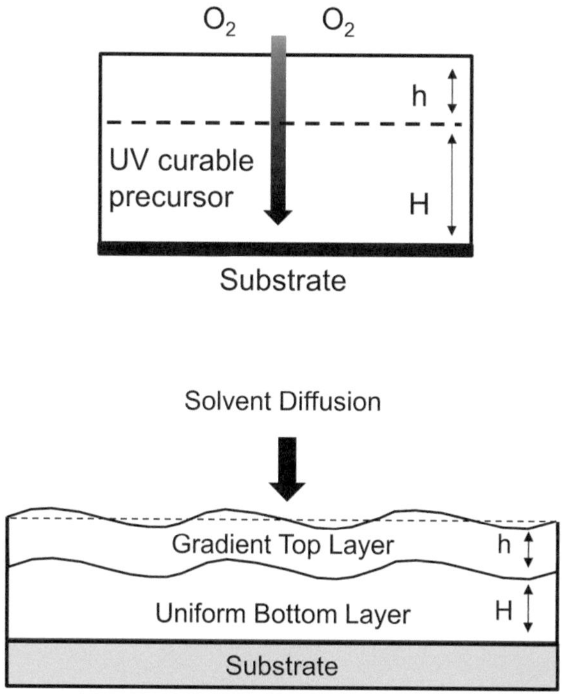

Fig. 9.3 (*Top*) Oxygen diffusion from the top of the UV curable precursor solution hinders crosslinking, which is more pronounced at the surface. (*Bottom*) Solvent diffusion is more pronounced at the top layer due to gradient crosslinking. This gradient creates surface instabilities

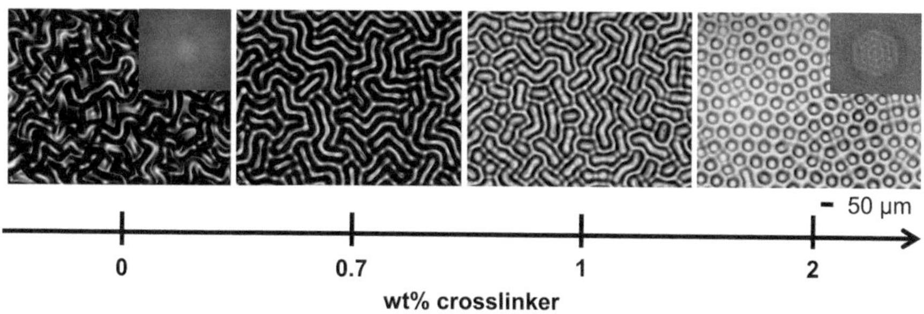

Fig. 9.4 A wide range of wrinkling patterns observed for PHEMA hydrogels with increasing crosslinker concentration. (Adapted with permission from [44]. Copyright 2009 WILEY-VCH Verlag GmbH & Co. KGaA, Weinheim)

the trichlorosilane groups with hydroxyl groups of the PHEMA led formation of a gradient crosslinking density with depth leading similar surface instabilities.

These surface instabilities are either in the form of shallow undulations, wrinkling, or in the form of sharp folds, creasing, which is determined by the linear expansion (α) of the confined film in a solvent. To develop a deeper understanding of the mechanism of wrinkling and creasing surface pattern formation, PHEMA

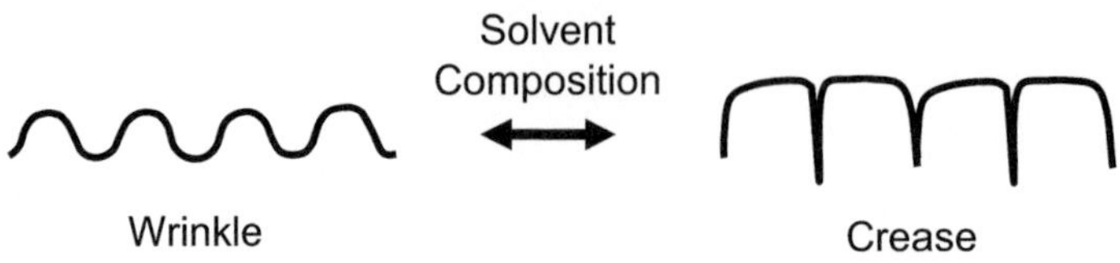

Fig. 9.5 Schematic of the wrinkling and the creasing instabilities (cross section of the film). Reversible transition of wrinkle to crease occurs by controlling solvent quality

films with depth-wise crosslinking gradient were used [34]. Films were swollen in a range of pure and mixed solvents to investigate the transition from wrinkling to creasing instabilities (Fig. 9.5). Note that the degree of the swelling and hence the magnitude of the compressive strain (or stress) is dependent on the polymer chemistry, the gel composition, the extent of the crosslinking, and the solvent quality, i.e., the affinity between the solvent molecules and the polymer chains. Therefore, it is possible to determine the critical condition for crease formation with respect to swelling. For a surface attached gel, the volumetric swelling ratio (V) is equal to the linear expansion (α), the ratio of the swollen film thickness to the dry thickness, as shown below [34].

$$\alpha = \frac{h}{h_0} = \frac{1}{(1 - \phi_s)}$$

In this equation, ϕ_s denotes the solvent fraction in the polymer network, which is dependent on the degree of crosslinking and solvent quality. To put this in perspective, the shear modulus (G) is proportional to crosslinking density (v_x) and the solvent fraction (ϕ_s), according to the Neo-Hookean model of rubber elasticity theory [62]

$$G \propto v_x \propto (1 - \phi_s)^{1/3}$$

On the other hand, the quality of the solvent or the solubility of the polymer in a solvent is determined by the solubility parameter (δ) and the Flory-Huggins polymer-solvent interaction parameter (χ). Solvating potential of a solvent can be written by using Hildebrand theory [34, 63, 64].

$$\frac{\alpha}{\alpha_{\max}} = \exp\left[a\alpha\left(\delta_s - \delta_p\right)^2\right]$$

In this equation, $\alpha_{\max}$ is the maximum linear expansion ratio, and a is a constant. δ_s and δ_p are the solubility parameters for the solvent and the polymer network, respectively. Therefore, the solubility of the polymer network is favored (i.e., the polymer network shows maximum linear swelling) when $\delta_s \approx \delta_p$. Therefore, it is possible to control the degree of the linear expansion ratio (or swelling) of the constraint polymer film by selecting suitable solvents (Fig. 9.6). For instance,

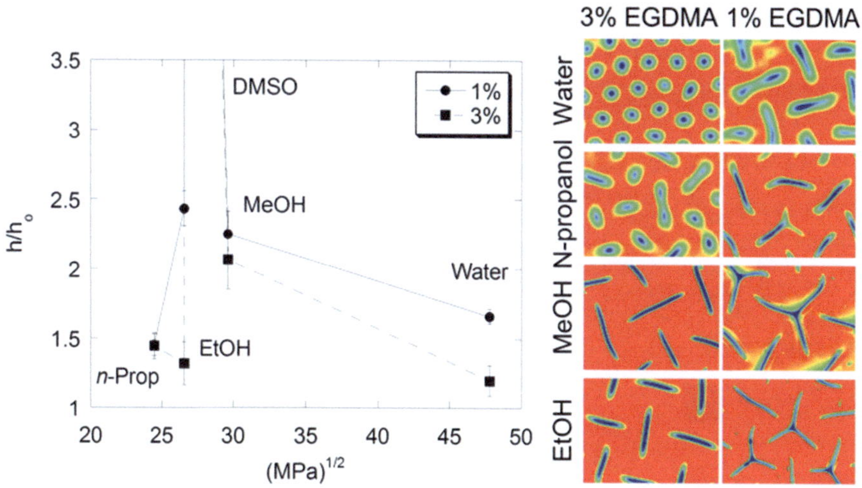

Fig. 9.6 Linear expansion with respect to the solvent solubility parameter (*left*), and optical images of the corresponding instabilities for PHEMA hydrogel with 3 and 1 % crosslinker (EGDMA) (*right*). (Reproduced from [34] with permission from The Royal Society of Chemistry)

PHEMA films with depth-wise gradient crosslinking allowed formation of wrinkling pattern in a poor solvent, such as water ($\delta_s = 47.8$), as wrinkling instability occurs in low compressive strains (see the discussion above). Large cracks and delimitations were observed for DMSO ($\delta_s = 26.7$), whereas creasing instabilities were observed in methanol ($\delta_s = 29.6$) and ethanol ($\delta_s = 26.5$). When solvent composition of ethanol-water mixture was dynamically tuned in situ, reversible transition of wrinkle to crease instabilities or vice versa was observed.

Another approach was reported to dynamically control the degree of swelling independent of solvent composition but by dynamically controlling the degree of crosslinking in the polymer gel. In this approach, polymers with photocleavable side groups were used to spatially and temporally alter hydrogel surface patterns via light exposure [65]. For this purpose, polymer films were fabricated via thermal polymerization of 2-hydroxyethylacrylate (TEA) and 2-nitrobenzyl acrylate (2-NBA) in the presence of ethylene glycol dimethacrylate and 2,2-azobisisobutyronitrile, at 65 °C for 60 min. Crosslinked polymer film was formed on methacrylated glass slides, which ensured covalent attachment of the polymer network to the glass slide. The film was exposed to UV light in DMSO, to cleave and remove the hydrophobic 2-NBA groups, forming free carboxylic acid side chains (Fig. 9.7), and thus, increasing hydrophilicity of the gel. The degree of the UV exposure was controlled to tune the water contact angle, the swelling ratio and the elastic modulus of the films, and thus, the display of swelling-induced surface patterns (Fig. 9.8). One of the advantages of this approach is the spatial control of the cleavage, which can be controlled by using a photomask.

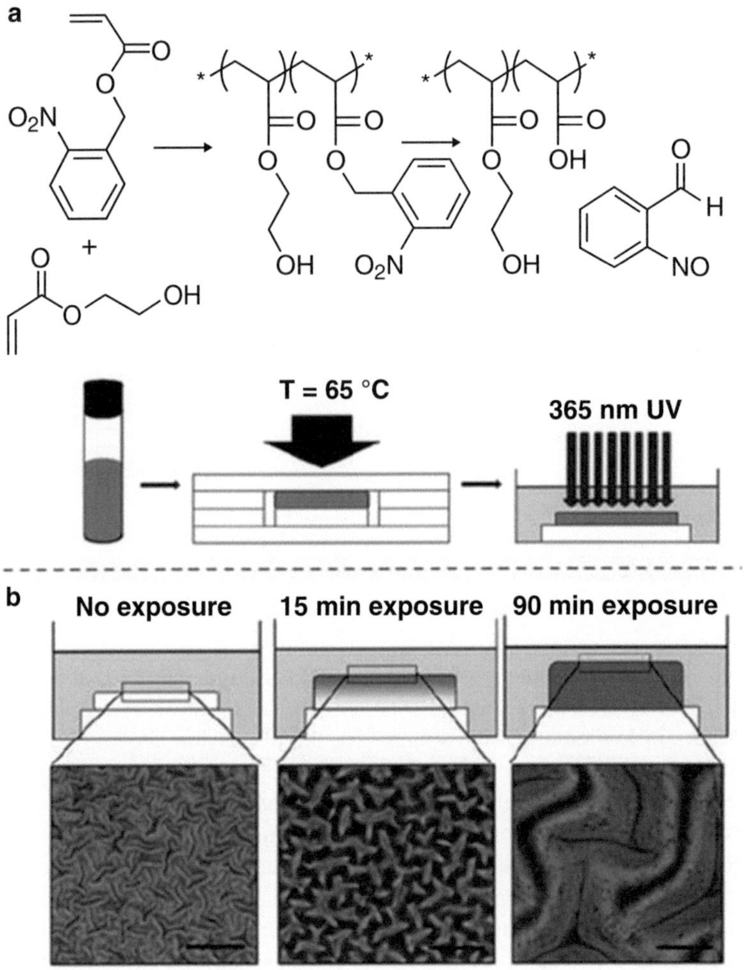

Fig. 9.7 (**a**) Thermal polymerization of HEA with 2-NBA followed by irradiation with UV light in DMSO to cleave 2-nitrobenzyl groups. (**b**) Swelling of the polymer, and hence the degree of surface instabilities, was controlled by UV exposure time. Scale bars are 100 μm. (Reproduced from [65] with permission from The Royal Society of Chemistry)

9.4 Self-Oscillating Dynamic Instabilities

Self-oscillatory patterns can be created by repeating swelling and de-swelling of the polymer network in a closed system, i.e., without any external stimuli. In comparison to stimuli-induced patterns (summarized above), self-oscillatory patterns have no on-off switch, and can repeat itself periodically, such as a beating heart (Fig. 9.9). The idea was based on Belousov-Zhabotinsky (BZ) type reaction [66] to induce spontaneous temporal change in the redox potential to control swelling

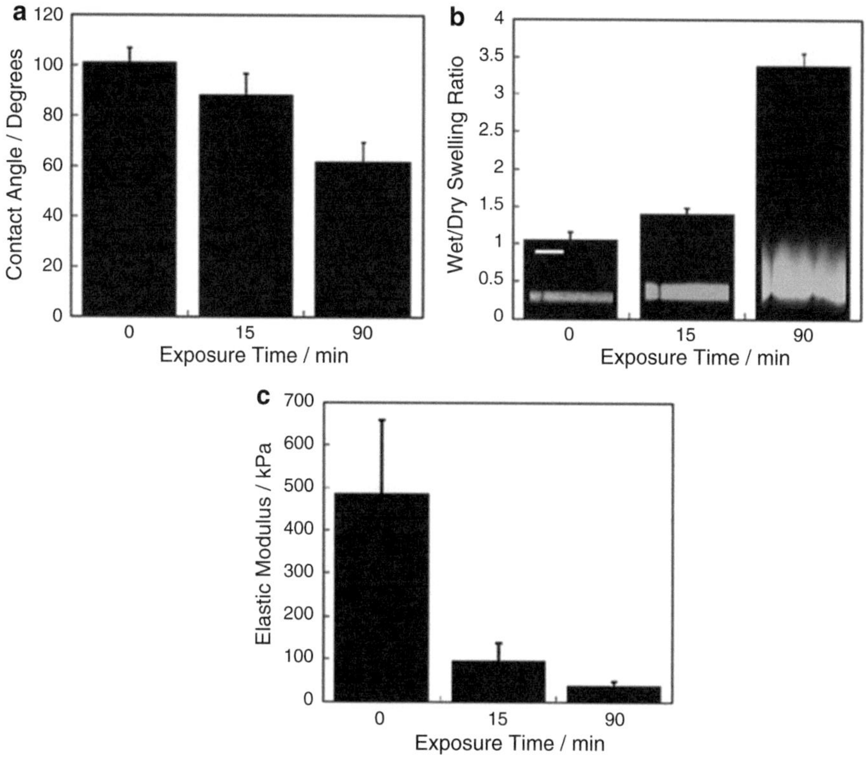

Fig. 9.8 (**a**) Contact angle, (**b**) swelling ratio, and (**c**) elastic modulus of the polymer films after 0, 15, and 90 min of UV exposure. The inset images in (**b**) are confocal microscopy images of the films (xz-plane), where gels are equilibrated in a fluorescent dye containing PBS solution to obtain contrast. (Reproduced from [65] with permission from The Royal Society of Chemistry)

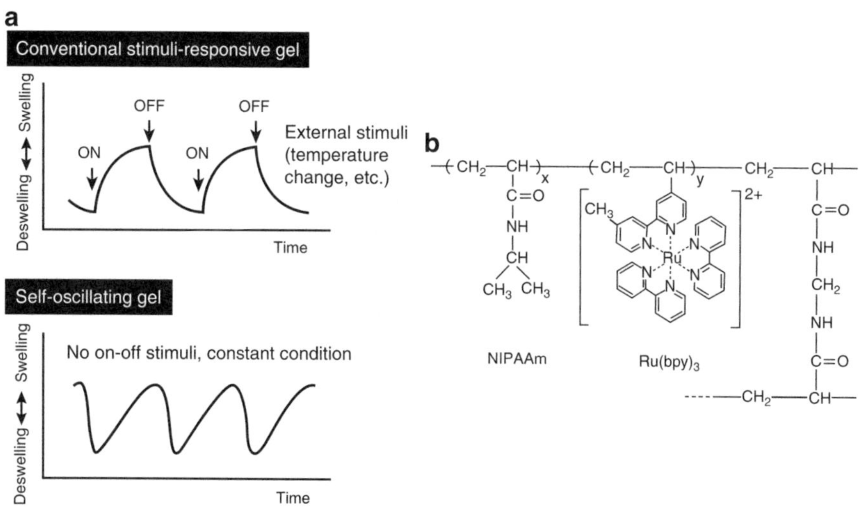

Fig. 9.9 (**a**) Plots indicating occurrence of swelling-deswelling for a stimuli responsive vs. a self-oscillating gel. (**b**) Chemical structure of poly(NIPAAM-co-Ru(bpy)$_3$) BZ gel. Reprinted with permission from [67]. (Copyright 1999 American Chemical Society)

and de-swelling of the polymer network periodically [67–70]. BZ reaction based patterns are first observed in homogenous solutions, then in many gel systems [66]. BZ reaction involves a metal catalyst that undergoes oscillatory reduction and oxidation, when exposed to the BZ reagents, such as $NaBrO_3$, HNO_3, and malonic acid [71]. BZ reaction transforms the chemical reaction into mechanical oscillation by contracting the gel in the reduced state and swelling the gel in the oxidized state, leading formation of the self-oscillating pattern.

Polymer gel systems gained special attention as they possessed many advantages, such as elimination of the hydrodynamic convection, ability to immobilize chemical groups including BZ catalysts, enabling the use of open systems without disturbing the pattern, possibility to explore water-soluble catalysts, and providing a bubble-free environment [66]. BZ catalyst can be free or tethered to the polymer network. For instance, ferroin-type catalyst [Fe(phen)$_3$] was used in soluble form for Millipore, Aerosil, agar-agar, and acrylamide systems. The use of highly hydrophobic and anionic BZ catalysis in tethered form were reported for silica gel, poly(acrylamide), calcium alginate, and anion-exchange resin (Fig. 9.10) [66]. These types of polymer gels undergoing BZ reaction are commonly referred to as BZ gels [67, 70, 72].

NIPAAm-based BZ gels are the most commonly used self-oscillating gels [67, 69, 70, 73]. For instance, the polymer network prepared from *N*-isopropylacrylamide (IPAAm) and copolymerized with ruthenium tris(2,2′-bipyridine) (Ru(bpy)$_3$), a catalyst for the BZ reaction, (poly(IPAAm-co-Ru(bpy)$_3$)) was shown to swell and de-swell at the oxidized (Ru(bpy)$_3^{3+}$) and reduced (Ru(bpy)$_3^{2+}$) states of the catalyst respectively, forming patterns with a wavelength equal to 50 μm [67]. BZ gels that are millimeters in size can self-oscillate in hours [74], and the system can be revived by refreshing the solution with reagents consumed in the reaction [75]. Velocity and the frequency of the chemical wave are shown to be controlled by controlling the

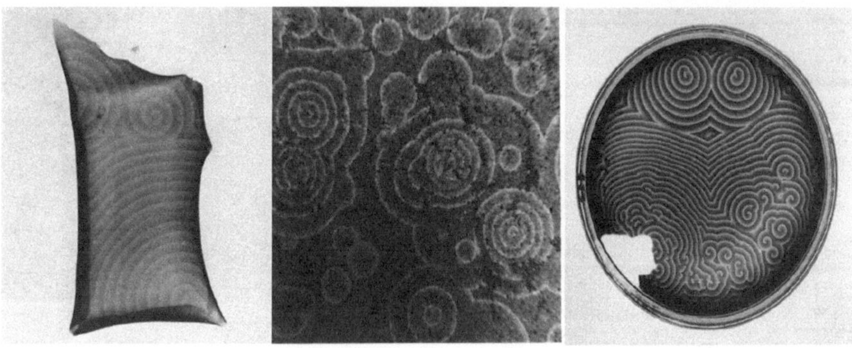

Fig. 9.10 Optical images of patterns formed on BZ gels. *From left to right*: Spiral BZ patterns (wavelength 0.82 mm) on calcium alginate gel; target patterns appeared in the discrete ion-exchange resin (wavelength of the patterns for the tight area on the left side of the image is about 1.2 mm); pattern on silica gel film. Reprinted with permission from [66]. (Copyright 1991 American Chemical Society.)

diffusion of the reacting species (such as $HBrO_2$) via temperature [74]. The size of the gel influences the oscillation behavior of the gel in such a way that duration of the oxidative state and the period of the oscillation are both reported to be increased with decreasing gel diameter (for a cylindrical geometry) [73, 76].

A wide range of BZ gels were developed based on polyacrylamide (PAAm), silica, poly(acrylamide-co-acrylate), and gelatin [77–79]. In addition, PAAm-based BZ gels were reported to de-swell when the catalyst was in the oxidized state [80], which was attributed to the formation of dynamic crosslinks at the oxidized state [77, 80]. Finally, postfunctionalization technique was employed by printing a reactive Ru catalyst ink onto a nonresponsive polymer substrate. The patterns were shown to form within the printed space, when the films were placed in the BZ solution medium [81]. The BZ gels with oscillatory behavior have many potentially useful applications including sensors [75], drug delivery applications [82], and robotics [83], but their use is currently limited due to availability of the fully developed systems.

9.4.1 Stationary Instabilities

It is possible to observe the spontaneous formation of stationary concentration instabilities in a closed reaction-diffusion system. These types of instabilities are referred to as Turing patterns. The theoretical model for Turing patterns was first developed by Turing [84], and the experimental evidence was obtained by Castets et al. for chlorite–iodide–malonic acid (CIMA) reaction in the presence of starch [85]. Later it was found in ferrocyanide-iodate-sulfite (FIS) reaction [86–89]. In order to form a Turing pattern a gradient of an activator and an inhibitor was generated within a gel film (or a membrane) in a reactor by hindering the diffusion of the activator to lower the diffusivity as compared to that of the inhibitor [90]. Gels are generally preferred as a reaction medium to restrict the mass transfer by convection. Transition from a uniform state to hexagonal and striped Turing patterns was also reported [90–92]. Polyacrylamide gels with tunable swelling ratio were used to investigate pattern formation in the FIS reaction [93]. The degree of swelling was shown to control the pattern formation in the FIS reaction for the poly (acrylamide) gels. At high swelling ratios oscillating spot patterns were observed, whereas lamellar patterns appeared at low swelling and self-replicating spot patterns were observed in between [93].

9.5 Summary

Polymeric systems displaying tunable and dynamic surface patterns are emerging as important platforms for many applications. Polymer networks swell and de-swell when exposed to a solvent. Controlling network heterogeneity enables to tune the reaction and the diffusion dynamics of species with the network. These systems enable display of transient to persistent creasing patterns, wrinkling to creasing transition, and self-oscillatory patterns can be created by repeating swelling and de-swelling of the polymer network without any external stimuli. The advances in our ability to control these surface instabilities will open new possibilities and applications for these systems.

References

1. Schaffer, E., et al.: Electrically induced structure formation and pattern transfer. Nature **403**(6772), 874–877 (2000)
2. Quake, S.R., Scherer, A.: From micro- to nanofabrication with soft materials. Science **290**(5496), 1536–1540 (2000)
3. Thorsen, T., et al.: Dynamic pattern formation in a vesicle-generating microfluidic device. Phys. Rev. Lett. **86**(18), 4163–4166 (2001)
4. Sidorenko, A., et al.: Reversible switching of hydrogel-actuated nanostructures into complex micropatterns. Science **315**(5811), 487–490 (2007)
5. Geim, A.K., et al.: Microfabricated adhesive mimicking gecko foot-hair. Nat. Mater. **2**(7), 461–463 (2003)
6. Lee, H., Lee, B.P., Messersmith, P.B.: A reversible wet/dry adhesive inspired by mussels and geckos. Nature **448**(7151), 338–341 (2007)
7. Lin, P.C., et al.: Mechanically tunable dry adhesive from wrinkled elastomers. Soft Matter **4**(9), 1830–1835 (2008)
8. Chan, E.P., Crosby, A.J.: Fabricating microlens arrays by surface wrinkling. Adv. Mater. **18**(24), 3238–3242 (2006)
9. Chandra, D., Yang, S., Lin, P.C.: Strain responsive concave and convex microlens arrays. Appl. Phys. Lett. **91**(25), 251912 (2007)
10. Yang, S., et al.: Functional biomimetic microlens arrays with integrated pores. Adv. Mater. **17**(4), 435–438 (2005)
11. Guvendiren, M., Burdick, J.A.: The control of stem cell morphology and differentiation by hydrogel surface wrinkles. Biomaterials **31**(25), 6511–6518 (2010)
12. Chen, C.S., et al.: Geometric control of cell life and death. Science **276**(5317), 1425–1428 (1997)
13. Stafford, C.M., et al.: A buckling-based metrology for measuring the elastic moduli of polymeric thin films. Nat. Mater. **3**(8), 545–550 (2004)
14. Hwa, T., Kardar, M.: Evolution of surface patterns on swelling gels. Phys. Rev. Lett. **61**(1), 106–109 (1988)
15. Sekimoto, K., Kawasaki, K.: Elastic instability of gels upon swelling. J. Phys. Soc. Jpn. **56**(9), 2997–3000 (1987)
16. Sekimoto, K., Kawasaki, K.: Elastic instabilities and phase coexistence of gels. Phys. A **154**(3), 384–420 (1989)
17. Tanaka, T.: Kinetics of phase-transition in polymer gels. Phys. A **140**(1–2), 261–268 (1986)
18. Tanaka, T., et al.: Phase-transitions in ionic gels. Phys. Rev. Lett. **45**(20), 1636–1639 (1980)
19. Tanaka, T., et al.: Mechanical instability of gels at the phase transition. Nature **325**(6107), 796–798 (1987)
20. Trujillo, V., Kim, J., Hayward, R.C.: Creasing instability of surface-attached hydrogels. Soft Matter **4**(3), 564–569 (2008)

21. Tanaka, H., Sigehuzi, T.: Surface-pattern evolution in a selling gel under a geometrical constraint - direct observation of fold structure and its coarsening dynamics. Phys. Rev. E **49**(1), R39–R42 (1994)
22. Tanaka, H., et al.: Morphological and kinetic evolution of surface patterns in gels during the swelling process - evidence of dynamic pattern ordering. Phys. Rev. Lett. **68**(18), 2794–2797 (1992)
23. Onuki, A.: Theory of pattern-formation in gels - surface folding in highly compressible elastic bodies. Phys. Rev. A **39**(11), 5932–5948 (1989)
24. Onuki, A.: Pattern-formation in gels. J. Phys. Soc. Jpn. **57**(3), 703–706 (1988)
25. Guvendiren, M., Burdick, J.A., Yang, S.: Kinetic study of swelling-induced surface pattern formation and ordering in hydrogel films with depth-wise crosslinking gradient. Soft Matter **6**(9), 2044–2049 (2010)
26. Toomey, R., Freidank, D., Ruhe, J.: Swelling behavior of thin, surface-attached polymer networks. Macromolecules **37**(3), 882–887 (2004)
27. Harmon, M.E., Kucking, D., Frank, C.W.: Photo-cross-linkable PNIPAAm copolymers. 5. Mechanical properties of hydrogel layers. Langmuir **19**(26), 10660–10665 (2003)
28. Harmon, M.E., Kuckling, D., Frank, C.W.: Photo-cross-linkable PNIPAAm copolymers. 2. Effects of constraint on temperature and pH-responsive hydrogel layers. Macromolecules **36**(1), 162–172 (2003)
29. Harmon, M.E., et al.: Photo-cross-linkable PNIPAAm copolymers. 4. Effects of copolymerization and cross-linking on the volume-phase transition in constrained hydrogel layers. Langmuir **19**(26), 10947–10956 (2003)
30. Basu, S.K., McCormick, A.V., Scriven, L.E.: Stress generation by solvent absorption and wrinkling of a cross-linked coating atop a viscous or elastic base. Langmuir **22**(13), 5916–5924 (2006)
31. Basu, S.K., et al.: Mechanism of wrinkle formation in curing coatings. Prog. Org. Coat. **53**(1), 1–16 (2005)
32. Gent, A.N., Cho, I.S.: Surface instabilities in compressed or bent rubber blocks. Rubber Chem. Technol. **72**(2), 253–262 (1999)
33. Hong, W., Zhao, X., Suo, Z.: Formation of creases on the surfaces of elastomers and gels. Appl. Phys. Lett. **95**(11), 111901 (2009)
34. Guvendiren, M., Burdick, J.A., Yang, S.: Solvent induced transition from wrinkles to creases in thin film gels with depth-wise crosslinking gradients. Soft Matter **6**(22), 5795–5801 (2010)
35. Weiss, F., et al.: Creases and wrinkles on the surface of a swollen gel. J. Appl. Phys. **114**(7), 073507 (2013)
36. Bowden, N., et al.: Spontaneous formation of ordered structures in thin films of metals supported on an elastomeric polymer. Nature **393**(6681), 146–149 (1998)
37. Groenewold, J.: Wrinkling of plates coupled with soft elastic media. Phys. A **298**(1–2), 32–45 (2001)
38. Stafford, C.M., et al.: Elastic moduli of ultrathin amorphous polymer films. Macromolecules **39**(15), 5095–5099 (2006)
39. Chan, E.P., Crosby, A.J.: Spontaneous formation of stable aligned wrinkling patterns. Soft Matter **2**(4), 324–328 (2006)
40. Efimenko, K., et al.: Nested self-similar wrinkling patterns in skins. Nat. Mater. **4**(4), 293–297 (2005)
41. Hayward, R.C., Chmelka, B.F., Kramer, E.J.: Template cross-linking effects on morphologies of swellable block copolymer and mesostructured silica thin films. Macromolecules **38**(18), 7768–7783 (2005)
42. Huang, R.: Kinetic wrinkling of an elastic film on a viscoelastic substrate. J. Mech. Phys. Solids **53**(1), 63–89 (2005)
43. Chen, X., Hutchinson, J.W.: Herringbone buckling patterns of compressed thin films on compliant substrates. J. Appl. Mech. **71**(5), 597–603 (2004)

44. Guvendiren, M., Yang, S., Burdick, J.A.: Swelling-induced surface patterns in hydrogels with gradient crosslinking density. Adv. Funct. Mater. **19**(19), 3038–3045 (2009)
45. Chung, J.Y., Nolte, A.J., Stafford, C.M.: Diffusion-controlled, self-organized growth of symmetric wrinkling patterns. Adv. Mater. **21**(13), 1358–1362 (2009)
46. Breid, D., Crosby, A.J.: Effect of stress state on wrinkle morphology. Soft Matter **7**(9), 4490–4496 (2011)
47. Cao, Y., Hutchinson, J.W.: From wrinkles to creases in elastomers: the instability and imperfection-sensitivity of wrinkling. Proc. Roy. Soc. A Math. Phys. Eng. Sci. **468**(2137), 94–115 (2012)
48. Wong, W.H., et al.: Surface instability maps for soft materials. Soft Matter **6**(22), 5743–5750 (2010)
49. Holmes, D.P., Crosby, A.J.: Snapping surfaces. Adv. Mater. **19**, 3589–3593 (2007)
50. Holmes, D.P., Ursiny, M., Crosby, A.J.: Crumpled surface structures. Soft Matter **4**(1), 82–85 (2008)
51. Zhang, Y., et al.: One-step nanoscale assembly of complex structures via harnessing of elastic instability. Nano Lett. **8**(4), 1192–1196 (2008)
52. Breid, D., Crosby, A.J.: Surface wrinkling behavior of finite circular plates. Soft Matter **5**, 425–431 (2009)
53. Sharp, J.S., Jones, R.A.L.: Swelling-induced morphology in ultrathin supported films of poly (d, l-lactide). Phys. Rev. E **66**(1), 011801 (2002)
54. Klein, Y., Efrati, E., Sharon, E.: Shaping of elastic sheets by prescription of non-Euclidean metrics. Science **315**(5815), 1116–1120 (2007)
55. Cerda, E., Mahadevan, L.: Geometry and physics of wrinkling. Phys. Rev. Lett. **90**(7), 4 (2003)
56. Huck, W.T.S., et al.: Ordering of spontaneously formed buckles on planar surfaces. Langmuir **16**(7), 3497–3501 (2000)
57. Sultan, E., Boudaoud, A.: The buckling of a swollen thin gel layer bound to a compliant substrate. J. Appl. Mech. Trans. ASME **75**(5), 051002 (2008)
58. Wu, Z.G., Bouklas, N., Huang, R.: Swell-induced surface instability of hydrogel layers with material properties varying in thickness direction. Int. J. Solids Struct. **50**(3–4), 578–587 (2013)
59. Lee, D., et al.: Surface instability of an elastic half space with material properties varying with depth. J. Mech. Phys. Solids **56**(3), 858–868 (2008)
60. Guvendiren, M., et al.: 9.22 - Photopolymerizable systems. In: Matyjaszewski, K., Möller, M. (eds.) Polymer Science: A Comprehensive Reference, pp. 413–438. Elsevier, Amsterdam (2012)
61. Li, Y.Y., et al.: Patterned polymer films via reactive silane infusion-induced wrinkling. Langmuir **29**(14), 4632–4639 (2013)
62. Flory, P.J., Rehner Jr., J.: Statistical mechanics of cross-linked polymer networks. II. Swelling. J. Chem. Phys. **11**, 521–526 (1943)
63. Barton, A.F.M.: Solubility parameters. Chem. Rev. **75**(6), 731–753 (1975)
64. Hildebrand, J.H., Prausnitz, J.M., Scott, R.L., Hildebrand, J.H., Prausnitz, J.M., Scott, R.L.: Regular and Related Solutions, 3rd edn. Van Nostrand-Reinhold, Princeton (1970)
65. Ramanan, V.V., et al.: Photocleavable side groups to spatially alter hydrogel properties and cellular interactions. J. Mater. Chem. **20**(40), 8920–8926 (2010)
66. Yamaguchi, T., et al.: Gel systems for the Belousov-Zhabotinskii reaction. J. Phys. Chem. **95**(15), 5831–5837 (1991)
67. Yoshida, R., Kokufuta, E., Yamaguchi, T.: Beating polymer gels coupled with a nonlinear chemical reaction. Chaos **9**(2), 260–266 (1999)
68. Yoshida, R., et al.: Self-oscillating gels. Adv. Mater. **9**(2), 175–178 (1997)
69. Yoshida, R., et al.: Self-oscillation of polymer gels coupled with the Belousov-Zhabotinsky reaction. ACH Model Chem. **135**(3), 409–416 (1998)
70. Yoshida, R., et al.: Self-oscillating gel. J. Am. Chem. Soc. **118**(21), 5134–5135 (1996)

71. Zaikin, A.N., Zhabotinsky, A.M.: Concentration wave propagation in two-dimensional liquid-phase self-oscillating system. Nature **225**(5232), 535–537 (1970)
72. Yashin, V.V., et al.: *Mechano-chemical oscillations and waves in reactive gels.* Rep. Prog. Phys. **75**(6), 066601 (2012)
73. Sasaki, S., et al.: Mechanical oscillation coupled with the Belousov–Zhabotinsky Reaction in Gel. Langmuir **19**(14), 5595–5600 (2003)
74. Miyakawa, K., et al.: Chemical waves in self-oscillating gels. Phys. Rev. E **62**(1), 793–798 (2000)
75. Chen, I.C., et al.: Mechanical resuscitation of chemical oscillations in Belousov–Zhabotinsky gels. Adv. Funct. Mater. **22**(12), 2535–2541 (2012)
76. Chen, I.C., et al.: Shape- and size-dependent patterns in self-oscillating polymer gels. Soft Matter **7**(7), 3141–3146 (2011)
77. Konotop, I.Y., et al.: Self-oscillatory systems based on polymer gels. Polym. Sci. Ser. B **51**(9–10), 383–388 (2009)
78. Konotop, I.Y., et al.: Chemomechanical oscillations in polymer gels: effect of the size of samples. Polym. Sci. Ser. B **53**(1–2), 26–30 (2011)
79. Smith, M.L., et al.: Autonomic hydrogels through postfunctionalization of gelatin. Chem. Mater. **24**(15), 3074–3080 (2012)
80. Yuan, P., et al.: UV patternable thin film chemistry for shape and functionally versatile self-oscillating gels. Soft Matter **9**(4), 1231–1243 (2013)
81. Kramb, R.C., et al.: Autonomic composite hydrogels by reactive printing: materials and oscillatory response. Soft Matter **10**(9), 1329–1336 (2014)
82. Yoshida, R., et al.: Pulsatile drug-delivery systems using hydrogels. Adv. Drug Deliv. Rev. **11**(1–2), 85–108 (1993)
83. Dayal, P., Kuksenok, O., Balazs, A.C.: Directing the behavior of active, self-oscillating gels with light. Macromolecules **47**(10), 3231–3242 (2014)
84. Turing, A.M.: The chemical basis of morphogenesis. Phil. Trans. R. Soc. Lond. B **237**(641), 37–72 (1952)
85. Castets, V., et al.: Experimental-evidence of a sustained standing Turing-type nonequilibrium chemical-pattern. Phys. Rev. Lett. **64**(24), 2953–2956 (1990)
86. Lee, K.J., et al.: Pattern-formation by interacting chemical fronts. Science **261**(5118), 192–194 (1993)
87. Lee, K.J., et al.: Experimental-observation of self-replicating spots in a reaction-diffusion system. Nature **369**(6477), 215–218 (1994)
88. Lee, K.J., Swinney, H.L.: Lamellar structures and self-replicating spots in a reaction-diffusion system. Phys. Rev. E **51**(3), 1899–1915 (1995)
89. Li, G., Qi, O.Y., Swinney, H.L.: Transitions in two-dimensional patterns in a ferrocyanide-iodate-sulfite reaction. J. Chem. Phys. **105**(24), 10830–10837 (1996)
90. Vigil, R.D., Ouyang, Q., Swinney, H.L.: Turing patterns in a simple gel reactor. Phys. A **188**(1–3), 17–25 (1992)
91. Ouyang, Q., Noszticzius, Z., Swinney, H.L.: Spatial bistability of 2-dimensional Turing patterns in a reaction-diffusion system. J. Phys. Chem. **96**(16), 6773–6776 (1992)
92. Ouyang, Q., Swinney, H.L.: Transition from a uniform state to hexagonal and striped Turing patterns. Nature **352**(6336), 610–612 (1991)
93. Ueno, T., Yoshida, R.: Effect of gel network on pattern formation in the ferrocyanide–iodate–sulfite reaction. J. Phys. Chem. A **115**(21), 5231–5237 (2011)

Chapter 10
Breath Figures: Fabrication of Honeycomb Porous Films Induced by Marangoni Instabilities

Alexandra Muñoz-Bonilla, Maud Save, Laurent Billon, and Juan Rodríguez-Hernández

10.1 Introduction

The creation of porous polymer surfaces is a center of interest in current research. Porous surfaces possess extremely high specific surface areas, thus allowing their employment in a large variety of applications including electronics, photonics or biotechnology [1, 2]. Pore size and distribution can play a major role in selective transportation or in insulation processes amongst others [3]. Those porous materials with cavities in the micrometer size range are interesting in catalysis, sensors, membrane preparation or as scaffolds for composite materials. Moreover porous materials with pore dimensions comparable to the wavelength of visible light are of interest as photonic band-gaps and optical stop-bands.

Structures with micrometer or submicrometer dimensions can be created using different templating methods [4, 5]. A large variety of approaches have been developed and employed to prepare microporous structured materials, including the use of templates such as ordered arrays of colloidal particles to produce inverse opal structures [6–9], from transformed polymeric sphere arrays [10, 11], using emulsion droplets as templates [12], employing natural biological templates [13–16], by

A. Muñoz-Bonilla
Departamento de Química Física Aplicada, Facultad de Ciencias, Universidad Autónoma de Madrid, C/Francisco Tomás y Valiente 7, Cantoblanco 28049, Madrid, Spain

M. Save • L. Billon (✉)
Equipe de Physique et Chimie des Polymères, CNRS, UMR 5254—IPREM, Universitè Pau & Pays Adour, 2 Avenue du Président Angot, Pau 64053, France
e-mail: laurent.billon@univ-pau.fr

J. Rodríguez-Hernández (✉)
Macromolecular Engineering Group, Institute of Polymer Science and Technology (ICTP-CSIC), C/Juan de la Cierva 3, 28006 Madrid, Spain
e-mail: rodriguez@ictp.csic.es

J. Rodríguez-Hernández, C. Drummond (eds.), *Polymer Surfaces in Motion*, DOI 10.1007/978-3-319-17431-0_10

phase inversion [17], self-organized surfactants [18], microphase separated or electric-field-induced block copolymers patterning [19–21], etc. Other alternatives include direct writing of polymer patterns [22], the use of photochemically or electrochemically polymerizable precursors [23] or soft lithographic methods [24].

Most of the above mentioned approaches resort to the use of templates that must be removed after the fabrication of the porous films. These templates are, in general, difficult to prepare or they are rather hard to be removed. This chapter is devoted to an alternative approach, i.e., the breath figures (BF) templating method [5, 25–29].

The breath figures technique is one of the most widely employed methods for the fabrication of organized porous polymer films [30, 31] and, as further depicted in detail, in this approach the template consists of an ordered array of water droplets that can be removed by simple evaporation. Indeed, the simultaneous evaporation of a volatile solvent and condensation of water vapor in combination with thermocapillary effects and Marangoni convection allow the formation and precise organization of water droplets at the polymer solution–air interface [30]. This array of water droplets will evaporate upon complete evaporation of the solvent of the polymeric solution, and the surface will reflect its presence in the form of pores.

Several significant advantages justify its extensive use. First, as mentioned above, the self-removal of the template favors a reduction on the production time and costs. Second, BF allows the employment of a large variety of materials ranging from polymers to hybrid nanocomposites thus leading to porous films with diverse properties. Finally, as will be analyzed throughout this chapter, the external parameters (temperature, air humidity, ...) and those related with the preparation procedure (solvent, polymer concentration, polymer features, ...) are directly related to the pore dimensions and shape obtained. Thus, by controlling these parameters the pore sizes can be easily manipulated.

10.1.1 Few Insights on the Breath Figures Mechanism

Water vapor condensation is a well-known phenomenon that leads to the formation of water droplets on solid surfaces. Pioneer works in this area focused on the study of the order that can be obtained by assembling water droplets on surfaces. This phenomenon was termed "breath figures" remembering their formation mechanism, i.e., breathing on cold surfaces [32–36]. Thus, breath figures are formed when a cold surface is brought in contact with moist air. Depending on the surface wettability we can find two situations. In a first situation, the condensed water vapor can wet the surface thus producing a thin layer. The second and most interesting case concerns the situation in which the surface is not wetted by the condensed vapor. In this case, moisture condenses on the cold surface forming water droplets that grow during the evaporation giving rise to distinct water droplet arrangements on the surface.

A particular case of breath figures formation concerns the use of polymeric solutions as cold surfaces that can produce highly ordered and functional porous materials. Thus, this approach to produce porous films is recognized as *breath figures templating method.*

The use of polymer solutions was first described by Francois et al. in 1994 that described how water droplets formed by condensation on top of an evaporating polymer solution can generate an imprint on polymer surfaces [31]. As a result of these exciting findings, this field received a new impulse, and it is during the last decade the breath figures mechanism has been extensively studied [25, 26, 29, 30, 37, 38]. The BF formation mechanism has been widely studied using solid and liquid surfaces [31, 39–44] and literature is today available concerning aspects such as droplet nucleation and growth [39, 42–47] or coalescence [39, 48–53]. On the contrary, far more limited literature can be found devoted to understanding the patterning formation using breath figures from polymer solutions [54–59].

According to the generally admitted mechanism, the formation of the breath figures pattern comprises several steps as depicted in Fig. 10.1 [41, 60]: in the initial stage (A), the endothermic evaporation of the solvent resulted in a decrease of the solvent temperature, thus triggering the water condensation. In the following stages the rising of temperature compensates the decrease of temperature of the initial stage [59]. Then, an initial heterogeneous nucleation of the water droplets formed on the surface and the growth of the isolated droplets occurs (B). Nucleation involves the formation of small liquid drops which are thermodynamically stable, and therefore, do not evaporate. During this period, the surface coverage, ε^2, which is related to the ratio of droplet area to substrate area, is rather low. Once a droplet of water has nucleated on the substrate, it grows at the expense of the vapor of the

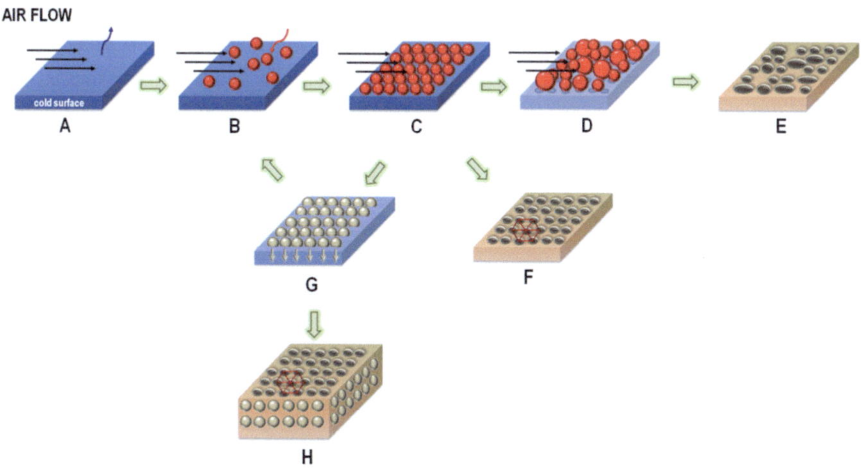

Fig. 10.1 Evolution of the water condensation during the evaporation of a polymer solution: breath figures formation following the mechanism of Srinivasarao and coworkers. This figure includes the formation of rather disordered structures (*E*), honeycomb structured films (*F*), and multilayers of pores (*H*). Reproduced with permission from ref. [27]

surrounding atmosphere. After the film returns to ambient temperature, the condensed water and residual solvent evaporate, leaving behind the honeycomb (HC) structure. It is important to consider that when both, the solvent vapor pressure and the air velocity across the surface are high, the surface cooling is more rapidly produced. In the intermediate stage, provided that the evaporation time was enough to induce growth of the water droplets and produce a complete coverage of the surface, a hexagonal closed packed array (C) can be obtained. Too short evaporation times will result in pores with a random distribution at the surface.

At this stage different situations can be found. In the case that solvent evaporation finishes before the coalescence begins, the film is formed by a monolayer of ordered pores with similar sizes (F). If the precipitating polymer employed stabilizes the condensed water droplets preventing the coagulation and there is still solvent available to be evaporated, the first layer of condensed droplets can sink into the solution (G). Besides, the density of the solvent used, if is more (i.e., carbon disulfide (CS_2)) or less (i.e., benzene, toluene) dense than water, can be also a key point on the development of monolayer or multilayered structures. As a consequence, the surface covered now with solvent can condense a new layer of water droplets. The result of this repetitive process is the formation of multilayers of pores (H).

Several models have been proposed in order to explain how droplets can be so close without coalescence. The models described resort to two different effects, i.e., thermocapillary effects and Marangoni convection. According to Srinivasarao and coworkers [30] coalescence is prevented by means of these two simultaneously occurring effects as follows: on the one hand, a thermocapillary effect which is a temperature-induced-gradient that drives the motion of the water droplets into the polymer solution. On the other hand, the Marangoni convection prevents coagulation between two different droplets (mechanism followed in Fig. 10.1). Alternatively, Fig. 10.2 illustrates the mechanism proposed by Shimomura et al. [60]. According to this group thermocapillary and Marangoni forces drives the movement of the water droplets inside of the polymeric solution. As a result and due to the drying process occurring from the edge to the center of the polymer solution, the thermocapillary and Marangoni forces submerged the droplets and

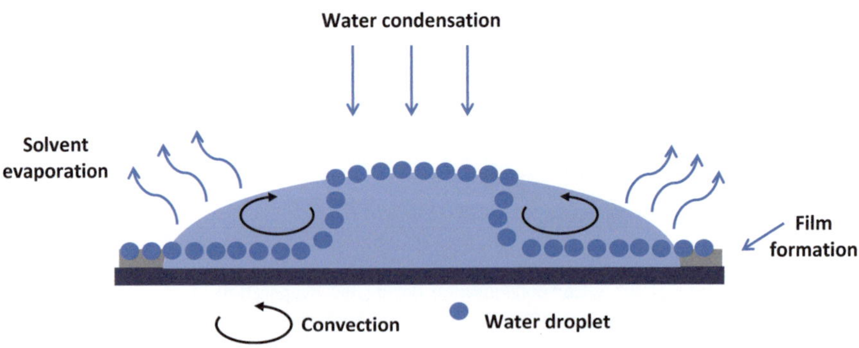

Fig. 10.2 Mechanism of BF formation according to Shimomura and coworkers [60]

direct the formation of hexagonal ordered porous surfaces [60]. Finally, accepting the two effects Francois et al. [31] evidenced that the use of polymers can prevent the coagulation to occur through the instant precipitation of a polymer layer around the water droplet.

If solvent evaporation proceeds and the polymer does not achieve the stabilization of the droplets, two drops can become in contact at their perimeter and interact by fusion or coalescence. After coalescence, a new drop is formed whose volume is the sum of the volumes of the initial drops. The formation of a new single drop is favorable in terms of energy since the surface energy is lowered by coalescence (D). As a result of the coalescence process and after evaporation of both the water droplets and the solvent employed, a disordered array of pores is formed (E).

10.2 Experimental Parameters to Induce Topographical Modifications Using the Breath Figures Approach

10.2.1 General Strategies: to Fabricate Porous Films by BFs

Several approaches have been developed to fabricate honeycomb porous films [61] (see Fig. 10.3). The most extended approach involves a casting process of polymer solution droplet on a solid support controlling the relative humidity of the chamber (i). In general, at least relative humidity (RH) higher than 50 % is necessary to induce the water condensation and produce highly ordered structures.

However it is alternatively allowed the preparation of porous films under dry conditions by introducing water directly into polymer solution by creation of homogeneous emulsions [62]. It is also reported the addition of certain amount of water in THF solution, a water miscible solvent [63].

Subsequent modifications of this basic technique, while maintaining the static conditions were introduced in order to enhance the order or tune the morphological characteristics of the structure. Among them, the control of the solution temperature by using a cold stage allows additional control over the condensation rate of water droplets and the evaporation of the solvent [28]. Indeed, a decrease of the temperature entails an increase in the condensation and therefore in the viscosity of the solution, lowering the solvent evaporation rate and improving the formation of highly ordered honeycomb structures [64, 65].

The spin coating method (ii) allows the preparation of films with uniform thickness and lower roughness [66–70]. Its use in combination with breath figures results in fabrication of large uniform areas of porous structures. In order to provide high humidity environment during the coating process, beakers containing hot water or an appropriate aqueous solution of a salt can be placed inside the spin-coater chamber [71]. In general, high spinning rates provide more regular porous structures whereas at low rotating speed the evaporation rate is lower and, thus, the

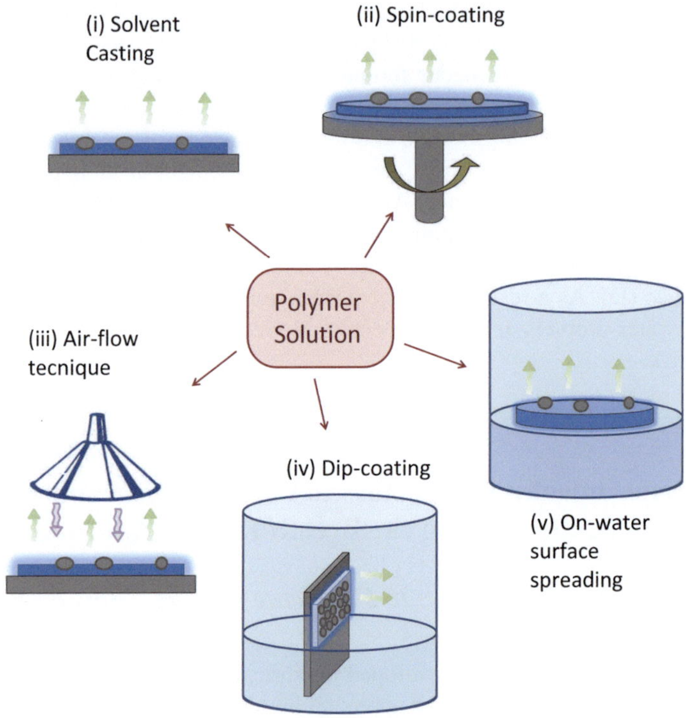

Fig. 10.3 Approaches available for the preparation of honeycomb-structured porous films via: (i) solvent casting; (ii) spin-coating; (iii) dynamic airflow technique; (iv) dip-coating, and (v) on-water surface spreading. Reprinted with permission from ref. [27]

probability to observe coalescence is higher [71]. Moreover, the pore size decreases with increasing spinning rate because the faster evaporation entails low time for water droplet growth although the nucleation of water droplets was favored. The use of spin-coating normally conducts to less ordered patterns as compared with drop-casting combined with airflow techniques and as a result of the rotation the shape of the pores obtained by this method is sometimes elongated rather than spherical.

Apart from the direct solvent evaporation using static conditions, several groups regulated the evaporation process by using a humid continuous airflow directed to the solution surface. The employment of airflow (iii) increases the solvent evaporation and improves the temperature gradient between the surface and the bulk. As a consequence, the airflow promotes convection in the solution and influences the arrangement of water droplets. Several variables need to be controlled including the flow speed or the humidity [72–75]. Moreover, the distance and the angle between the surface and the air current is also an additional key parameter to control the pore characteristics and optimize the quality of the film. As an example, low airflow rate allows water droplets to grow for longer time giving place to larger pores. On the other hand, an excessive airflow causes too fast evaporation and prevents water droplets both to grow and to organize into a regular array.

Dip-coating is also an alternative methodology successfully employed to prepare regular arrays of micrometer pores [76, 77]. As depicted in (iv), in this case, due to the vertical position of the glass slide and the liquid surface, the mechanism is slightly different. For instance, the sinking of the droplets may not be due to a gravity process and it is more probable that the droplets touch the surface, and interfacial forces control the movement of the droplets and the polymer solution.

Finally, whereas the honeycomb structures have been mainly obtained on solid supports (glass and silicon wafers) there are few studies devoted to the preparation of porous films on liquid interfaces resulting in the formation of self-supporting films (v). As an example, Nishikawa et al. reported the preparation of honeycomb films at the air–water interface [61, 78, 79]. The volume of cast solution and the temperature of the water bath allowed them to control the parameters of the microstructure including pore size of the mesh and film thickness. This approach, called "on-water spreading" method, has been also employed by other groups [80, 81].

10.2.2 Role of the Experimental Conditions and Polymers Employed to Control the Features of the Ordered Pattern of Pores

10.2.2.1 Pore Dimensions and Morphology

The patterning of porous polymer films by the breath figure approach involves two liquid phases in a dynamic process and the features of the solvents are a key point of the process to control the quality of the pore ordering. The film formation is driven by a rapid solvent evaporation while the droplet stabilization is promoted by the polymer precipitation at the liquid–liquid interface preventing their coalescence [82]. The arrangement of the droplets into the thermodynamically preferred porous hexagonal pattern is favored by the use of a volatile organic solvent fulfilling the following criteria: high vapor pressure, low boiling point, low solubility in water and preferentially higher density than water. The features of different organic solvents gathered in Table 10.1 show that carbon disulfide and chloroform fulfill the four criteria. These solvents are indeed the most widely used solvents for breath figure process [25–27, 29, 83].

In the early work of François and coworkers, water was used for the fabrication of the pores in honeycomb-structured porous polystyrene films, the pores being the replica of the organized pattern of condensed water droplets [31]. Water is still the most efficient and widely used solvent for breath figure process as it is environmentally friendly and easy to process as static humid atmosphere or dynamic humid

Table 10.1 Data of solvents used for honeycomb film fabrication [84]

	Chloroform	Dichloromethane	Carbon disulfide	Toluene	Tetrahydrofurane	Pentane	Diethyl ether	Ethyl acetate
Density (g/cm^3)	1.48	1.32	1.26	0.86	0.89	0.62	0.71	0.90
Vapor pressure (kPa)	21.3	46.5	39.7	2.9	21.6	56.8	58.6	9.7
Solubility in water (g/L)	8.2	20	2.9	0.5	Miscible	0.4	69	87
Boiling point (°C)	61	39	46	110	65	36	34	77

airflow. Moreover, water is favorably condensed at the surface of the organic solvent as the endothermic evaporation process induces surface cooling. Indeed, the evolution of the solution temperature was explored during solvent (carbon disulfide) evaporation and it was shown that a noticeable temperature variation occurred during the solvent evaporation. The first stage showed a sharp drop in temperature down to 10 °C for the first 100 s and the second stage showed a slow rising of temperature up to 25 °C [59]. The moist atmosphere, used either in static or dynamic conditions, usually produces periodic structures within a pore size range of 1–20 μm [25, 27, 28]. Formation of pore patterns with smaller pores is more rarely observed. Few examples of polymers such as poly(vinyldiphenylquinoline) [85], fluorinated acrylic copolymer [86], *p*-toluenesulfonate piperidinium acid salt-terminated polystyrene [87], polystyrene-*b*-poly(4-vinylpyridine) [81], or porphyrin-core star polystyrene-*b*-poly(2-hydroxyethyl methacrylate) [88], showed the ability to produce pore diameter below 1 μm under humid atmosphere. A study of Ding et al. [89] reported the possibility to prepare porous honeycomb polymer films using breath figure method under nonaqueous atmospheres. The nature of the vapors clearly influenced both size and shape of pores for films prepared by casting a solution of polystyrene-*b*-poly(dimethylsiloxane) (PS-*b*-PDMS) in carbon disulfide with a concentration of 80 mg mL^{-1}. An ordered array of cylindrical pores with large section of 9.6 μm was observed with methanol vapor while the diameter decreased to 7.0 μm for ellipsoidal pores and 4.7 μm for spheroidal pores under the ethanol and water atmospheres, respectively (Fig. 10.4) [89].

The difference in pore shape was explained by the difference of interfacial tension between the polymer solution and the solvent or water droplet. Indeed, spreading water droplets over a unit area of a polymer solution can be determined by the spreading coefficient (S) as follows: $S = \gamma_P - (\gamma_W + \gamma_{W/P})$, where γ_P is the surface tension of the polymer solution, γ_W is the surface tension of the water droplet, and $\gamma_{W/P}$ is the interfacial tension between the polymer solution and the water droplets in this case [90]. Water has the largest surface tension compared with the two alcohols, thus the water droplet spreading is reduced and water droplets maintain their spherical shape [89]. On the other hand, the value of $\gamma_{W/P}$ is assumed to be very low for alcohol droplets because both methanol and ethanol are miscible with carbon disulfide. It should be noticed that under such alcoholic vapors, the polymer concentration should be higher than under aqueous atmosphere as no regular patterns were formed for polymer concentrations less than 10 g L^{-1}.

Apart from thermodynamic (interfacial/surface tensions, vapor pressures) which plays a role in tuning the size and shape of pores in honeycomb films, the final array of pores is also regulated by kinetics as breath figure is a dynamic process driven by solvent evaporation and polymer precipitation. As summarized in different reviews, the pore size are also strongly impacted by experimental parameters such as the level of relative humidity, the volume of cast solution or the concentration of the polymer solution [25–29, 83]. Breath figure process is actually a tradeoff between the fast solvent evaporation required to promote surface cooling inducing water droplet condensation, arrangement and absence of droplet coalescence by thermocapillary effects, and the sufficient time allowing water droplet growth and

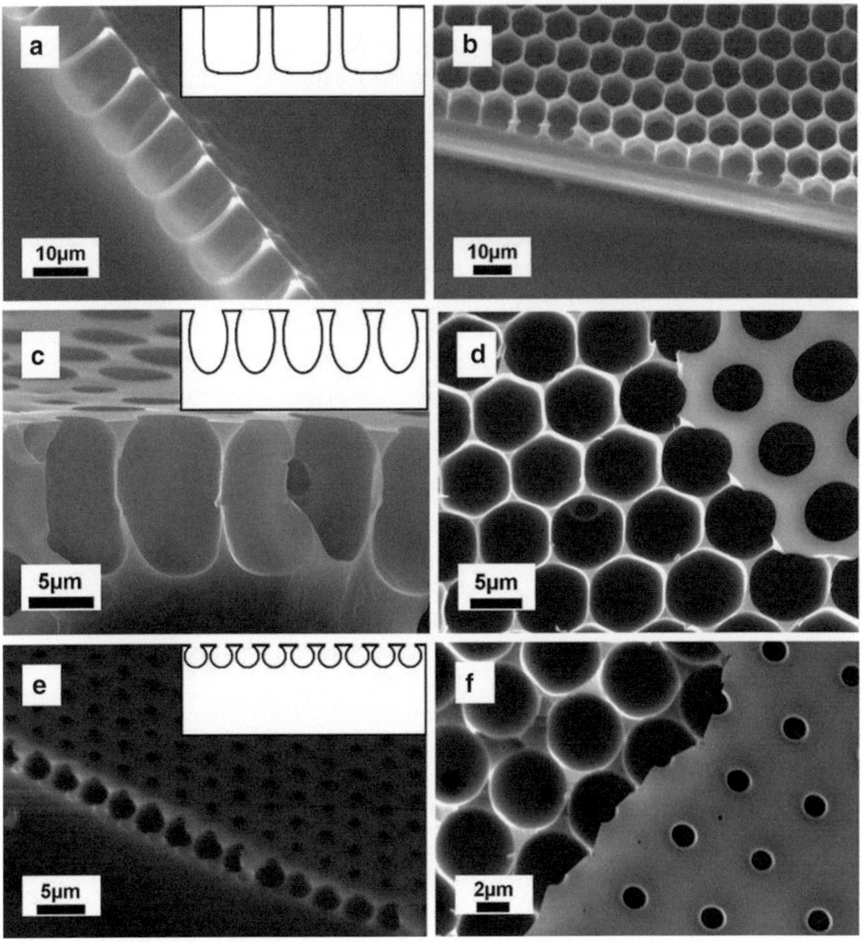

Fig. 10.4 SEM images of PS-*b*-PDMS/CS_2 films prepared under different solvent vapors: methanol (**a**, **b**), ethanol (**c**, **d**), and water (**e**, **f**). The *right column* corresponds to the HC film after peeling the *top layer* (from ref. [89])

arrangement. The time of the overall rapid process is only few minutes. Breath figures generally do not form in an atmosphere that has less than 40–50 % relative humidity. For an established polymer/solvent system, increasing the level of relative humidity in the range of 40–90 % induces an increase of the pore size [25, 27–29, 91–97]. Under dynamic conditions, the humid airflow rate and the distance between airflow and surface both influence the rate of solvent evaporation and the droplet condensation [65, 72–75]. A slow solvent evaporation allows water droplets to grow on the surface producing large pores while a too fast solvent evaporation limits the water droplet growth and the pore organization via the Bénard–Marangoni convection arising in a solution presenting a thermal gradient. The solvent evaporation time is also dependent of the volume of polymer solution cast onto the substrate. A higher casting volume reduces the evaporation time and provokes an increase of the average pore diameter [98, 99]. Likewise, the faster solvent evaporation occurring from the edge of the honeycomb film to the center induced the presence of smaller pores at the edge of the film [86, 100]. The polymer concentration also plays an important role on the control of pore size. As reported by several studies, for a given polymer/solvent system, the increase of the concentration of polymer solution systematically led to lower pore diameter [93, 97, 98, 101–108]. For higher polymer concentration, the precipitation of the thin polymer layer at the interface of polymer solution and water droplet occurs in earlier stage, hence limiting droplet growth.

Finally, another way to tune the pore shape is to apply directional mechanical forces such as shrinking [109] or stretching [73] to viscoelastic honeycomb films or to orientate the humid airflow [110]. In addition peeling-off the top surface of honeycomb films, exhibiting an organized hexagonal pattern of spherical pores, produced pincushion structures at the origin of surface hydrophobicity [86, 111, 112].

10.2.2.2 Pore Ordering at the Surface and in the Volume of Film

Breath figure process is a simple and robust method based on the fast endothermic solvent evaporation upon casting a polymer solution under humid atmosphere. The water condensation occurring on the top surface produces an ordered array of hexagonally packed water droplets. The surface of the final polymer film is structured by a hexagonal array of pores which are structured into either mono- or three-dimensional structure. Observation of the cross section of honeycomb films shows that pores can be connected in the longitudinal direction [113–115] (Fig. 10.5).

The regular stacking of the water droplets on the surface and inside the polymer solution is favored by the Bénard–Marangoni convection arising in a solution presenting a thermal gradient [30, 116, 117]. Research teams attempted to understand the parameters driving the formation of monolayer or multiple layers of pores. A first report related the presence of pore multilayer to the lower density of the water-immiscible solvent (toluene, benzene) in comparison with water as three-dimensional pattern pores was assigned to successive sinking of condensed water droplets into the polymer solution [30]. However, multilayer structures have also

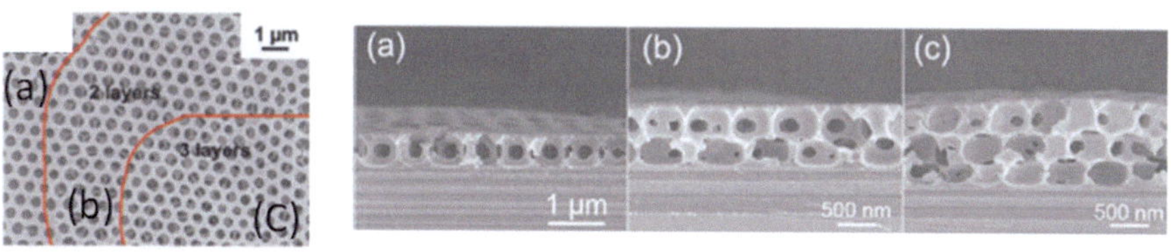

Fig. 10.5 *Top* and *side view* of single layer film (near the *edge*) and multilayer film (in the *center*) with connected pores (from ref. [115])

been observed from solvent exhibiting a higher density than water (carbon disulfide and chloroform) [31, 101, 106, 115, 118]. Bolognesi et al. [119] proposed a prediction model showing that the formation of monolayer or multilayer structures can be tuned by values of surface tensions of polymer solution and water but also by the interfacial tension between water and the polymer solution. It is worthwhile to remark that during solvent evaporation, the interfacial tension between water and polymer solution continuously changes as the concentration of the polymer solution raises. This dynamic phenomenon makes the model difficult to systematically predict the final structure. For instance, a comparison between the theoretical predictions of the model based on interfacial interactions with the experimental data for poly(D,L-lactide) honeycomb films suggested that interfacial interaction alone cannot explain the formation of either monolayered or multilayered porous films [104]. In this system, when choosing solvents with interfacial tension suitable to promote multilayer film, a single layer of pores was observed [104].

As mentioned for experimental parameters influencing the pore size, both volume and concentration of the cast polymer solution can also modify the three-dimensional organization of the pore array [79, 101, 102, 106]. Indeed, for a given polymer/solvent system, low volume or concentration of polymer solution cast under humid atmosphere led to a single layer of pores while increasing the wet coating thickness by increasing either cast volume or polymer concentration resulted to multiple layers of pores ordered in three dimensions [88, 101, 106]. Heterogeneity in a single film can be observed with a monolayer of pores at the outer region of the film where a rapid solvent evaporation occurs while towards the center of film, several layers of pores are present due to a slower solvent evaporation [79, 115]. Such results confirmed that values of interfacial tensions are unlikely to predict the number of pore layers created during the dynamic BF process.

10.2.2.3 Effect of Chemical Nature and Topology of Polymers on Pore Ordering

Honeycomb porous films were prepared from a wide range of polymers with different chemical structures, topologies or microstructures. However, the polymer should fulfill some prerequisites to favor the formation of honeycomb film by breath figures. First, the chemical composition of the polymer should be hydrophobic enough to promote the fast precipitation of the polymer at the organic solvent–water interface which maintains the droplet shape. The results of literature highlight that polymers able to self-assemble into supramolecular structures or polymers with complex chemical topologies are more robust to form highly structured films under variable conditions [25, 27–29]. The first study on honeycomb-patterned film prepared by breath figure reported casting of polystyrene-*b*-poly(paraphenylene) block copolymer in carbon disulfide [31]. The poly(paraphenylene) block being insoluble in carbon disulfide, the copolymer self-assembled into aggregates to form star-like topologies [31, 120]. Alpha-telechelic polystyrenes with limited molecular weight (polymers also called ionomers) proved to be good candidates to prepare well-organized honeycomb porous films as the polar chain end induces self-assembly of polymer chains in the organic solvent [30, 87, 106, 121]. The later examples rely on polymer self-assembly induced by the insolubility of either one block or a polar group in the initial diluted solution of organic solvent. The immiscibility between two blocks of soluble hydrophobic rod–coil or coil–coil block copolymers is also at the origin of copolymer self-assembly into nanostructures inside the HC film [81, 101, 111, 122]. Block copolymer self-assembly occurring simultaneously with both solvent evaporation and water condensation promotes the formation of highly structured hexagonal pattern of pores showing iridescence properties [111, 122]. Other examples of honeycomb films based on coil–coil [123, 124] or rod–coil [125–128] block copolymers were reported without focusing on the ability of the block copolymers to self-assemble into nanostructures. Well-organized honeycomb films were prepared from star polymers [88, 129, 130], miktoarm star polymer [131], core cross-linked star polymers [79, 95, 132, 133], comb-like polymers [134], hyperbranched polymers [98], or dendritic copolymers [135]. Core cross-linked star polymers were synthesized by "arm first" strategy using atom transfer radical polymerization [136–138]. Interestingly, star polymers were able to produce organized porous honeycomb films by BF method whereas none of the original linear precursors (poly(dimethylsiloxane) PDMS, poly(ethyl acrylate), poly(methyl acrylate), poly(tert-butyl acrylate), poly (methyl methacrylate)) with low glass transition temperature were able to form stable ordered porous films [133]. Recently, a library of biocompatible linear–dendritic amphiphilic copolymers based on linear hydrophilic poly(ethylene glycol) (PEG) block and dendritic hydrophobic poly(ε-caprolactone) (PCL) were synthesized in order to assess how the dendritic linker affects the pore formation [135]. Nevertheless, images of the recovered porous films highlighted that these amphiphilic copolymers can create a second smaller porosity in between the main pores (Fig. 10.6). This phenomenon was previously described with other amphiphilic block copolymers, which was ascribed to the water condensation into inverse micelles formed in the organic solvent [139–141]. Despite the low amount of 2-hydroxyethyl methacrylate (HEMA) hydrophilic units ($<$16 mol.%), the final structure of a star (PS-*b*-PHEMA)$_4$ porous film fabricated by the breath figure method was greatly different from the four-arm star polystyrene (PS)$_4$ [88]. The former copolymer resulted in multilayer disordered structure exhibiting a second smaller porosity while monolayer structured films were recovered from (PS)$_4$ polymer [88]. In order to investigate how the copolymer microstructure can affect the honeycomb porosity, both block and gradient amphiphilic copolymers based on poly(2-2(2′, 3′, 4′, 6′-tetra-*O*-acetyl-*β*-D-galactosyloxy)ethyl acrylate) and polystyrene were synthesized with a similar molar fraction of hydrophilic content [141]. It was shown that gradient copolymers were suitable candidates to prevent the formation of the disordered second porosity observed for amphiphilic block copolymer [141].

It was reported the use of fluorinated copolymers for the preparation of honeycomb films by the breath figure process in order to increase the hydrophobic character [95, 99, 142], indeed even superhydrophobic behavior [143]. Some studies

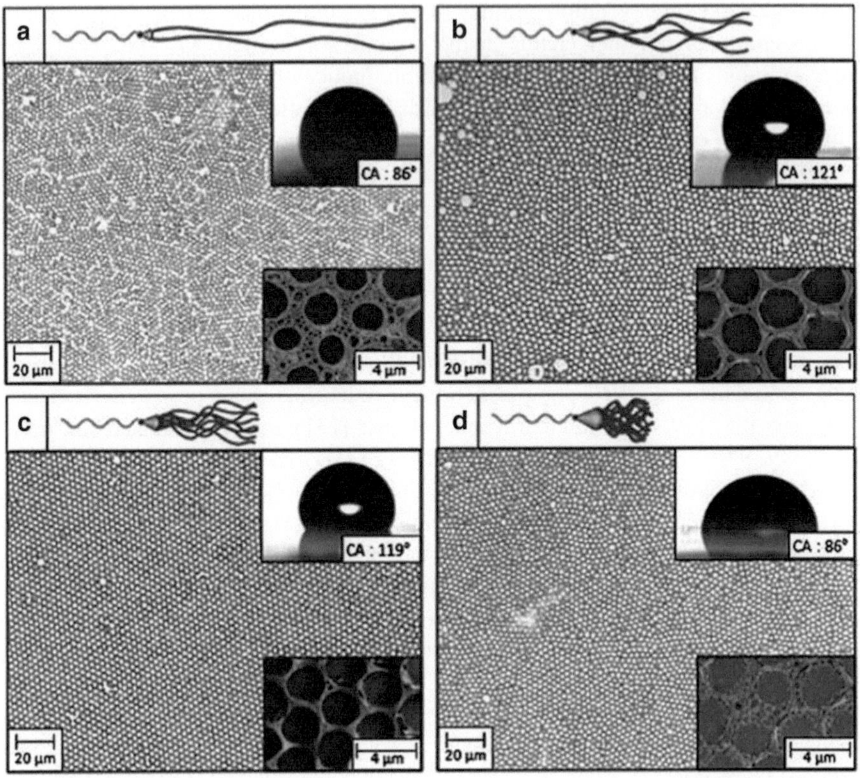

Fig. 10.6 Optical microscopy images of films cast from the dendritic PEG–PCL amphiphilic copolymers. Insets: image of the contact angle taken directly after deposition of a water droplet and the SEM micrograph of the film (from ref. [135])

highlighted the influence of the fluorine content on the quality of pore ordering [95, 142], showing for instance that the optimum pore ordering was observed for molar content of pentafluoropropyl acrylate below 35 % in poly((^{1}H,^{1}H-pentafluoropropyl acrylate)-*ran*-(methyl methacrylate)) star polymer films [95].

10.3 Simultaneous Functionalization and Patterning of the Surfaces

10.3.1 Directing the Position of the Functional Groups

Besides the control of the surface topography, the breath figures approach offers the possibility to control the chemical functionality within the patterned film directed by the formation mechanism. The versatility of the method enables the preparation of

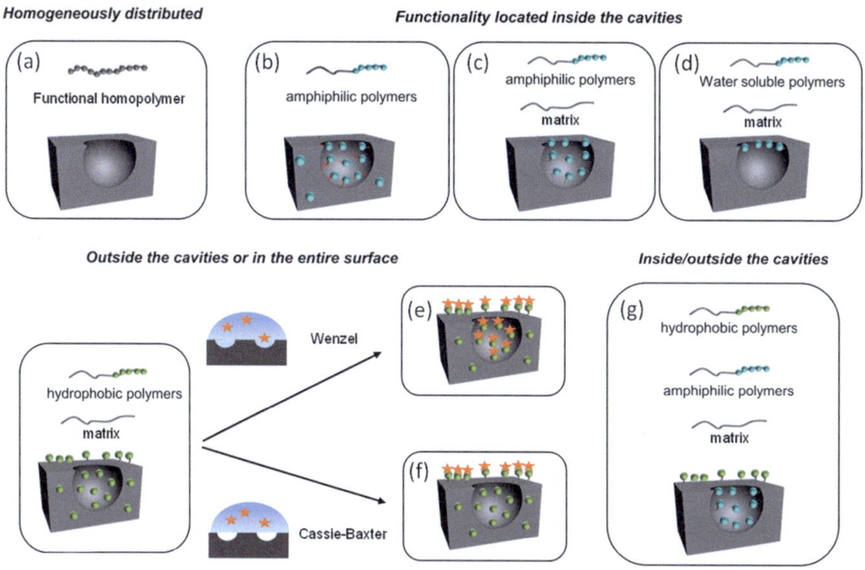

Scheme 10.1 Illustration of the different approaches to control the chemical functionality in breath figures films

functional honeycomb films with advanced properties from a great variety of materials. In addition, as will be described as follows, the distribution of the functional groups within the film can be designed and specifically located in different positions of the pattern. Thus, with an appropriate design of the polymer, a desired functionality can be homogeneously distributed in the entire film, exclusively inside the cavities, both inside/outside the pores or on the top surface (Scheme 10.1). A variety of techniques have been employed to analyze the chemical distribution of the surfaces, such as fluorescence microscopy [114, 140], transmission electron microscopy [75], atomic force microscopy [144], time-of-flight secondary-ion mass spectrometry [145], X-ray photoelectron spectroscopy [146], energy-dispersive X-ray (EDX) measurements [147] and Raman microscopy [148].

10.3.1.1 Functionality Homogeneously Distributed

The simple case is the use of single functional homopolymers that produce honeycomb films with the functionality homogeneously distributed (Scheme 10.1a). The first report of breath figures films was based on polystyrene [31], then others polymeric structures have been progressively used to introduce physical properties or more advanced characteristics to the whole film such as optical grade poly (methyl methacrylate) [149], flexible polydimethylsiloxane [150], biocompatible and biodegradable polycaprolactone [73], thermally stable polyimide [151], or even conjugated polymers [152] with electronic and optical properties useful for solar and optoelectronic devices.

10.3.1.2 Functionality Located Inside the Cavities

In order to localize the functionality preferentially inside the cavities amphiphilic macromolecules either end-functionalized polymers or copolymers are typically employed in the BF process. Besides the well-known capability of the polymers containing a low amount of polar groups to form honeycomb films much easier than nonpolar polymers, they self-assemble in organic solvent and the hydrophilic moieties tend to segregate at the water–solution interface. More precisely, the condensation of water droplets during the BF progress implies the reorientation of the polymers allowing the interaction of the polar groups and the water droplets. Therefore, at the end of the process pores are enriched in polar groups (Scheme 10.1b). As commented, the simplest amphiphilic structures used in this processes are linear polymers, mainly polystyrene, with polar end-groups usually synthetized by living polymerization techniques. For instance polystyrene with *p*-toluenesulfonate piperidinium acid salt end groups [87], hydroxyl-terminated, carboxy-terminated [30, 106] and also PS with a polar head of glucose [153] have been successfully employed to form porous films with functional groups preferentially inside the holes. Nevertheless, amphiphilic copolymers are the materials mostly used to create honeycomb films with this particular chemical distribution. This is because of their enhanced capability to self-assemble in solution and also their higher content of functional groups per macromolecule. A large variety of functional amphiphilic copolymers have been used in the BF process in order to direct the functionality toward the holes, from linear statistical and block copolymers to star, graft copolymers or comb-like structures [98, 140, 154]. Of all, amphiphilic glycopolymers have received a particular interest because they provide honeycomb structured films with bioactive sites for protein patterning [141, 155]. The preferential functionality of the structures and their biological activity are usually tested by the study of the specific recognition to fluorescent-labeled lectins.

In spite of this favored distribution of the polar moieties within the pores when amphiphilic polymers are used, these polar functional groups are also located in the rest of the film. Alternatively, the use of blends allows the specific distribution of the different components within or out of the holes (Scheme 10.1c). The condensed water droplets direct the phase separation of the blend components during the BF process, conducting the more hydrophilic component exclusively toward the holes. Russell et al. reported the first work of BF using blends [75]. They created honeycomb films with pores decorated with tri-*n*-octylphosphine oxide-stabilized CdSe nanoparticles that were able to segregate to water–polymer solution interface. Afterward, other blends based not only on polymer/inorganic compounds [156, 157] but also in all polymeric blends were employed. In the latter case, typically a small amount of amphiphilic copolymer bearing specific functional groups is mixed with a hydrophobic polymer matrix as major component [114, 144, 158]. This strategy also offers a great versatility of functional groups due to the small amount that is required to functionalize the interior of the cavities. A particular case is the use of water soluble amphiphilic copolymer as minor component of the blend. The copolymer can migrate inside the condensed water droplet and a coffee-stain phenomenon could takes place during the water droplet evaporation, conducting to a ring-like deposition on the top edge of the cavities (Scheme 10.1d) [96, 159].

Equally, blends of incompatible homopolymers, usually hydrophilic/hydrophobic, can be microstructured into honeycomb arrays by combining the BF process and phase separation. Cui et al. have worked on blends of PS and poly (2-vinylpyridine) (P2VP) that lead to ordered holes enriched in the more hydrophilic P2VP polymer [160].

As described above, the functionalization of the pores with polar moieties as a consequence of self-assembly process during the BF formation is a very straightforward and rapid strategy to control the chemical distribution of these surfaces. However, this is limited to polymers or amphiphilic structures which are compatible with the breath figure formation. Alternatively, the porous films can be functionalized by post-modification reactions giving the possibility to introduce a wide range of functionalities into the pores. Taking the advantage of the in-situ functionalization, BF patterns can be obtained from materials containing ligands [75], reactive groups [161] or initiator sites [162] located within in the cavities for further chemical reactions or polymerization step.

10.3.1.3 Functionality Located Outside the Cavities or at the Entire Surface

In contrast, the exclusive functionalization of the surfaces outside the cavities is not as direct and obvious as previous strategies described above. All the reported approaches involve a post-modification step and make use of the particular wettability behavior of the honeycomb film. Two different situations can occur: (a) when a droplet of a liquid with the reagents is placed on the porous surface and it is able to wet the whole surface, penetrating inside the pores of the surface (Wenzel model [163]), simultaneous modification of the interior of pores and external surface can be obtained (Scheme 10.1e); (b) on the other hand, if the droplets cannot penetrate in the cavities because of the air bubbles entrapped (Cassie–Baxter model [164]) only the external surface is modified (Scheme 10.1f). Wan et al. [165, 166] prepared films from polystyrene-*b*-poly(*N*,*N*-dimethylaminoethyl methacrylate) amphiphilic copolymers which were further modified by electrostatic interactions across the whole surface or selectively at the external surface. However, an intermediate state of wettability in which the holes are partially filled was observed as a consequence of the hydrophilic character of the pores and the functionalization of the external surface was not exclusive. To avoid the partial filling of the holes and obtain a total Cassie-Baxter behavior, hydrophobic copolymers were used in another approach to fabricate the honeycomb films [167]. In particular, regular porous patterns were obtained from blends of PS and a hydrophobic block copolymer of polystyrene-*b*-poly(2,3,4,5,6-pentafluorostyrene) (PS-*b*-P5FS) that tends to segregate to the air interface. The wettability behavior and the presence of the fluorinated segment on the surface allowed the chemical modification of the surface outside the pores by the thiol-para fluorine "click" reaction based on a nucleophilic substitution reaction.

10.3.1.4 Varies Functionalities Inside/Outside the Cavities

The use of more complex systems such as ternary blends allows the functionalization of the surfaces with varies chemical functionalities. For instance a PS matrix was mixed with two block copolymers, a hydrophobic (PS-*b*-P5FS) and an amphiphilic polystyrene-*b*-poly[poly(ethylene glycol) methyl ether methacrylate] (PS-*b*-P(PEGMA)) copolymer [96]. The chemical distribution of the resultant surface pattern implies an enrichment of the holes in the amphiphilic copolymer with an external surface mainly functionalized in the fluorinated copolymer with low surface energy (Scheme 10.1g). Other ternary blends combining incompatible copolymers and homopolymers have been reported leading to more complex topographies and chemical distributions [148].

10.3.1.5 Overview of the Functional Building Blocks Introduced in Honeycomb Films

Due to the simplicity of fabrication, the breath figure method has stimulated a wide interest, and a variety of materials have been also exploited as building blocks, such as different polymers architectures, amphiphilic polyion complex, metal and hybrid nanoparticles. The concept of using materials of different chemical natures and assembling them to obtain a new one that exhibits enhanced properties can be applied to many different fields. Nevertheless, to obtain interesting properties, a simple mixture of organic and inorganic materials is not enough, one has to organize them in a specific way.

10.3.1.6 Nanoparticles and Others Hybrid Materials

Hybrid honeycomb polymer films can be elaborated through either a simultaneous self-assembly of polymer and nanoparticles or by the in situ formation of hybrid particles from precursor or by using HC film as a template.

As early as 2004, Russell et al. reported honeycomb films where the pore walls are decorated with CdSe nanoparticles [75]. Since this first study, several other metallic nanoparticles such as Fe_2O_3 [168, 169], Fe_3O_4 [169], ZnO [170], gold [156, 169, 171–173], or silver [169, 171] nanoparticles and CdSe/CdS QDs [169] have been used to create highly ordered hybrid honeycomb films. The nanoparticles which are immiscible with the polymer matrix showed interfacial activities driving the film formation by stabilizing the water droplets during the BF process. The self-organized macroporous honeycomb films showed the combined properties of both the NPs and the ordered inverse opal structures leading to new photonic band gap materials [169], light-emitting device [174], or magnetic patterned surfaces [168, 169]. The design of organic electronic devices can be achieved by mixing organic conductive NPs as carbon nanotubes CNT [156, 175], fullerene C60 [176] but also graphene [177]. Moreover, a study of Ji et al. describes the possibility to tune the localization of silica NPs inside a highly structured honeycomb film depending on the NP surface functionalization [178]. Using the particle-assisted fabrication of honeycomb-structured hybrid films, both hydrophilic and hydrophobic SiO_2 NPs can be directed either on the pore surface or inside the walls, respectively.

An alternative to form honeycomb-patterned films with embedded inorganic NPs involves their in-situ preparation from inorganic precursors dispersed in a polymer matrix. The polymer governs the honeycomb film formation with micron-sized pores while the walls are simultaneously filled with the precursors of NPs. Such a process has been employed to generate cadmium sulfide CdS nanocrystals in PMMA matrix [179], Ag NPs in polyurethane (PU) where the formation of regular pore arrays is controlled by the humidity levels, the content of Ag NPs and polymer [157]. From TiO_2 precursor located inside the condensed water droplets which acted as "microreactors" for the $TiCl_4$ hydrolysis, Li et al. fabricated patterned composite film with hemispherical TiO_2 microparticles lying in the holes of a honeycomb polystyrene film [180]. Following this approach, non-polymeric honeycomb films can be achieved by further burning of the organic phase and thus ZnO or carbon nanotubes honeycomb were achieved (Fig. 10.7) [181].

Polymer honeycomb films can be also used as templates to produce micropatterned inorganic/organic structures by external deposition of NPs or chemical vacuum deposition CVD. The selective assembly of the SiO_2 NPs into the pores is strongly impacted by the wettability and thus the Cassie–Baxter/ Wenzel transition was demonstrated as the key factor of such an approach [182]. Cadmium tellurium quantum dots were deposited onto polystyrene honeycomb film leading to red emitting films [183]. Recently, surface-enhanced Raman Scattering SERS substrates were also prepared by Ag NPs adsorption onto poly (*N*,*N*-dimethylaminoethyl methacrylate) PDMAEMA surface [184].

Fig. 10.7 (*Left*) SEM image of hydrothermal ZnO nanorod arrays grown from $Zn(acct)_2$ honeycomb structured pattern. (*Right*) SEM image of carbon nanotube arrays grown from ferrous honeycomb structured pattern (from ref. [181])

10.3.1.7 Stimuli-Responsive Polymers

Stimuli-responsive surfaces have received considerable attention because of their applications, in drug and gene delivery/complexation, sensing, wettability switching, … under different external stimuli such as temperature, solvent, pH, light, or mechanical stress by changing back and forth the surface properties. The use of stimuli-responsive block copolymers offers unique possibilities to achieve such behavior [185–187]. Based on honeycomb films, some interesting results that are different from those for flat films have been reported. Stenzel et al. used a polystyrene-*b*-poly(*N*-isopropyl acrylamide) (PS-*b*-PNIPAM) block copolymer and Yabu et al. a blend of PS and a copolymer containing PNIPAM, the both for the preparation of thin dense films, honeycomb films, and the corresponding pincushion-like films after removing the top surfaces [187, 188]. They found that the thin dense film and pincushion-like film show obvious thermo-responsive property whereas the honeycomb film does not. This is because the PNIPAM block that is hydrophobic at room temperature mainly segregates at the interface of the water–polymer solution, forming a hydrophobic top surface and hydrophilic pores. In both cases, only the pincushion-like films show thermoresponsiveness. To overcome this behavior, Stenzel and coworkers also fabricated thermo-responsive honeycomb films by grafting PNIPAM from the P(S-*co*-HEMA) film surface where the grafting mainly takes place in the pores [162]. Therefore, the honeycomb films grafted with PNIPAM display switchable hydrophilic/hydrophobic characteristics and if a copolymer of NIPAM and *N*-acryloyl glucosamine is grafted, it shows selective recognition of Con A [189]. pH-sensitive honeycomb films have also been investigated by Escalé et al. [81, 111]. They presented interesting hierarchical surfaces obtained by direct self-assembly of block copolymers and composed of pores at the micrometer scale and pH sensitive phase separated structures at the nanometer scale based on PS-*b*-P4VP and PS-*b*-PAA (PAA from thermal hydrolysis of poly(ethoxyethyl acrylate) (PEEA)). Considering this hierarchical structure with segregated carboxyl groups, it may provide not only a pH-responsive surface but also a platform for functional films through site specific post-modification.

Responsive polymers surfaces can be also produced by exposure to an external environment, as a selective solvent atmosphere or relative humidity atmosphere to turn the topography of the surface. Han et al. described the formation of an ordered array of holes combining the BF approach and the phase separation of polymer blends composed of P2VP and PS [160]. The surface topography changes in a reversible manner between an ordered honeycomb structure and hexagonal islands when treated with different solvents. Munoz-Bonilla et al. described the changes on the surface topography by changing the relative humidity of the environment [70, 144, 158].

Moreover, some investigations have dealt with the combination of the photonic properties and the ordered porous structure for light-harvesting or emitting. In both cases, light-sensitive honeycomb films based on semi-conductive NPs or conjugated polymers were used to improve the opto-electronic properties of organic thin-film devices [126, 161, 174, 176, 190–192].

In summary, the responsive behaviors on honeycomb film surfaces are dependent on their unique structure, which makes it possible to design new smart surfaces.

10.3.1.8 Trends for New Functional Building Blocks

If formation of macroporous honeycomb structures using polymers, nanocomposites, and hybrids have been described, only few examples of honeycomb films based on new building blocks introduced via the different methods described above have been recently achieved as functional materials. Indeed, functional honeycomb film with self-assembled Horseradish peroxidase (HRP) enzyme nanogels at the pore walls for biocatalysis can be mentioned for clinical diagnostic kits and for immunoassays [193].

Kim and coworkers reported the application of breath figure method to a small photo-responsive molecule, i.e., a organogelator self-assembled into supramolecular fibrillar networks and further ordered in a hierarchically honeycomb structure [194]. Recently, a new organogelator was synthesized and large-scale ordered honeycomb patterns were also observed [195]. Moreover, Babu et al. also reported the formation of hierarchical macroporous structures from an amino acid linked *p*-conjugated organogelator [196].

We can also mention the use of bio-sourced building blocks based on cellulose or dextran. Kadla et al. described value-added materials from naturally abundant polymers for system that may serve as a platform for the design and development of biosensors [197]. A hierarchically structured honeycomb film from dextran-*b*-PS amphiphilic linear diblock copolymers has also been described by Chen et al. leading to ordered porous bio-hybrid films. [198] Honeycomb patterned surfaces functionalized with biomolecules for specific recognition of proteins or bacteria have been also achieved either by self-assembly of amphiphilic copolymers based on galactose moieties [155] or by post-modification with peptide sequences [199].

10.4 Complex Porous Surface Patterns: Ordered Multiscale Structures

Multiscale structuration of polymeric materials is challenging to provide a wide range of properties. For instance, iridescence phenomenon is created by a periodic structure of pores (or particles) organized at the micron-scale interacting with sunlight wavelength. Functionality can be introduced by nanoscopic domains. Self-assembly of block copolymers, two chemically dissimilar polymers joined together, leads to phase separation at the nanometer scale as the covalent linkage between blocks restricts the phase separation to the length scale of each block [200, 201]. Diblock copolymers are able to produce four main morphologies (BCC spheres, hexagonally packed cylinders,

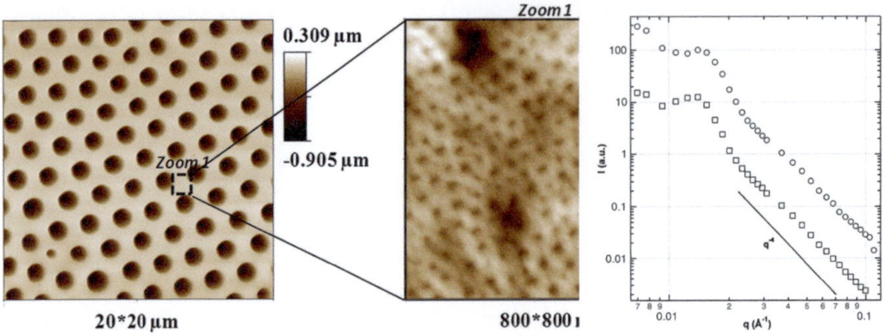

Fig. 10.8 Example of hierarchical structuration in polystyrene-*b*-poly(ethoxy ethyl acrylate) diblock copolymer honeycomb film (AFM topography images). *Left*: hexagonal-pattern of pore; *Middle*: zoom of honeycomb film showing diblock copolymer structuration into nanodomains; *Right*: SANS curves measured on the deprotected HC film soaked to D_2O (from ref. [111])

gyroid and lamellae) at the equilibrium ordered state. For sufficiently high Flory–Huggins parameter values and degree of polymerization, the different morphologies depend on the volume fraction of each block [201]. This spontaneous thermodynamic nanophase segregation proved to be a straightforward method to hierarchically structure porous films [81, 101, 111, 122, 144]. Indeed, self-assembly of suitable rod–coil [101], or coil–coil diblock copolymers [81, 111, 122] occurring simultaneously to BF process, offers the possibility to create nanodomains in the walls and on the surface of honeycomb-structured films (see Fig. 10.8). The hydrophobic polystyrene block is required to favor the honeycomb film formation via BF process while the other chemically different block provides additional properties to the hierarchically structured film via the intrinsic properties of the mesophase (adhesive properties of poly(*n*-butyl acrylate and pH-responsive properties of poly (4-vinylpyridine) or poly(acrylic acid)) [81, 111, 122]. The organized pattern of pores is observed at the micron-length scale either by optical microscopy or by atomic force microscopy. The presence of a sub-structuration of diblock copolymer was demonstrated by different techniques probing the interfaces at the nanometer scale: small angle neutron scattering (SANS) of a stack of honeycomb films providing the average distance of the nanodomains [81, 111, 122], transmission electron microscopy of a perpendicularly cross-sectioned film showing the nanostructuration inside de film [101] and atomic force microscopy to probe the nanostructure morphology at the film surface [81, 111, 122] or in the bottom of pores [144].

Apart from the use of block copolymers providing nanoscale features other methodologies alone or in combination can provide hierarchically structured patterns. For instance taking advantage of the template assisted structuration, demixing of polymer blend solutions on structured substrates can also be employed to create multiscale ordered surfaces. Boneberg et al. [202] employed this approach and blended two components, PVP and PS, already known to form ordered porous films during phase separation [203]. The authors used a micrometer patterned (hydrophilic/hydrophobic areas) substrate and they prepared the films under a high relative humidity (~50 % R.H.) that favor both the water vapor condensation and simultaneously improves the demixing of the components [204]. As a result depending on their affinity for the substrate patterns, PVP and PS are distributed in specific areas. In addition, water condensation occurs both in PVP and PS domains producing pores with an average size below 30 nm. As a result of these two processes and, eventually after selective removal of PVP, the surface reveals a polymer pattern formed by alternating micrometer size layers of polymer with nanopores.

Films formed from blends can also produce hierarchical ordered materials. Raczkowska et al. [66] prepared films from blends of polar PMMA and nonpolar PS. Spin-coating of the blends under an appropriate relative humidity produced macrophase separated domains and simultaneously droplet-like patterns selectively in the domains rich in the polar polymer. Submicrometer pores decorated large PMMA-rich surface regions thus resulting in hierarchical film morphology.

Hierarchical structures in the micrometer and/or submicrometer scale from blends can also be prepared by combination of the BF and a different patterning technique, in general, lithographic techniques. For instance, Ge and Lu [205] prepared multiscale structured films by combining the breath figures formation with embossing techniques (Fig. 10.9). More precisely the authors prepared porous

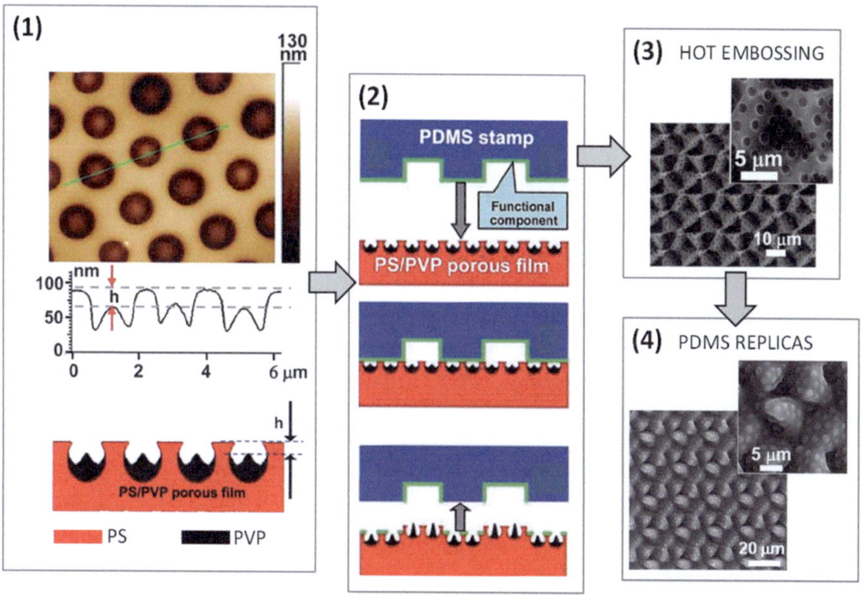

Fig. 10.9 Formation of multiscale structures by combination of the breath figures approach and embossing techniques. The immiscibility between both homopolymers added (PS and PVP) due to the different hydrophilicity produces hierarchical ordered patterns where the PVP is located inside the pore and the PS at the pore wall. In addition, embossing produces an additional level of order at the micrometer length-scale as well as replicas. Reproduced with permission from ref. [205]

surfaces from PS/PVP blend in which the pores are filled at the bottom with the hydrophilic homopolymer. Due to its hydrophobicity, PS forms the wall of the pores while the hygroscopic PVP and the hydrophilic polymer enrich the water droplet–polymer solution interface. Thus, upon solvent evaporation the PVP is deposited within the pore. Using this porous structure, the hierarchical structures with well-defined multi-scale features were obtained upon embossing the films using a PDMS stamp. In addition, these porous structures can be employed as template to prepare microstructures of negative PDMS replicas.

Similarly, two levels of order were obtained by Connal et al. combining the breath figures approach with photolithographic cross-linking strategies [206]. The authors employed a cross-linked star (CCS) based on poly(methyl acrylate) modified at the chain ends with 9-anthracene carbonyl chloride. This polymer was cast under flow of humid air forming porous, microstructured films. The 9-anthracene groups introduce the ability to reversibly cross-link the microstructured film when irradiated with long wave UV (>350 nm) and the reaction can be reversed upon exposure to short-wave UV light (<300 nm). Hence, in addition to the structural micrometer size pores, larger scale (several hundreds of microns) surface patterns due to the mask employed (TEM grids) were obtained.

The use of TEM grids offers additional possibilities since they can be employed to induce surface features in nonplanar surfaces. In this sense, Connal and Qiao [132] prepared core CCS polymers and used them for the preparation of macroporous materials with two levels of ordered surface features by casting the star PDMS solution onto the TEM grid [206]. The first level of ordered surface features is originated from the contours of the TEM grid while the second level is obtained by the condensation of water vapor. The concept was later extended to other types of substrates thus giving a variety of porous surfaces with doughnut, golf balls and hollow porous pockets shapes [150].

Whereas the polymers employed by Qiao et al. [133, 206] concerned only star polymers with low T_g (below 48 °C), Li and coworkers [207] studied the construction of macroporous polymeric films on various nonplanar substrates with static breath figures technique, using linear polymers with high T_g. For this purpose, two kinds of linear polymers with high T_g, polystyrene-*b*-poly(acrylic acid) and polystyrene without polar end groups, were employed to prepare three-dimensional macroporous films on different nonplanar substrates.

Other examples of nonplanar hierarchically structured porous films are for instance the preparation of membranes for advanced filtration (also known as microsieves) composed by a hierarchical pore structure [208] or the construction of hierarchical structures by combination of electrospinning or electrospraying and breath figures [209–211].

Finally, nanoscale features can be also obtained by using nanoparticles. For instance, hierarchical hybrid nanoparticle-polymer porous assemblies have been obtained by self-assembly of nanoparticles at the polymer solution–water droplet interface. The first example was reported by Russell et al. [75] who prepared honeycomb structured interfaces by casting a chloroform solution of PS using TOPO-stabilized CdSe nanoparticles. As a result of the breath figures mechanism, the TOPO-CdSe nanoparticles segregated to the pore interface and form a uniform layer in the inner part of the pore that will be stabilized upon complete evaporation of both water droplet and solvent. Apart from CdSe nanoparticles, other groups employed alternative metals. This is the case of Hao's group [147, 171] who described how gold nanoparticles accumulate at the water–chloroform interface or metal oxide nanoparticles such as ZnO or SiO_2 previously modified by surface initiated polymerizations [170]. The inorganic particles require an organic coating, mostly a polymer layer, to be stable in solution.. Thus, the pore sizes of the films can be tuned depending on the polymer attached to the nanoparticles.

10.5 Applications

Porous materials prepared by the BF approach have shown great potential for application in many fields. Among all, this section is limited only to some of the most promising areas.

10.5.1 Superhydrophobic Surfaces

Superhydrophobic surfaces with contact angles greater than 150° have found applications as self-cleaning surface, to prevent oxidation or even snow sticking. In general, a combination of highly hydrophobic chemical functionality and rough structure is necessary to obtain such a surface. The breath figures method is a very simple approach to combine both requisites that allows the use of a wide variety of hydrophobic polymer and creates surface with high porosity. Besides, this versatile technique offers additional possibilities to introduce further roughness, for instance in the preparation of hierarchical surfaces as described in previous sections. Superhydrophobic films have been obtained from porous surfaces made of fluorinated polymer [143] or waxy-dendron-grafted polymers [212] by peeling off the top layer. The resultant pincushion structures exhibit very high contact angles. Decreasing the pore size is another strategy that leads to superhydrophobic surfaces [86]. High water contact angle values have been also obtained from a polybutadiene film by CF_4 plasma-chemical fluorination that cross-links the polymer introducing an additional texturing at the surface [213].

10.5.2 Templating

This technique is also suitable for templating and microfabrication and offers new prospects in the fields of soft lithography and micropatterning. As an advantage the polymeric templates can be easily removed by solution or pyrolysis. Various metals

[214, 215], carbon nanotubes [216], SiO_2, TiO_2, and CdS [217], etc. have been already microstructured. Besides, these honeycomb films obtained via BF can be used as template to construct pillars or inverse pores of PDMS by replica molding [119].

10.5.3 Biomaterials

Moreover, the BF technique is very attractive to develop cell culture scaffolds with a great interest in biomedical applications including tissue engineering and diagnosis. In general scaffolds should present high porosity as requirement in addition to other properties such as biocompatibility, biodegradability or mechanical properties. The depth, the pore size and the pore density are factors that strongly influence the attachment of cells to solid substrate, and these parameters can be modulated using the breath figures approach. A wide variety of polymers can be successfully used such as poly(L-lactic acid) [78] or poly(ε-caprolactone) [218]. Bacterial cells have been micropatterned by a selective modification of the pore in polystyrene film using appropriate polypeptide sequences. The design of the pore sizes allowed the immobilization of single bacteria with similar dimensions at the inner pore interface [199].

10.5.4 Filtration and Separation Processes

Microporous membranes are often used in many processes to remove impurities or contaminants through size-selective filtration. The breath figures method also finds application in this field, specially the approaches that facilitate the easy transfer to other supports. Another prerequisite is the formation of through pores that penetrate from the top of the layer to the bottom and the use of ice support favors this fact. For example, highly uniform membranes of PS-*b*-PDMAEMA have been prepared with pores on the micrometer scale for size-selective separation. The films were prepared by casting at an air–ice interface and easily transferred onto other supports [219]. Miktoarm star copolymers with proper water wettability and mechanical stability have been used to fabricate separation membranes also using ice substrate [131]. Moreover, the breath figures approach has been employed to build polymer membranes on structured substrates in order to obtain hierarchically structured microsieves [208].

10.5.5 Surface Enhanced Raman Scattering (SERS) Substrates

Nowadays SERS is a very attractive spectroscopic technique able to detect trace-level substances including contaminants, biomolecules and other chemical analytes. Periodically arranged metallic nano/microstructures, especially gold and silver nanoparticles, have proven to be good candidates for SERS substrates. Breath figures technique offers the possibility to simply prepare microstructured surfaces with metal nanoparticles. Combining the BF and vapor deposition processes, silver spikes arrays with triangular sharp have been fabricated showing strong enhanced Raman scattering of rhodamine 6G [220]. Other strategies involve the electrodeposition of gold nanoparticles [221] or the incorporation of the nanoparticles in the polymeric solution [173]. Recently, SERS substrates have been created by in situ generation and immobilization of Ag NPs nanoparticles at the surface of honeycomb films based on PDMAEMA [184].

10.5.6 Optical Materials

BF technique permits the fabrication of three dimensional ordered porous structured with a wide range of pore size, thus the obtained materials can operate as "inverse opals" at infrared, visible, ultraviolet, and even soft X-ray wavelengths. Many optical uses have been proposed for these materials, based on the optical interferences of the light with the periodic structures, for example as antireflective coatings at near-infrared wavelength [67]. The use of polymers with special optical properties opens more the field of application. Honeycomb films were obtained from light-emitting poly(distyryldimethylbenzene-*co*-triethyleneglycol) rod–coil copolymer. The introduction of microstructure enhances photocurrent generation as compared to flat surfaces as a result of the improvement of the light-harvesting efficiency and also the charge separation and charge transfer [126]. Another reported example is the utilization of small molecule tetraphenylethene derivatives for the fabrication of highly emissive BF film, as a consequence of an aggregation-induced emission phenomenon [190].

References

1. Ostuni, E., Chen, C.S., Ingber, D.E., Whitesides, G.M.: Selective deposition of proteins and cells in arrays of microwells. Langmuir **17**, 2828–2834 (2001)
2. Joannopoulos, J.D., Johnson, S.G., Winn, J.N., Meade, R.D.: Photonic Crystals: Molding the Flow of Light, 2nd edn. Princeton University Press, Princeton (2008)
3. Yoshida, M.: Novel thin film with cylindrical nanopores that open and close depending on temperature: first successful synthesis. Macromolecules **29**, 8987–8989 (1996)

4. Hoa, M.L.K., Lu, M.H., Zhang, Y.: Preparation of porous materials with ordered hole structure. Adv. Colloid Interface Sci. **121**, 9–23 (2006)
5. Wu, D., Xu, F., Sun, B., Fu, R., He, H., Matyjaszewski, K.: Design and preparation of porous polymers. Chem. Rev. **12**(7), 3959–4015 (2012)
6. Cassagneau, T., Caruso, F.: Conjugated polymer inverse opals for potentiometric biosensing. Adv. Mater. **14**, 1837–1841 (2002)
7. Hong, J.C., Park, J.H., Chun, C., Kim, D.Y.: Photoinduced tuning of optical stop bands in azopolymer based inverse opal photonic crystals. Adv. Funct. Mater. **17**, 2462–2469 (2007)
8. You, B., Shi, L., Wen, N., Liu, X., Wu, L., Zi, J.: A facile method for fabrication of ordered porous polymer films. Macromolecules **41**, 6624–6626 (2008)
9. Li, C., Hong, G.S., Yu, H., Qi, L.M.: Facile fabrication of honeycomb-patterned thin films of amorphous calcium carbonate and mosaic calcite. Chem. Mater. **22**, 3206–3211 (2010)
10. Xue, M.J., Xiao, W.T., Zhang, Z.J.: Porous films from transformation of polymeric sphere arrays. Adv. Mater. **20**, 439–442 (2008)
11. Santos, L., Martin, P., Ghilane, J., Lacaze, P.C., Randriamahazaka, H., Abrantes, L.M., et al.: Electrosynthesis of well-organized nanoporous poly(3,4-ethylenedioxythiophene) by nanosphere lithography. Electrochem. Commun. **12**, 872–875 (2010)
12. Pietsch, T., Gindy, N., Fahmi, A.: Nano- and micro-sized honeycomb patterns through hierarchical self-assembly of metal-loaded diblock copolymer vesicles. Soft Matter **5**, 2188–2197 (2009)
13. Shin, Y., Li, X.S., Wang, C., Coleman, J.R., Exarhos, G.J.: Synthesis of hierarchical titanium carbide from titania-coated cellulose paper. Adv. Mater. **16**, 1212–1215 (2004)
14. Shin, Y., Wang, C., Exarhos, G.J.: Synthesis of SiC ceramics by the carbothermal reduction of mineralized wood with silica. Adv. Mater. **17**, 73–77 (2005)
15. Huang, J., Kunitake, T.: Nano-precision replication of natural cellulosic substances by metal oxides. J. Am. Chem. Soc. **125**, 11834–11835 (2003)
16. Meldrum, F., Seshadri, R.: Porous gold structures through templating by echinoid skeletal plates. Chem. Commun. **29–30** (2000)
17. Luo, R.L., Young, T.H., Sun, Y.M.: Structure formation and characterization of EVAL membranes with cosolvent of isopropanol and water. Polymer **44**, 157–166 (2003)
18. Zhang, Y.L., Ding, H., Wei, S., Liu, S., Wang, Y.P., Xiao, F.S.: Hierarchical macroporous epoxy resin templated from single semi-fluorinated surfactant. J. Por. Mater **17**, 693–698 (2010)
19. Tang, C.B., Bang, J., Stein, G.E., Fredrickson, G.H., Hawker, C.J., Kramer, E.J., et al.: Square packing and structural arrangement of ABC triblock copolymer spheres in thin films. Macromolecules **41**, 4328–4339 (2008)
20. Xiang, H., Lin, Y., Russell, T.P.: Electrically induced patterning in block copolymer films. Macromolecules **37**, 5358–5363 (2004)
21. Morikawa, Y., Nagano, S., Watanabe, K., Kamata, K., Iyoda, T., Seki, T.: Optical alignment and patterning of nanoscale microdomains in a block copolymer thin film. Adv. Mater. **18**, 883–886 (2006)
22. Li, G., Burggraf, L.W.: Controlled patterning of polymer films using an AFM tip as a nano-hammer. Nanotechnology **18**, 245302 (2007)
23. Behl, M., Seekamp, J., Zankovych, S., Sotomayor Torres, C.M., Zentel, R., Ahopelto, J.: Towards plastic electronics: patterning semiconducting polymers by nanoimprint lithography. Adv. Mater. **14**, 588–591 (2002)
24. Campbell, M., Sharp, D.N., Harrison, M.T., Denning, R.G., Turberfield, A.J.: Fabrication of photonic crystals for the visible spectrum by holographic lithography. Nature **404**, 53–56 (2000)
25. Hernandez-Guerrero, M., Stenzel, M.H.: Honeycomb structured polymer films via breath figures. Polym. Chem **3**, 563–577 (2012)
26. Bunz, U.H.F.: Breath figures as a dynamic templating method for polymers and nanomaterials. Adv. Mater. **18**, 973–989 (2006)

27. Muñoz-Bonilla, A., Fernández-García, M., Rodríguez-Hernández, J.: Towards hierarchically ordered functional porous polymeric surfaces prepared by the breath figures approach. Prog. Polym. Sci. **39**, 510–554 (2014)
28. Stenzel, M.H., Barner-Kowollik, C., Davis, T.P.: Formation of honeycomb-structured, porous films via breath figures with different polymer architectures. J. Polym. Sci. A Polym. Chem **44**, 2363–2375 (2006)
29. Escale, P., Rubatat, L., Billon, L., Save, M.: Recent advances in honeycomb-structured porous polymer films prepared via breath figures. Eur. Polym. J. **48**, 1001–1025 (2012)
30. Srinivasarao, M., Collings, D., Philips, A., Patel, S.: Three-dimensionally ordered array of air bubbles in a polymer film. Science **292**, 79–83 (2001)
31. Widawski, G., Rawiso, M., François, B.: Self-organized honeycomb morphology of star-polymer polystyrene films. Nature **369**, 387–389 (1994)
32. Aitkek, J.: Breath figures. Nature **86**, 516–517 (1911)
33. Aitkek, J.: Breath figures. Nature **90**, 619–621 (1912)
34. Giltay, J.W.: Breath figures. Nature **86**, 585–586 (1911)
35. Rayleigh, L.: Breath figures. Nature **86**, 416–417 (1911)
36. Rayleigh, L.: Breath figures. Nature **90**, 436–438 (1912)
37. Jenekhe, S.A., Chen, X.L.: Self-assembly of ordered microporous materials from rod-coil block copolymers. Science **283**, 372–375 (1999)
38. Sun, H., Wu, L.X.: Ordered honeycomb-patterned films via breath figures. Prog. Chem **22**, 1784–1798 (2010)
39. Beysens, D., Steyer, A., Guenoun, P., Fritter, D., Knobler, C.M.: How does dew form? Phase Trans **31**, 219–246 (1991)
40. Beysens, D.: The formation of dew. Atmos. Res **39**, 215–237 (1995)
41. Beysens, D.: Dew nucleation and growth. Compt. Rend. Phys **7**, 1082–1100 (2006)
42. Fritter, D., Knobler, C.M., Beysens, D.A.: Experiments and simulation of the growth of droplets on a surface (breath figures). Phys. Rev. A **43**, 2858–2869 (1991)
43. Family, F., Meakin, P.: Scaling of the droplet-size distribution in vapor-deposited thin films. Phys. Rev. Lett. **61**, 428–431 (1988)
44. Fritter, D., Knobler, C.M., Roux, D., Beysens, D.: Computer simulations of the growth of breath figures. J. Stat. Phys **52**, 1447–1459 (1988)
45. Briscoe, B.J., Galvin, K.P.: Breath figures. J. Phys. D Appl. Phys. **23**, 1265 (1990)
46. Narhe, R.D., Khandkar, M.D., Shelke, P.B., Limaye, A.V., Beysens, D.A.: Condensation-induced jumping water drops. Phys. Rev. E Stat. Nonlin. Soft Matter Phys. **80**(3 Pt 1), 031604 (2009)
47. Rogers, T.M., Elder, K.R., Desai, R.C.: Droplet growth and coarsening during heterogeneous vapor condensation. Phys. Rev. A **38**, 5303–5309 (1988)
48. Beysens, D., Knobler, C.M.: Growth of breath figures. Phys. Rev. Lett. **57**, 1433–1436 (1986)
49. Blaschke, J., Lapp, T., Hof, B., Vollmer, J.: Breath figures: nucleation, growth, coalescence, and the size distribution of droplets. Phys. Rev. Lett. **109**, 068701 (2012)
50. Gau, H., Mönch, W., Herminghaus, S.: Coalescence dynamics of ordered breath figures. Prog. Coll. Polym. Sci **110**, 34–36 (1998)
51. Gau, H., Herminghaus, S.: Ripening of ordered breath figures. Phys. Rev. Lett. **84**, 4156 (2000)
52. Lopez, G., Biebuyck, H., Frisbie, C., Whitesides, G.: Imaging of features on surfaces by condensation figures. Science **260**, 647–649 (1993)
53. Knobler, C.M., Beysens, D.: Growth of breath figures on fluid surfaces. Europhys. Lett **6**, 707–712 (1988)
54. Bormashenko, E., Pogreb, R., Stanevsky, O., Bormashenko, Y., Stein, T., Gengelman, O.: Mesoscopic patterning in evaporated polymer solutions: new experimental data and physical mechanisms. Langmuir **21**, 9604–9609 (2005)
55. Bormashenko, E., Pogreb, R., Musin, A., Stanevsky, O., Bormashenko, Y., Whyman, G., et al.: Patterning in rapidly evaporated polymer solutions: formation of annular structures under evaporation of the poor solvent. J. Colloid Interface Sci. **300**, 293–297 (2006)

56. Bormashenko, E., Musin, A., Pogreb, R., Bormashenko, Y., Gendelman, O.: Self-assembled patterns obtained with evaporated polymer solutions and pre-stretched polymer substrates. Colloids Surf. A Physicochem. Eng. Asp. **303**, 253–256 (2007)
57. Bormashenko, E., Balter, S., Pogreb, R., Bormashenko, Y., Gendelman, O., Aurbach, D.: On the mechanism of patterning in rapidly evaporated polymer solutions: is temperature-gradient-driven Marangoni instability responsible for the large-scale patterning? J. Colloid Interface Sci. **343**, 602–607 (2010)
58. Barrow, M.S., Jones, R.L., Park, J.O., Srinivasarao, M., Williams, P.R., Wright, C.J.: Physical characterisation of microporous and nanoporous polymer films by atomic force microscopy, scanning electron microscopy and high speed video microphotography. Spectroscopy **18**, 577–585 (2004)
59. Kuo, C.T., Lin, Y.S., Liu, T.K., Liu, H.C., Hung, W.C., Jiang, I.M., et al.: Dynamics of single-layer polymer breath figures. Opt. Express **18**, 18464–18470 (2010)
60. Maruyama, N., Koito, T., Nishida, J., Sawadaishi, T., Cieren, X., Ijiro, K., et al.: Mesoscopic patterns of molecular aggregates on solid substrates. Thin Solid Films **327**, 854–856 (1998)
61. Wong, K., Hernández-Guerrero, M., Granville, A., Davis, T., Barner-Kowollik, C., Stenzel, M.: Water-assisted formation of honeycomb structured porous films. J. Por. Mater **13**, 213–223 (2006)
62. Kasai, W., Kondo, T.: Fabrication of honeycomb-patterned cellulose films. Macromol. Biosci. **4**, 17–21 (2004)
63. Wang, Y., Liu, Z., Huang, Y., Han, B., Yang, G.: Micropatterned polymer surfaces induced by nonsolvent. Langmuir **22**, 1928–1931 (2006)
64. Govor, L.V., Bashmakov, I.A., Kiebooms, R., Dyakonov, V., Parisi, J.: Self-organized networks based on conjugated polymers. Adv. Mater. **13**, 588–591 (2001)
65. Angus, S.D., Davis, T.P.: Polymer surface design and informatics: facile microscopy/image analysis techniques for self-organizing microporous polymer film characterization. Langmuir **18**, 9547–9553 (2002)
66. Madej, W., Budkowski, A., Raczkowska, J., Rysz, J.: Breath figures in polymer and polymer blend films spin-coated in dry and humid ambience. Langmuir **24**, 3517–3524 (2008)
67. Park, M.S., Kim, J.K.: Broad-band antireflection coating at near-infrared wavelengths by a breath figure. Langmuir **21**, 11404–11408 (2005)
68. Park, M.S., Joo, W., Kim, J.K.: Porous structures of polymer films prepared by spin coating with mixed solvents under humid condition. Langmuir **22**, 4594–4598 (2006)
69. Orlov, M., Tokarev, I., Scholl, A., Doran, A., Minko, S.: pH-Responsive thin film membranes from poly(2-vinylpyridine): water vapor-induced formation of a microporous structure. Macromolecules **40**, 2086–2091 (2007)
70. Muñoz-Bonilla, A., Ibarboure, E., Papon, E., Rodriguez-Hernandez, J.: Engineering polymer surfaces with variable chemistry and topography. J. Polym. Sci. A Polym. Chem **47**, 2262–2271 (2009)
71. Pilati, F., Montecchi, M., Fabbri, P., Synytska, A., Messori, M., Toselli, M., et al.: Design of surface properties of PET films: effect of fluorinated block copolymers. J. Colloid Interface Sci. **315**, 210–222 (2007)
72. Yabu, H., Tanaka, M., Ijiro, K., Shimomura, M.: Preparation of honeycomb-patterned polyimide films by self-organization. Langmuir **19**, 6297–6300 (2003)
73. Nishikawa, T., Nonomura, M., Arai, K., Hayashi, J., Sawadaishi, T., Nishiura, Y., et al.: Micropatterns based on deformation of a viscoelastic honeycomb mesh. Langmuir **19**, 6193–6201 (2003)
74. Song, L., Bly, R.K., Wilson, J.N., Bakbak, S., Park, J.O., Srinivasarao, M., et al.: Facile microstructuring of organic semiconducting polymers by the breath figure method: hexagonally ordered bubble arrays in rigid rod-polymers. Adv. Mater. **16**, 115–118 (2004)
75. Boker, A., Lin, Y., Chiapperini, K., Horowitz, R., Thompson, M., Carreon, V., et al.: Hierarchical nanoparticle assemblies formed by decorating breath figures. Nat. Mater. **3**, 302–306 (2004)

76. Roszol, L., Lawson, T., Koncz, V., Noszticzius, Z., Wittmann, M., Sarkadi, T., et al.: Micropatterned polyvinyl butyral membrane for acid-base diodes. J. Phys. Chem. B **114**, 13718–13725 (2010)
77. Hiwatari, K., Serizawa, T., Seto, F., Kishida, A., Muraoka, Y., Akashi, M.: Graft copolymers having hydrophobic backbone and hydrophilic branches XXXIV. Fabrication and control of honeycomb structure prepared from amphiphilic graft copolymers. Polym J **33**, 669–675 (2001)
78. Nishikawa, T., Ookura, R., Nishida, J., Arai, K., Hayashi, J., Kurono, N., et al.: Fabrication of honeycomb film of an amphiphilic copolymer at the air–water interface. Langmuir **18**, 5734–5740 (2002)
79. Connal, L.A., Gurr, P.A., Qiao, G.G., Solomon, D.H.: From well defined star-microgels to highly ordered honeycomb films. J. Mater. Chem. **15**, 1286–1292 (2005)
80. Cheng, C.X., Tian, Y., Shi, Y.Q., Tang, R.P., Xi, F.: Porous polymer films and honeycomb structures based on amphiphilic dendronized block copolymers. Langmuir **21**, 6576–6581 (2005)
81. Escalé, P., Rubatat, L., Derail, C., Save, M., Billon, L.: pH sensitive hierarchically self-organized bioinspired films. Macromol. Rapid Commun **32**, 1072–1076 (2011)
82. Karthaus, O., Maruyama, N., Cieren, X., Shimomura, M., Hasegawa, H., Hashimoto, T.: Water-assisted formation of micrometer-size honeycomb patterns of polymers. Langmuir **16**, 6071–6076 (2000)
83. Bai, H., Du, C., Zhang, A., Li, L.: Breath figure arrays: unconventional fabrications, functionalizations, and applications. Angew. Chem. Int. Ed. Engl. **52**, 12240–12255 (2013)
84. Lide, D.R.: Handbook of Chemistry and Physics, 80th edn. CRC, Boca Raton (1999)
85. Barner-Kowollik, C., Dalton, H., Davis, T.P., Stenzel, M.H.: Nano- and micro-engineering of ordered porous blue-light-emitting films by templating well-defined organic polymers around condensing water droplets. Angew. Chem. Int. Ed. **42**, 3664–3668 (2003)
86. Yabu, H., Shimomura, M.: Single-step fabrication of transparent superhydrophobic porous polymer films. Chem. Mater. **17**, 5231–5234 (2005)
87. Yunus, S., Delcorte, A., Poleunis, C., Bertrand, P., Bolognesi, A., Botta, C.: A route to self-organized honeycomb microstructured polystyrene films and their chemical characterization by ToF-SIMS imaging. Adv. Funct. Mater. **17**, 1079–1084 (2007)
88. Zhu, L.W., Wan, L.S., Jin, J., Xu, Z.K.: Honeycomb porous films prepared from porphyrin-cored star polymers: submicrometer pores induced by transition of monolayer into multilayer structures. J. Phys. Chem. C **117**, 6185–6194 (2013)
89. Ding, J., Zhang, A., Bai, H., Li, L., Li, J., Ma, Z.: Breath figure in non-aqueous vapor. Soft Matter **9**, 506–514 (2013)
90. Harkins, W.D., Feldman, A.: Films the spreading of liquids and the spreading coefficient. J. Am. Chem. Soc. **44**, 2665–2685 (1922)
91. Jiang, X., Gu, J., Shen, Y., Wang, S., Tian, X.: Formation of honeycomb-patterned micro-porous films based on a fluorinated poly(siloxane imide) segmented copolymer. J. Appl. Polym. Sci. **119**, 3329–3337 (2011)
92. Li, J., Zhao, Q.L., Chen, J.Z., Li, L., Huang, J., Ma, Z., et al.: Highly ordered microporous films containing a polyolefin segment fabricated by the breath-figure method using well-defined polymethylene-b-polystyrene copolymers. Polym. Chem **1**, 164–167 (2010)
93. Wang, C.Y., Mao, Y.D., Wang, D.Y., Qu, Q.S., Yang, G.J., Hu, X.Y.: Fabrication of highly ordered microporous thin films by PS-b-PAA self-assembly and investigation of their tunable surface properties. J. Mater. Chem. **18**, 683–690 (2008)
94. Wu, X.J., Jones, M.D., Davidson, M.G., Chaudhuri, J.B., Ellis, M.J.: Surfactant-free poly (lactide-co-glycolide) honeycomb films for tissue engineering: relating solvent, monomer ratio and humidity to scaffold structure. Biotechnol. Lett. **33**, 423–430 (2011)
95. Zhang, Z., Hughes, T.C., Gurr, P.A., Blencowe, A., Uddin, H., Hao, X., et al.: The behaviour of honeycomb film formation from star polymers with various fluorine content. Polymer **54**, 4446–4454 (2013)

96. de León, A.S., del Campo, A., Fernández-García, M., Rodríguez-Hernández, J., Muñoz-Bonilla, A.: Hierarchically structured multifunctional porous interfaces through water templated self-assembly of ternary systems. Langmuir **28**, 9778–9787 (2012)
97. Chen, S., Alvès, M., Save, M., Billon, L.: Synthesis of amphiphilic diblock copolymers derived from renewable dextran by nitroxide mediated polymerization: towards hierarchically structured honeycomb porous films. Polym. Chem. **5**(18), 53105 (2014). doi:10.1039/C4PY00390J
98. Dong, W.Y., Zhou, Y.F., Yan, D.Y., Mai, Y.Y., He, L., Jin, C.Y.: Honeycomb-structured microporous films made from hyperbranched polymers by the breath figure method. Langmuir **25**, 173–178 (2009)
99. Qin, S., Li, H., Yuan, W.Z., Zhang, Y.M.: Fabrication of polymeric honeycomb microporous films: breath figures strategy and stabilization of water droplets by fluorinated diblock copolymer micelles. J. Mater. Sci. **47**, 6862–6871 (2012)
100. de Leon, A.S., Del Campo, A., Cortajarena, A.L., Fernandez-Garcia, M., Munoz-Bonilla, A., Rodriguez-Hernandez, J.: Formation of multigradient porous surfaces for selective bacterial entrapment. Biomacromolecules **15**, 3338–3348 (2014)
101. Hayakawa, T., Horiuchi, S.: From angstroms to micrometers: self-organized hierarchical structure within a polymer film. Angew. Chem. Int. Ed. Engl. **42**, 2285–2289 (2003)
102. Ejima, H., Iwata, T., Yoshie, N.: Morphology-retaining carbonization of honeycomb-patterned hyperbranched poly(phenylene vinylene) film. Macromolecules **41**, 9846–9848 (2008)
103. Tian, Y., Liu, S., Ding, H.Y., Wang, L.H., Liu, B.Q., Shi, Y.Q.: Formation of honeycomb-patterned polyetherketone cardo (PEK-C) films in a highly humid atmosphere. Macromol. Chem. Phys **207**, 1998–2005 (2006)
104. Servoli, E., Ruffo, G.A., Migliaresi, C.: Interplay of kinetics and interfacial interactions in breath figure templating - A phenomenological interpretation. Polymer **51**, 2337–2344 (2010)
105. Xiong, X.P., Lin, M.F., Zou, W.W., Liu, X.Y.: Kinetic control of preparing honeycomb patterned porous film by the method of breath figure. React. Funct. Polym. **71**, 964–971 (2011)
106. Billon, L., Manguian, M., Pellerin, V., Joubert, M., Eterradossi, O., Garay, H.: Tailoring highly ordered honeycomb films based on ionomer macromolecules by the bottom-up approach. Macromolecules **42**, 345–356 (2009)
107. Cong, H.L., Wang, J.L., Yu, B., Tang, J.G.: Preparation of a highly permeable ordered porous microfiltration membrane of brominated poly(phenylene oxide) on an ice substrate by the breath figure method. Soft Matter **8**, 8835–8839 (2012)
108. Honglawan, A., Yang, S.: Evaporative assembly of ordered microporous films and their hierarchical structures from amphiphilic random copolymers. Soft Matter **8**, 11897–11904 (2012)
109. Yabu, H., Jia, R., Matsuo, Y., Ijiro, K., Yamamoto, S., Nishino, F., et al.: Preparation of highly oriented nano-pit arrays by thermal shrinking of honeycomb-patterned polymer films. Adv. Mater. **20**, 4200–4204 (2008)
110. Li, J., Peng, J., Huang, W.H., Wu, Y., Fu, J., Cong, Y., et al.: Ordered honeycomb-structured gold nanoparticle films with changeable pore morphology: from circle to ellipse. Langmuir **21**, 2017–2021 (2005)
111. Escale, P., Van Camp, W., Du Prez, F., Rubatat, L., Billon, L., Save, M.: Highly structured pH-responsive honeycomb films by a combination of a breath figure process and in situ thermolysis of a polystyrene-block-poly(ethoxy ethyl acrylate) precursor. Polym. Chem **4**, 4710–4717 (2013)
112. Yabu, H., Shimomura, M.: Mesoscale pincushions, microrings, and microdots prepared by heating and peeling of self-organized honeycomb-patterned films deposited on a solid substrate. Langmuir **22**, 4992–4997 (2006)

113. Saito, Y., Kawano, T., Shimomura, M., Yabu, H.: Fabrication of mussel-inspired highly adhesive honeycomb films containing catechol groups and their applications for substrate-independent porous templates. Macromol. Rapid Comm **34**, 630–634 (2013)
114. Pessoni, L., Lacombe, S., Billon, L., Brown, R., Save, M.: Photoactive, porous honeycomb films prepared from rose bengal-grafted polystyrene. Langmuir **29**, 10264–10271 (2013)
115. Dong, R.H., Yan, J.L., Ma, H.M., Fang, Y., Hao, J.C.: Dimensional architecture of ferrocenyl-based oligomer honeycomb-patterned films: from monolayer to multilayer. Langmuir **27**, 9052–9056 (2011)
116. Limaye, A.V., Narhe, R.D., Dhote, A.M., Ogale, S.B.: Evidence for convective effects in breath figure formation on volatile fluid surfaces. Phys. Rev. Lett. **76**, 3762–3765 (1996)
117. Xu, S., Li, M., Mitov, Z., Kumacheva, E.: Surface textures induced by convection in thin films of polymeric and polymerizable fluids. Prog. Org. Coat. **48**, 227–235 (2003)
118. Sun, H., Li, H.L., Bu, W.F., Xu, M., Wu, L.X.: Self-organized microporous structures based on surfactant-encapsulated polyoxometalate complexes. J. Phys. Chem. B **110**, 24847–24854 (2006)
119. Bolognesi, A., Mercogliano, C., Yunus, S., Civardi, M., Comoretto, D., Turturro, A.: Self-organization of polystyrenes into ordered microstructured films and their replication by soft lithography. Langmuir **21**, 3480–3485 (2005)
120. Pitois, O., Francois, B.: Formation of ordered micro-porous membranes. Eur. Phys. J. B **8**, 225–231 (1999)
121. Ghannam, L., Manguian, M., Francois, J., Billon, L.: A versatile route to functional biomimetic coatings: ionomers for honeycomb-like structures. Soft Matter **3**, 1492–1499 (2007)
122. Escalé, P., Save, M., Lapp, A., Rubatat, L., Billon, L.: Hierarchical structures based on self-assembled diblock copolymers within honeycomb micro-structured porous films. Soft Matter **6**, 3202–3210 (2010)
123. Li, L., Chen, C.K., Zhang, A.J., Liu, X.Y., Cui, K., Huang, J., et al.: Fabrication of robust honeycomb polymer films: a facile photochemical cross-linking process. J. Colloid Interface Sci. **331**, 446–452 (2009)
124. Du, C., Zhang, A., Bai, H., Li, L.: Robust microsieves with excellent solvent resistance: cross-linkage of perforated polymer films with honeycomb structure. ACS Macro Lett **2**, 27–30 (2013)
125. de Boer, B., Stalmach, U., van Hutten, P.F., Melzer, C., Krasnikov, V.V., Hadziioannou, G.: Supramolecular self-assembly and opto-electronic properties of semiconducting block copolymers. Polymer **42**, 9097–9109 (2001)
126. Heng, L.P., Zhai, J., Zhao, Y., Xu, J.J., Sheng, X.L., Jiang, L.: Enhancement of photocurrent generation by honeycomb structures in organic thin films. Chemphyschem **7**, 2520–2525 (2006)
127. Lin, C.L., Tung, P.H., Chang, F.C.: Synthesis of rod-coil diblock copolymers by ATRP and their honeycomb morphologies formed by the 'breath figures' method. Polymer **46**, 9304–9313 (2005)
128. Chiu, Y.C., Kuo, C.C., Lin, C.J., Chen, W.C.: Highly ordered luminescent microporous films prepared from crystalline conjugated rod-coil diblock copolymers of PF-b-PSA and their superhydrophobic characteristics. Soft Matter **7**, 9350–9358 (2011)
129. Francois, B., Pitois, O., Francois, J.: Polymer films with a self-organized honeycomb morphology. Adv. Mater. **7**, 1041–1044 (1995)
130. Hsu, J.C., Sugiyama, K., Chiu, Y.C., Hirao, A., Chen, W.C.: Synthesis of new star-shaped polymers with styrene-fluorene conjugated moieties and their multicolor luminescent ordered microporous films. Macromolecules **43**, 7151–7158 (2010)
131. Zhang, C., Wang, X., Min, K., Lee, D., Wei, C., Schulhauser, H., et al.: Developing porous honeycomb films using miktoarm star copolymers and exploring their application in particle separation. Macromol. Rapid Comm **35**, 221–227 (2014)
132. Connal, L.A., Qiao, G.G.: Preparation of porous poly(dimethylsiloxane)-based honeycomb materials with hierarchal surface features and their use as soft-lithography templates. Adv. Mater. **18**, 3024–3028 (2006)

133. Connal, L.A., Vestberg, R., Gurr, P.A., Hawker, C.J., Qiao, G.G.: Patterning on nonplanar substrates: flexible honeycomb films from a range of self-assembling star copolymers. Langmuir **24**, 556–562 (2008)
134. Hernandez-Guerrero, M., Davis, T.P., Barner-Kowollik, C., Stenzel, M.H.: Polystyrene comb polymers built on cellulose or poly(styrene-co-2-hydroxyethylmethacrylate) backbones as substrates for the preparation of structured honeycomb films. Eur. Polym. J. **41**, 2264–2277 (2005)
135. Walter, M.V., Lundberg, P., Hult, D., Hult, A., Malkoch, M.: A one component methodology for the fabrication of honeycomb films from biocompatible amphiphilic block copolymer hybrids: a linear-dendritic-linear twist. Polym. Chem **4**, 2680–2690 (2013)
136. Blencowe, A., Tan, J.F., Goh, T.K., Qiao, G.G.: Core cross-linked star polymers via controlled radical polymerisation. Polymer **50**, 5–32 (2009)
137. Gao, H.F., Ohno, S., Matyjaszewski, K.: Low polydispersity star polymers via cross-linking macromonomers by ATRP. J. Am. Chem. Soc. **128**, 15111–15113 (2006)
138. Helms, B., Guillaudeu, S.J., Xie, Y., McMurdo, M., Hawker, C.J., Frechet, J.M.J.: One-pot reaction cascades using star polymers with core-confined catalysts. Angew. Chem. Int. Ed. **44**, 6384–6387 (2005)
139. Beattie, D., Wong, K.H., Williams, C., Poole-Warren, L.A., Davis, T.P., Barner-Kowollik, C., et al.: Honeycomb-structured porous films from polypyrrole-containing block copolymers prepared via RAFT polymerization as a scaffold for cell growth. Biomacromolecules **7**, 1072–1082 (2006)
140. Wong, K.H., Davis, T.P., Bamer-Kowollik, C., Stenzel, M.H.: Honeycomb structured porous films from amphiphilic block copolymers prepared via RAFT polymerization. Polymer **48**, 4950–4965 (2007)
141. Escalé, P., Ting, S.R.S., Khoukh, A., Rubatat, L., Save, M., Stenzel, M.H., et al.: Synthetic route effect on macromolecular architecture: from block to gradient copolymers based on acryloyl galactose monomer using RAFT polymerization. Macromolecules **44**, 5911–5919 (2011)
142. Xue, Y., Lu, H.-C., Zhao, Q.-L., Huang, J., Xu, S.-G., Cao, S.-K., et al.: Polymethylene-b-poly(styrene-co-2,3,4,5,6-pentafluoro styrene) copolymers: synthesis and fabrication of their porous films. Polym. Chem **4**, 307–312 (2013)
143. Yabu, H., Takebayashi, M., Tanaka, M., Shimomura, M.: Superhydrophobic and lipophobic properties of self-organized honeycomb and pincushion structures. Langmuir **21**, 3235–3237 (2005)
144. Muñoz-Bonilla, A., Ibarboure, E., Papon, E., Rodriguez-Hernandez, J.: Self-organized hierarchical structures in polymer surfaces: self-assembled nanostructures within breath figures. Langmuir **25**, 6493–6499 (2009)
145. Zhu, L.-W., Ou, Y., Wan, L.-S., Xu, Z.-K.: Polystyrenes with hydrophilic end groups: synthesis, characterization, and effects on the self-assembly of breath figure arrays. J. Phys. Chem. B **118**, 845–854 (2014)
146. Stenzel, M.H., Davis, T.P.: Biomimetic honeycomb-structured surfaces formed from block copolymers incorporating acryloyl phosphorylcholine. Aust. J. Chem **56**, 1035–1038 (2003)
147. Samanta, S., Chatterjee, D.P., Layek, R.K., Nandi, A.K.: Multifunctional porous poly(vinylidene fluoride)-graft-poly(butyl methacrylate) with good Li + ion conductivity. Macromol. Chem. Phys **212**, 134–149 (2011)
148. De León, A.S., Del Campo, A., Fernández-García, M., Rodríguez-Hernández, J., Muñoz-Bonilla, A.: Fabrication of structured porous films by breath figures and phase separation processes: tuning the chemistry and morphology inside the pores using click chemistry. ACS Appl. Mater. Interfaces **5**, 3943–3951 (2013)
149. Peng, J., Han, Y., Fu, J., Yang, Y., Li, B.: Formation of regular hole pattern in polymer films. Macromol. Chem. Phys **204**, 125–130 (2003)
150. Connal, L.A., Qiao, G.G.: Honeycomb coated particles: porous doughnuts, golf balls and hollow porous pockets. Soft Matter **3**, 837–839 (2007)

151. Bormashenko, E., Schechter, A., Stanevsky, O., Stein, T., Balter, S., Musin, A., et al.: Free-standing, thermostable, micrometer-scale honeycomb polymer films and their properties. Macromol. Mater. Eng **293**, 872–877 (2008)
152. Tsai, H., Xu, Z., Pai, R.K., Wang, L., Dattelbaum, A.M., Shreve, A.P., et al.: Structural dynamics and charge transfer via complexation with fullerene in large area conjugated polymer honeycomb thin films†. Chem. Mater. **23**, 759–761 (2010)
153. Stenzel, M.H., Davis, T.P., Fane, A.G.: Honeycomb structured porous films prepared from carbohydrate based polymers synthesized via the RAFT process. J. Mater. Chem. **13**, 2090–2097 (2003)
154. Ke, B.B., Wan, L.S., Zhang, W.X., Xu, Z.K.: Controlled synthesis of linear and comb-like glycopolymers for preparation of honeycomb-patterned films. Polymer **51**, 2168–2176 (2010)
155. Ting, S.R.S., Min, E.H., Escale, P., Save, M., Billon, L., Stenzel, M.H.: Lectin recognizable biomaterials synthesized via nitroxide-mediated polymerization of a methacryloyl galactose monomer. Macromolecules **42**, 9422–9434 (2009)
156. Nurmawati, M.H., Ajikumar, P.K., Renu, R., Valiyaveettil, S.: Hierarchical self-organization of nanomaterials into two-dimensional arrays using functional polymer scaffold. Adv. Funct. Mater. **18**, 3213–3218 (2008)
157. Jiang, X., Zhou, X., Zhang, Y., Zhang, T., Guo, Z., Gu, N.: Interfacial effects of in situ-synthesized Ag nanoparticles on breath figures. Langmuir **26**, 2477–2483 (2010)
158. Muñoz-Bonilla, A., Ibarboure, E., Bordegé, V., Fernández-García, M., Rodríguez-Hernández, J.: Fabrication of honeycomb-structured porous surfaces decorated with glycopolymers. Langmuir **26**, 8552–8558 (2010)
159. De León, A.S., Del Campo, A., Fernández-García, M., Rodríguez-Hernández, J., Muñoz-Bonilla, A.: Tuning the pore composition by two simultaneous interfacial self-assembly processes: breath figures and coffee stain. Langmuir **30**, 6134–6141 (2014)
160. Cui, L., Peng, J., Ding, Y., Li, X., Han, Y.C.: Ordered porous polymer films via phase separation in humidity environment. Polymer **46**, 5334–5340 (2005)
161. Galeotti, F., Calabrese, V., Cavazzini, M., Quici, S., Poleunis, C., Yunus, S., et al.: Self-functionalizing polymer film surfaces assisted by specific polystyrene end-tagging. Chem. Mater. **22**, 2764–2769 (2010)
162. Hernandez-Guerrero, M., Min, E., Barner-Kowollik, C., Muller, A.H.E., Stenzel, M.H.: Grafting thermoresponsive polymers onto honeycomb structured porous films using the RAFT process. J. Mater. Chem. **18**, 4718–4730 (2008)
163. Wenzel, R.N.: Resistance of solid surfaces to wetting by water. Ind. Eng. Chem. **28**, 988–994 (1936)
164. Cassie, A.B.D., Baxter, S.: Wettability of porous surfaces. Trans. Faraday Soc. **40**, 546–551 (1944)
165. Wan, L.-S., Lv, J., Ke, B.-B., Xu, Z.-K.: Facilitated and site-specific assembly of functional polystyrene microspheres on patterned porous films. ACS Appl. Mater. Interfaces **2**, 3759–3765 (2010)
166. Ke, B.-B., Wan, L.-S., Li, Y., Xu, M.-Y., Xu, Z.-K.: Selective layer-by-layer self-assembly on patterned porous films modulated by Cassie-Wenzel transition. Phys. Chem. Chem. Phys. **13**, 4881–4887 (2011)
167. De León, A.S., Campo, A.D., Labrugère, C., Fernández-García, M., Muñoz-Bonilla, A., Rodríguez-Hernández, J.: Control of the chemistry outside the pores in honeycomb patterned films. Polym. Chem **4**, 4024–4032 (2013)
168. Sun, H., Li, H.L., Wu, L.X.: Micro-patterned polystyrene surfaces directed by surfactant-encapsulated polyoxometalate complex via breath figures. Polymer **50**, 2113–2122 (2009)
169. Ma, H., Cui, J., Chen, J., Hao, J.: Self-organized polymer nanocomposite inverse opal films with combined optical properties. Chemistry **17**, 655–660 (2011)
170. Deleuze, C., Derail, C., Delville, M.H., Billon, L.: Hierarchically structured hybrid honeycomb films via micro to nanosized building blocks. Soft Matter **8**, 8559–8562 (2012)

171. Ma, H., Cui, J., Song, A., Hao, J.: Fabrication of freestanding honeycomb films with through-pore structures via air/water interfacial self-assembly. Chem. Commun. **47**, 1154–1156 (2011)
172. Wong, K.H., Davis, T.P., Barner-Kowollik, C., Stenzel, M.H.: Gold-loaded organic/inorganic nanocomposite honeycomb membranes. Aust. J. Chem **59**, 539–543 (2006)
173. Kong, L., Dong, R., Ma, H., Hao, J.: Au NP honeycomb-patterned films with controllable pore size and their surface-enhanced Raman scattering. Langmuir **29**, 4235–4241 (2013)
174. Vohra, V., Bolognesi, A., Calzaferri, G., Botta, C.: Multilevel organization in hybrid thin films for optoelectronic applications. Langmuir **25**, 12019–12023 (2009)
175. Zou, J., Chen, H., Chunder, A., Yu, Y., Huo, Q., Zhai, L.: Preparation of a superhydrophobic and conductive nanocomposite coating from a carbon-nanotube-conjugated block copolymer dispersion. Adv. Mater. **20**, 3337 (2008)
176. Tsai, H., Xu, Z., Pai, R.K., Wang, L., Dattelbaum, A.M., Shreve, A.P., et al.: Structural dynamics and charge transfer via complexation with fullerene in large area conjugated polymer honeycomb thin films. Chem. Mater. **23**, 759–761 (2011)
177. Ma, H., Gao, P., Fan, D., Du, B., Hao, J., Wei, Q.: Assembly of graphene nanocomposites into honeycomb-structured macroporous films with enhanced hydrophobicity. N. J. Chem **37**, 1307–1311 (2013)
178. Sun, W., Shao, Z., Ji, J.A.: Particle-assisted fabrication of honeycomb-structured hybrid films via breath figures method. Polymer **51**, 4169–4175 (2010)
179. Wang, J., Wang, C.F., Shen, H.X., Chen, S.: Quantum-dot-embedded ionomer-derived films with ordered honeycomb structures via breath figures. Chem. Commun. **46**, 7376–7378 (2010)
180. Li, X.F., Zhang, L.A., Wang, Y.X., Yang, X.L., Zhao, N., Zhang, X.L., et al.: A bottom-up approach to fabricate patterned surfaces with asymmetrical TiO_2 microparticles trapped in the holes of honeycomblike polymer film. J. Am. Chem. Soc. **133**, 3736–3739 (2011)
181. Li, L., Zhong, Y.W., Ma, C.Y., Li, J., Chen, C.K., Zhang, A.J., et al.: Honeycomb-patterned hybrid films and their template applications via a tunable amphiphilic block polymer/inorganic precursor system. Chem. Mater. **21**, 4977–4983 (2009)
182. Ke, B.B., Wan, L.S., Chen, P.C., Zhang, L.Y., Xu, Z.K.: Tunable assembly of nanoparticles on patterned porous film. Langmuir **26**, 15982–15988 (2010)
183. Galeotti, F., Mroz, W., Bolognesi, A.: CdTe nanocrystal assemblies guided by breath figure templates. Soft Matter **7**, 3832–3836 (2011)
184. Ou, Y., Wang, L.-Y., Zhu, L.-W., Wan, L.-S., Xu, Z.-K.: In-situ immobilization of silver nanoparticles on self-assembled honeycomb-patterned films enables surface-enhanced Raman scattering (SERS) substrates. J. Phys. Chem. C **118**, 11478–11484 (2014)
185. Mendes, P.M.: Stimuli-responsive surfaces for bio-applications. Chem. Soc. Rev. **37**, 2512–2529 (2008)
186. Chen, T., Ferris, R., Zhang, J., Ducker, R., Zauscher, S.: Stimulus-responsive polymer brushes on surfaces: transduction mechanisms and applications. Prog. Polym. Sci. **35**, 94–112 (2010)
187. Yabu, H., Hirai, Y., Kojima, M., Shimomura, M.: Simple fabrication of honeycomb- and pincushion-structured films containing thermoresponsive polymers and their surface wettability. Chem. Mater. **21**, 1787–1789 (2009)
188. Nygard, A., Davis, T.P., Barner-Kowollik, C., Stenzel, M.H.: A simple approach to micropatterned surfaces by breath figures with internal structure using thermoresponsive amphiphilic block copolymers. Aust. J. Chem **58**, 595–599 (2005)
189. Min, E.H., Ting, S.R.S., Billon, L., Stenzel, M.H.: Thermo-responsive glycopolymer chains grafted onto honeycomb structured porous films via RAFT polymerization as a thermo-dependent switcher for lectin concanavalin a conjugation. J. Polym. Sci. A Polym. Chem **48**, 3440–3455 (2010)
190. Wang, J., Shen, H.-X., Wang, C.-F., Chen, S.: Multifunctional ionomer-derived honeycomb-patterned architectures and their performance in light enhancement of light-emitting diodes. J. Mater. Chem. **22**, 4089–4096 (2012)

191. Yabu, H., Akagi, K., Shimomura, M.: Micropatterning of liquid crystalline polyacetylene derivative by using self-organization processes. Synth. Met. **159**, 762–764 (2009)
192. Bolognesi, A., Galeotti, F., Giovanella, U., Bertini, F., Yunus, S.: Nanophase separation in polystyrene-polyfluorene block copolymers thin films prepared through the breath figure procedure. Langmuir **25**, 5333–5338 (2009)
193. Wan, L.-S., Li, Q.-L., Chen, P.-C., Xu, Z.-K.: Patterned biocatalytic films via one-step self-assembly. Chem. Commun. **48**, 4417–4419 (2012)
194. Kim, J.H., Seo, M., Kim, S.Y.: Lithographically patterned breath figure of photoresponsive small molecules: dual-patterned honeycomb lines from a combination of bottom-up and top-down lithography. Adv. Mater. **21**, 4130 (2009)
195. Zhang, M., Sun, S., Yu, X., Cao, X., Zou, Y., Yi, T.: Formation of a large-scale ordered honeycomb pattern by an organogelator via a self-assembly process. Chem. Commun. **46**, 3553–3555 (2010)
196. Babu, S.S., Mahesh, S., Kartha, K.K., Ajayaghosh, A.: Solvent-directed self-assembly of pi gelators to hierarchical macroporous structures and aligned fiber bundles. Chem. Asian J. **4**, 824 (2009)
197. Xu, W.Z., Zhang, X., Kadla, J.F.: Design of functionalized cellulosic honeycomb films: site-specific biomolecule modification via "click chemistry". Biomacromolecules **13**, 350–357 (2011)
198. Chen, S., Alves, M.-H., Save, M., Billon, L.: Synthesis of amphiphilic diblock copolymers derived from renewable dextran by nitroxide mediated polymerization: towards hierarchically structured honeycomb porous films. Polym. Chem. **5**, 5310–5319 (2014)
199. Sanz de Leon, A., Rodriguez-Hernandez, J., Cortajarena, A.L.: Honeycomb patterned surfaces functionalized with polypeptide sequences for recognition and selective bacterial adhesion. Biomaterials **34**, 1453–60 (2013)
200. Leibler, L., Benoit, H.: Theory of correlations in partly labeled homopolymer melts. Polymer **22**, 195–201 (1981)
201. Matsen, M.W., Bates, F.S.: Origins of complex self-assembly in block copolymers. Macromolecules **29**, 7641–7644 (1996)
202. Geldhauser, T., Walheim, S., Schimmel, T., Leiderer, P., Boneberg, J.: Influence of the relative humidity on the demixing of polymer blends on prepatterned substrates. Macromolecules **43**, 1124–1128 (2009)
203. Cui, L., Xuan, Y., Li, X., Ding, Y., Li, B.Y., Han, Y.C.: Polymer surfaces with reversibly switchable ordered morphology. Langmuir **21**, 11696–11703 (2005)
204. Hecht, U., Schilz, C.M., Stratmann, M.: Influence of relative humidity during film formation processes on the structure of ultrathin polymeric films. Langmuir **14**, 6743–6748 (1998)
205. Ge, W., Lu, C.: Hierarchical honeycomb patterns with tunable microstructures: controllable fabrication and application as replication templates. Soft Matter **7**, 2790–2796 (2011)
206. Connal, L.A., Vestberg, R., Hawker, C.J., Qiao, G.G.: Fabrication of reversibly crosslinkable, 3-dimensionally conformal polymeric microstructures. Adv. Funct. Mater. **18**, 3315–3322 (2008)
207. Ding, J., Gong, J., Bai, H., Li, L., Zhong, Y., Ma, Z., et al.: Constructing honeycomb micropatterns on nonplanar substrates with high glass transition temperature polymers. J. Colloid Interface Sci. **380**, 99–104 (2012)
208. Greiser, C., Ebert, S., Goedel, W.A.: Using breath figure patterns on structured substrates for the preparation of hierarchically structured microsieves. Langmuir **24**, 617–620 (2008)
209. Kwak, G., Fukao, S., Fujiki, M., Sakaguchi, T., Masuda, T.: Nanoporous, honeycomb-structured network fibers spun from semiflexible, ultrahigh molecular weight, disubstituted aromatic polyacetylenes: superhierarchical structure and unique optical anisotropy. Chem. Mater. **18**, 5537–5542 (2006)
210. Zheng, J., Zhang, H., Zhao, Z., Han, C.C.: Construction of hierarchical structures by electrospinning or electrospraying. Polymer **53**, 546–554 (2012)

211. Fashandi, H., Karimi, M.: Pore formation in polystyrene fiber by superimposing temperature and relative humidity of electrospinning atmosphere. Polymer **53**, 5832–5849 (2012)
212. Ting, W.-H., Chen, C.-C., Dai, S.A., Suen, S.-Y., Yang, I.K., Liu, Y.-L., et al.: Superhydrophobic waxy-dendron-grafted polymer films via nanostructure manipulation. J. Mater. Chem. **19**, 4819–4828 (2009)
213. Brown, P.S., Talbot, E.L., Wood, T.J., Bain, C.D., Badyal, J.P.S.: Superhydrophobic hierarchical honeycomb surfaces. Langmuir **28**, 13712–13719 (2012)
214. Yabu, H., Hirai, Y., Shimomura, M.: Electroless plating of honeycomb and pincushion polymer films prepared by self-organization. Langmuir **22**, 9760–9764 (2006)
215. Ishii, D., Yabu, H., Shimomura, M.: Selective metal deposition in hydrophobic porous cavities of self-organized honeycomb-patterned polymer films by all-wet electroless plating. Colloids Surf. A Physicochem. Eng. Asp. **313–314**, 590–594 (2008)
216. Ma, C.Y., Zhong, Y.W., Li, J., Chen, C.K., Gong, J.L., Xie, S.Y., et al.: Patterned carbon nanotubes with adjustable array: a functional breath figure approach. Chem. Mater. **22**, 2367–2374 (2010)
217. Sakatanì, Y., Boissière, C., Grosso, D., Nicole, L., Soler-Illia, G.J.A.A., Sanchez, C.: Coupling nanobuilding block and breath figures approaches for the designed construction of hierarchically templated porous materials and membranes. Chem. Mater. **20**, 1049–1056 (2008)
218. Sato, T., Tanaka, M., Yamamoto, S., Ito, E., Shimizu, K., Igarashi, Y., et al.: Effect of honeycomb-patterned surface topography on the function of mesenteric adipocytes. J. Biomater. Sci. Polym. Ed. **21**, 1947–1956 (2010)
219. Wan, L.S., Li, J.W., Ke, B.B., Xu, Z.K.: Ordered microporous membranes templated by breath figures for size-selective separation. J. Am. Chem. Soc. **134**, 95–98 (2012)
220. Hirai, Y., Yabu, H., Matsuo, Y., Ijiro, K., Shimomura, M.: Arrays of triangular shaped pincushions for SERS substrates prepared by using self-organization and vapor deposition. Chem. Commun. **46**, 2298–2300 (2010)
221. Tang, P., Hao, J.: Directionally electrodeposited gold nanoparticles into honeycomb macropores and their surface-enhanced Raman scattering. N. J. Chem **34**, 1059–1062 (2010)

Chapter 11
Spontaneous Structuration of Hydrophobic Polymer Surfaces in Contact with Salt Solutions

Igor Siretanu, Hassan Saadaoui, Jean-Paul Chapel, and Carlos Drummond

It has been described in previous chapters how spontaneous instabilities related to interfacial phenomena can be used to produce controlled patterns on polymer surfaces. Strategies of polymer patterning assisted by dewetting or water drop condensation were described. In this chapter we present a waterborne process based on the interaction between ions in water and hydrophobic polymer surfaces, modulated by the gases dissolved in the aqueous phase. We show how by controlling this interaction the polymer surface can be conveniently modified. In the first section of the chapter we describe some aspects of the interface between water and a hydrophobic surface. We then describe how the composition of the aqueous phase can have important consequences on the morphology of the hydrophobic surface, and then illustrate how this process can be conveniently used to modify the morphology of a hydrophobic polymer in a controlled manner.

11.1 Water-Hydrophobic Interfaces, WHI: A Deceptively Simple System

The interface between water and hydrophobic surfaces plays a central role in a number of important subjects. Proteins folding, self-assembly, detergency, or oil recovery are just few examples of important problems involving the contact between water and a hydrophobic surface. Nevertheless, a complete description

I. Siretanu (✉)
Physics of Complex Fluids, MESA Institute for Nanotechnology, University of Twente, Post Office Box 217, 7500 AE Enschede, The Netherlands
e-mail: i.siretanu@utwente.nl

H. Saadaoui • J.-P. Chapel • C. Drummond
CNRS, Centre de Recherche Paul Pascal (CRPP), UPR 8641, 33600 Pessac, France

Centre de Recherche Paul Pascal, Université de Bordeaux, 33600 Pessac, France

J. Rodríguez-Hernández, C. Drummond (eds.), *Polymer Surfaces in Motion*,
DOI 10.1007/978-3-319-17431-0_11

of WHI is yet to be reached. In particular, two aspects related to the process discussed in this chapter have been the subject of some controversy: the presence of a layer of reduced water density near the WHI and the adsorption of ions on them.

When water is in contact with air in standard conditions, oxygen and nitrogen are dissolved in concentrations of 0.2 mM and 0.5 mM, respectively. There is a general consensus that when a hydrophobic surface is in contact with water, the fluid density is substantially reduced in the vicinity of the surface. On the contrary, the reasons behind this depleted layer and its actual structure are still subjects of debate. It has been reported that dissolved gases can adsorb on hydrophobic surfaces in the form of small spherical caps (nanobubbles) [1] or flattened objects [2, 3] (micro-pancakes; it has been argued that these are just bubbles with a large but well-defined radius of curvature in order to satisfy Laplace's equation [4]) or as a reduced-density continuous interfacial layer. After they were first proposed about 20 years ago [1], many experimental studies have reported the existence of nanobubbles on hydrophobic surfaces. In most of them, tapping-mode atomic force microscope has been used to image the bubbles under gentle perturbation [5]. An example of nanobubble formation on a hydrophobic surface is presented in Fig. 11.1a. It has been suggested that the high-frequency oscillatory motion of the AFM tip may be responsible for nanobubble nucleation. However, other techniques (infrared spectroscopy [6], surface plasmon resonance [7], optical interference-enhanced reflection microscopy [8], and rapid cryofixation/freeze fracture [9]) have reinforced the idea of their existence.

The mere existence of gaseous nanostructures (GNs) on surfaces has been the subject of scorching debates in the literature. This is partially due to the fact that their materialization and stability appear to be extremely sensitive to the experimental conditions, as was recently shown by Seddon and coworkers [10]. However, the existence of GN on hydrophobic surfaces is gaining consensus in the scientific

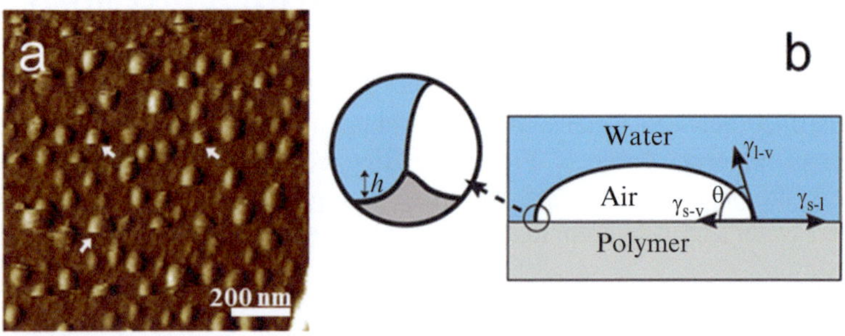

Fig. 11.1 (**a**) Contact-mode AFM deflection images of PS in water. The presence of nanobubbles is observed. Occasionally the bubbles are removed by the effect of the tip; only a portion of the nanobubble appears in the image (*white arrows*). (**b**) Schematic representation of a nanobubble in a water/polymer interface. The contact angle θ is determined by the equilibrium between the horizontal forces in the triple solid–liquid–vapor contact line: liquid–vapor $\gamma_{l\text{-}v}$, solid–liquid $\gamma_{s\text{-}l}$, and solid–vapor $\gamma_{s\text{-}v}$ interfacial tensions. The vertical component of the liquid–vapor interfacial tension, $\gamma_{l\text{-}v}\sin(\theta)$, is equilibrated by a deformation of the substrate, as described in the text

community, after a large number of papers have been published in that subject. On the contrary, the reasons for their stability remain unclear. It has often been argued that because of their small radii of curvature, the excess Laplace pressure inside the nanobubbles ($\Delta P = 2\gamma/R$, where γ is the water-air interfacial tension and R the GN radius of curvature) should lead to their immediate disappearance: the bubbles are diffusively unstable. Even though R of the GN is larger than it may appear from their typical nanometric size height (because of the unusually low contact angle of the GN on hydrophobic surfaces), the excess Laplace pressure inside the GN is of the order of few atmospheres, which leads to estimations of lifetimes of the order of microseconds [5]. Nevertheless, it has been repeatedly observed that adsorbed nanobubbles can be stable for as long as few days, suggesting that they are thermodynamically stable.

Several theories have been advanced to explain this anomaly. Spurious contamination has been evoked several times. As everybody recognizes, it is extremely difficult to keep water and hydrophobic surfaces contamination free. Minute amounts of surface-active contaminants could be responsible for the unexpected stability of the GN [11]. However, the widespread presence of GN hints for more fundamental reasons behind their stability. Dynamic stationary models based on continuous gas flow between the bubbles and the surroundings have also been proposed [12]. However it is unclear what would be the driving force for the continuous gas flow. As some ions naturally adsorb on hydrophobic surfaces in contact with water, models based on the pressure due to the presence of charges adsorbed at the interface have also been discussed [13]. Attard recently proposed that gas supersaturation on the aqueous phase can explain both the stability and the anomalous contact angle, due to reduction in the water vapor interfacial tension [4]. This model conciliates the existence of the GN with Laplace's equation. However, it imposes supersaturation as a *sine qua non* condition for the existence of GN, which has been contested by other groups [10]. In addition, it is not clear why a long-lasting vapor supersaturation is observed in many different experimental situations. In summary, the reasons for the stability of GN are yet to be soundly established; for our purposes, it suffices to recognize their existence and stability. As discussed in next chapter and mentioned in next section, this may have important implications in the morphology of polymer surfaces in contact with water.

Another puzzling aspect of WHI is the fact that they are often electrically charged. In the absence of dissociating groups, this can only be due to ion adsorption, even though ions should strongly partition into the aqueous phase of higher dielectric constant [14]. Different mechanisms—entropy changes [15], ionic asymmetry [16], ionic polarizability [17], and ionic induced decrement of water polarization fluctuations [18, 19]—have been evoked to explain this fact. A satisfactory description of a charged interface at low-salt concentrations can be achieved by mean field theories, developed after the work of Gouy and Chapman, which is the starting point for the celebrated Poisson-Boltzmann equation [20]. One aspect that cannot be accounted for by this simple description is the distinct effect of particular ions, because it only considers the ionic charge, and does not take into account the discrete nature of ions or solvent molecules, or the possibility of correlation between the charges. The error introduced by considering ions as indistinct point charges, neglecting the dispersion forces involving the ions and their disruptive effect on the solvent, becomes unacceptable as ionic strength increases. Another source of error that is frequently overlooked is related to the deformability of the boundary: the walls are often considered to be unalterable and non-deformable. This assumption is adequate for hard solid walls, but is certainly not so for soft boundaries, e.g., charged membranes or polymer substrates. In this sense, we have observed that smooth, hydrophobic polymer surfaces can be substantially deformed due to the effect of adsorbing ions. We have called this phenomenon ion-induced polymer nanostructuration, IPN [21]. We describe the IPN process in detail in the rest of this chapter. Some examples of how this method can be used to control the morphology of polymer surfaces are advanced in the last section.

11.2 Ionic Solutions in Contact with Hydrophobic Polymers: Ion-Induced Polymer Nanostructuration

We discuss now how the morphology of a hydrophobic polymer can be modified by the contact with water. For the reasons described in previous section, this transformation depends on the amount of gases dissolved and the ionic species present. The typical size and nature of formed pattern can then be tuned if these variables are precisely controlled. Two processes must be considered. First, the GNs have a transforming effect, as described in next chapter. Second, the specific adsorption of ions may induce spontaneous surface deformation.

Reducing the concentration of the gases dissolved in water may deeply affect many interfacial properties. Despite the fact that oil and water do not mix, spontaneous emulsification of hydrocarbons occurs after water degassing [22–24]. Colloid stability [25], the range of the hydrophobic interaction [26], the conductivity of salt solutions [27], the efficiency of flotation process [28], and the slippage boundary conditions in flowing water [29] are all affected by the amount of gas in solution. The amount of gas dissolved has also significant consequences on the morphology of a surface in contact with the aqueous phase. Several workers have shown how the presence of GNs can induce surface modification. Wang and coworkers [30] reported the formation of nanoindents in very thin films of polystyrene (<5 nm) (PS) after few tens of minutes of exposure to water. Similar results were reported by Mazumder and Bhushan [13] after exposing PS films (25 nm thick) to deionized water. Janda and coworkers [31] reported the exfoliation and indentation of highly ordered pyrolytic graphite in contact with nanobubbles. Tarábková and Janda [32] reported a netlike nanopatterning of thin PS films (around 10 nm) in contact with nanobubbles. A complete description of this subject is discussed in next chapter.

Even in the absence of dissolved gas, polymer surfaces may be modified in contact with water. As advanced above, we have observed that the adsorption of specific ions on hydrophobic polymers can induce important surface

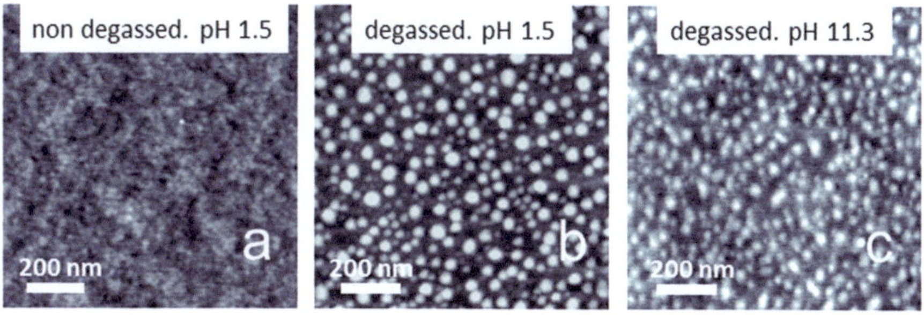

Fig. 11.2 1 μm × 1 μm height tapping-mode AFM micrographs taken in air of 300 nm thick 250 kDa polystyrene films after exposure to a non-degassed (**a**) and degassed solution of nitric acid at pH 1.5 (**b**) and sodium hydroxide at pH 11.3 (**c**) in double-distilled water and room temperature. The presence of asperities of regular nanometric size is observed on the surface exposed to the degassed solutions for 5 min. On the contrary, no modification was detected when an identical film was exposed to the same solution under identical conditions (or much longer times) before removing the dissolved gases. XPS tests showed that no chemical modification of the films has occurred

transformations. For this to happen, the intimate contact between the aqueous phase and the polymer has to be favored by the removal of gases dissolved in the water. Water degassing avoids formation of GNs, allowing the direct adsorption of ions to the interface. As an example, when polystyrene films are put in contact with degassed water solutions of acidic or basic pH for a few minutes, a long-lasting nanostructuration spontaneously forms on the solid surface, as illustrated in Fig. 11.2. This only happens in the presence of certain ions; otherwise no modification of the polymer film is observed. The typical size of the self-assembled patterns depends on pH, type, and concentration of dissolved salts; temperature and amount of dissolved gas in the aqueous phase; as well as mobility of the surface, which can be affected by the interaction with the supporting substrate.

The examples shown in Fig. 11.2 correspond to thin films of polystyrene in contact with acidic or basic water. The surface structuration has no effect on the wettability of the surfaces: the contact angle of water on the substrate is not changed by the process. The morphology change is not accompanied by any chemical modification. However, the hydrophobicity of the surface seems to determine the outcome of the process: the structuration of the film is not observed if the polymer surface is rendered hydrophilic ($\theta = 30°$) before the water treatment by partial UV oxidation. Two remarkable conclusions can be drawn from this interesting effect. First, the interaction between the aqueous phase and the hydrophobic substrate is fundamentally modified by the gases dissolved which shelter the hydrophobic substrate from the aqueous phase. Second, there is a mobile surface layer on the polymer which can be restructured under external stimuli, even though we are working at temperatures largely below the glass transition temperature T_g of PS. This indicates that macromolecular chains near the interface have properties which deviate significantly from their bulk counterparts. Indeed, no structuration

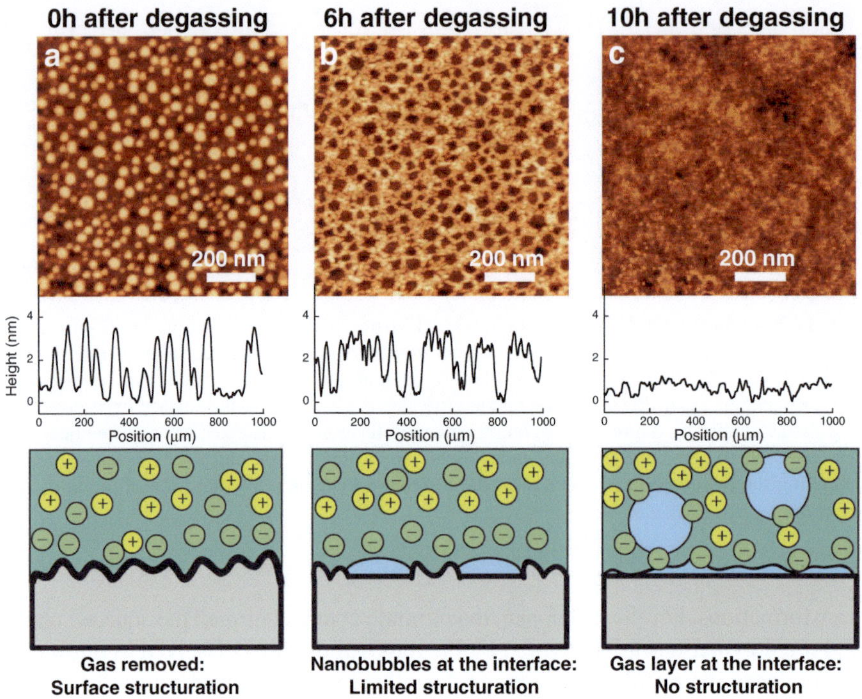

Fig. 11.3 The amount of dissolved gas determines the nature and extent of the surface structuration. 300 nm thick 250 kDa PS films were immersed during 10 min in water at pH 1.5 and then studied by AFM in air. Before treating the surfaces, the aqueous solution was in contact with air during different periods of time after degassing as indicated, to change the amount of gas dissolved. (**a**) Shortly after degassing the preferential adsorption of ions at the water/polymer interface is likely to be responsible for the observed self-assembled nanostructure. Increasing amounts of gas in the solution move the ions away from the polymer surface limiting the structuration below the bubbles (**b**), or the low-density layer (**c**). Adapted with permission from [21]. Copyright (2011) American Chemical Society

was observed in films of UV cross-linkable PS after UV curing, evidencing the importance of surface mobility.

It is very revealing to explore the effect of the amount of gas dissolved on IPN. As can be observed in Fig. 11.3, the size of the observed bumps is larger at short times after degassing. Ten hours after degassing water has regained its equilibrium with the atmosphere, no structuring effect is detected. On the contrary, at intermediate times a seemingly different structuration effect is observed: the contact with the aqueous phase induces the formation of a regular array of shallow holes/rims (nanoindents) in the polymer film, with a typical size of 30 nm and depth of 2 nm. The thickness of the structured region (between 2 and 3 nm) is consistent with recent reports of the presence of a thin mobile layer on the surface of glassy PS films [33].

Our understanding of the fundamental reasons behind these phenomena is far from being complete. Some preliminary and speculative ideas waiting to be validated are described here. At low concentration of dissolved gas (cf. Fig. 11.3a), the preferential adsorption of ions at the interface induces the formation of bumps. As mentioned above, it has long being recognized that pristine water/oil interfaces can become charged due to adsorption of ions. It is only in the presence of adsorbing ions that IPN is observed: we believe that the observed surface structuration is related to the electric field originated by their presence. As mentioned above, the discontinuity in dielectric constant conspires against ion adsorption on the interface [14]. However, if other forces promote ion adsorption (e.g., ion-dispersion forces [34], "cavity" forces, due to perturbation of the hydrogen bond network of water [35]), subsequent reconstruction of the interface, driven by electrostatic, can then arise to reduce the total energy of the system. Some examples related to this phenomena can be found in the literature; membrane deformation by interaction with ions was discussed few decades ago by Parsegian [36]; water-air interfacial deformation in the presence of adsorbing ions has been predicted by molecular dynamic simulations [37]. The complete description of this problem requires integrating the viscoelastic and plastic properties of the polymer film. In addition, the film shape and the electrostatic boundary conditions are intermingled and cannot be treated independently: the presence of the ions distorts the shape of the surface, changing the boundary conditions of the electrostatic problem that determines the ionic distribution.

At large gas concentration larger bubbles or even a continuous low-density layer is present at the water/polymer interface and no structuration effect is detected (Fig. 11.3c). Finally, at intermediate gas concentrations a layer of nanobubbles nucleate on the surface, which is responsible for the netlike pattern observed (Fig. 11.3b). The formation of a similar pattern has been reported by Wang and coworkers [30], and is certainly related to the process of nanobubble-assisted nanostructuration described in next chapter. They suggested that the combination of increased Laplace pressure and interfacial tension is at the origin of the nanoindents observed. The problem of the elastic deformation of soft surfaces due to the vertical component of the interfacial tension when a liquid drop (or a gas bubble) is posed in the surface has been discussed in the literature. The surface modification shown in Fig. 11.3 is long lasting, indicating an irreversible (plastic) deformation. However, we can try to estimate the order of magnitude of the deformation ridge h using the elastic model described by Shanahan and Carré [38]. By equating the vertical component of the interfacial tension to the elastic stress field on the solid surface, one obtains $h \approx \gamma_{LV} \cdot \sin\theta/G$, where G is the shear modulus of the solid and γ_{LV} the water-air interfacial tension (cf. Fig. 11.1b). If G and θ for bulk polystyrene and γ_{LV} for air-water are considered, h values of the order of 10^{-12} m are obtained, much smaller than the deformation observed experimentally. The inconsistency is even more important if experimental values of the contact angle θ typically observed for nanobubbles are considered. Several reasons could be evoked to account for this discrepancy. The value of G of the polymer surface is probably smaller than the bulk value. Recently, by studying the relaxation of nanometric deformations on films of PS, Fakhraai and Forrest showed that the mobility of polymer chains is dramatically accelerated nearby the surfaces with respect to the bulk [33]. Although it is still poorly understood, several explanations have been advanced for this enhanced dynamic, related to the reduced entanglement density, enrichment of chain ends, and increased free volume for the polymer molecules at the surface. Another possible explanation is that water may have a plasticizing effect on the polymer effectively reducing *G*. However, it is difficult to conceive a reduction of *G* of three orders of magnitude, which is necessary to account for the discrepancy observed. The effect of adsorbing ions may also be important. As can be observed in the AFM micrographs, small bumps are still formed in the region between or at the rim of the nanobubbles. Hence, the observed deformation at intermediate concentrations may be related to the transforming effect of the adsorbing ions, as in the case of complete degassing, as illustrated in Fig. 11.3.

For the rest of this chapter we describe the IPN process: the influence of some variables that determine the outcome of the process and how it can be used to achieve easy polymer patterning in a single step.

11.3 IPN: Ionic Specificity

We found that the IPN process is ion specific: in the absence of adsorbing ions in the aqueous phase no structuring effect was observed. Some examples of structuring and non-structuring ions are presented in Fig. 11.4. In all cases, the IPN was correlated with ion adsorption, which was indirectly determined by measuring the electrokinetic zeta potential of PS films in contact with salt solutions. From our studies, several general trends were recognized:

- Substantial PS structuration is observed in the presence of water ions ($pH < 2.5$ or $pH > 10$). The deformation of the hydrophobic polymer surface at different pH values indicates a greater influence of the presence of hydroxide ions compared with protons: for example, similar surface patterning is obtained at pH 1.5 and 11, even though the $[H_3O^+]$ in the former case is much larger than the $[OH^-]$ in the latter. This result suggests a preferential adsorption of OH^- with respect to H_3O^+ on the hydrophobic substrate. It has been reported many times that hydrophobic interfaces and even gas bubbles are negatively charged in contact with water at neutral or basic pH and become positively charged in acidic solutions [39]. The origin of this charge is hotly debated. The main disagreement comes from the fact that the majority of experiments suggest a preferential adsorption of hydroxide ions [39], while the opposite picture has emerged from reports of surface-sensitive spectroscopic techniques and MD simulations [40]. Our results point in the direction of preferential OH^- adsorption.
- Degassed solutions of most inorganic electrolytes evaluated did not induce large structuration on PS surfaces. Some examples are presented in Fig. 11.4. With this kind of salts, the observed mild surface deformation is similar to the one

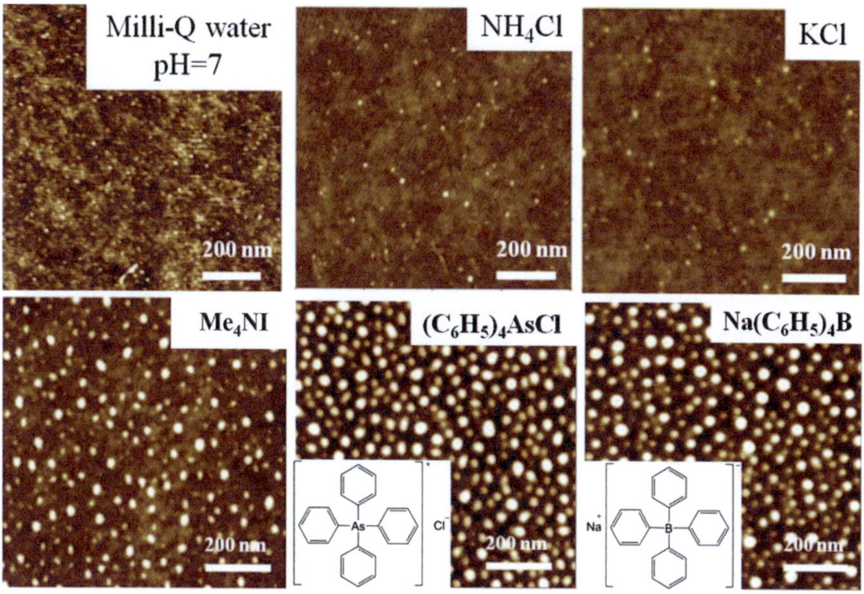

Fig. 11.4 AFM micrographs of 300 nm thick PS films after 5-min exposure to water or 0.03 M degassed aqueous solutions of ammonium chloride, potassium chloride, tetramethyl ammonium iodide, tetraphenyl arsonium chloride, Ph_4AsCl, and sodium tetraphenylborate, $NaPh_4B$

obtained by treatment with degassed water at neutral pH. As a remarkable exception to this rule, relatively large bumps (although at a lesser density) are observed after treatment of PS surfaces with degassed solutions of some lithium salts (see below).

- Degassed aqueous solutions of amphiphilic ions (e.g., sodium dodecyl sulfate, SDS, dodecyl trimethyl ammonium bromide, DTAB) or hydrophobic ions (e.g., halides of tetra-alkyl-ammonium quaternary, Ph_4BNa or Ph_4AsCl) induce substantial deformation of the PS surfaces. The hydrophobicity of the ion seems to have a distinctive influence on the nanostructuration. The longer is the chain length of tetra-alkyl-ammonium iodide salts the larger is the typical size of the induced nanostructure. Furthermore, ions with similar molecular structure seem to induce similar nanostructuration effect, regardless of the sign of the charge. For example, the salt Ph_4BNa induces the same effect as Ph_4AsCl (Fig. 11.4), probably due to the similar effect of the tetraphenyl ions of opposite charge.
- The observed nanostructuration is not just due to the effect of a particular ion, but to the combined influence of the different ions present in solution. For example, the structuration effect of tetrabutylammonium salts is different for different counterions. The size of the bumps was larger for the case of iodide or bromide compared with nitrate or chloride. A more marked difference is observed for the case of lithium salts. While LiCl and $LiSO_4$ did not induce extensive surface structuration, substantial effect of LiI and LiBr solutions was

observed. As a comparison, sodium or potassium salts with similar counterions did not produce any noticeable surface modification.

- For salts that do not provoke extensive structuration of PS surfaces, there is no appreciable effect of salt concentration up to 0.1 M. This indicates that the small structuration observed in these cases is mainly determined by the small amount of water ions present in the solutions. On the contrary, for salts with larger structuring influence, the resulting morphology depends on salt concentration. In general, higher concentrations result in higher surface charge density and in a more important modification of the polymer surface. Typically, the surface structuration was no longer enhanced for concentrations above 0.03 M.

An extensive report on ionic specificity in IPN was recently published [41].

11.4 Thermal Effects in IPN

As can be observed in Fig. 11.5, the typical size of the self-assembled pattern is strongly modified by the temperature during treatment. We can distinguish two different scenarios. At low temperature ($T < 30$ °C) a monomodal bump size

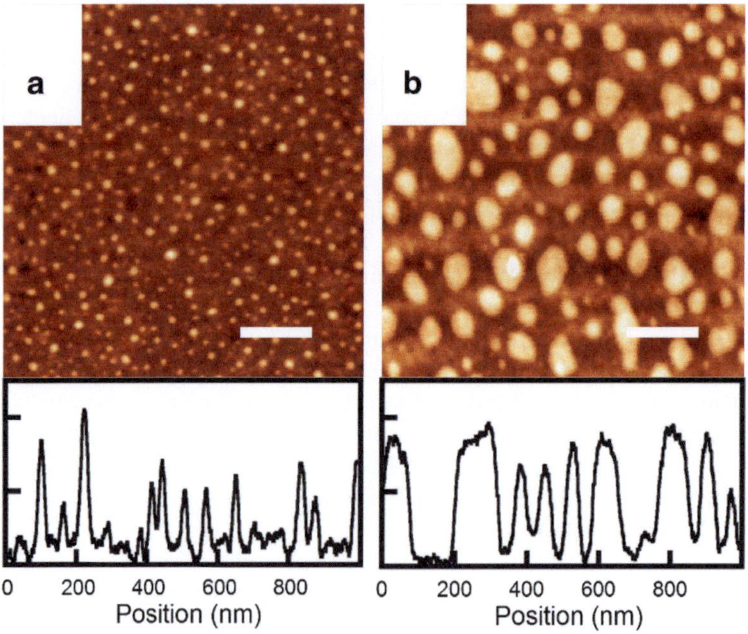

Fig. 11.5 1 μm × 1 μm height tapping-mode AFM micrographs taken in air of 300 nm thick 250 kDa PS films after exposure to degassed solutions of nitric acid in double-distilled water at pH 1.5. The temperature during treatment was (**a**) 25 °C and (**b**) 70 °C. A typical height profile for each condition is presented. The presence of self-assembled asperities (*bumps*) is observed only on the surfaces exposed to the degassed solution for 1 min. The scale bars correspond to 200 nm

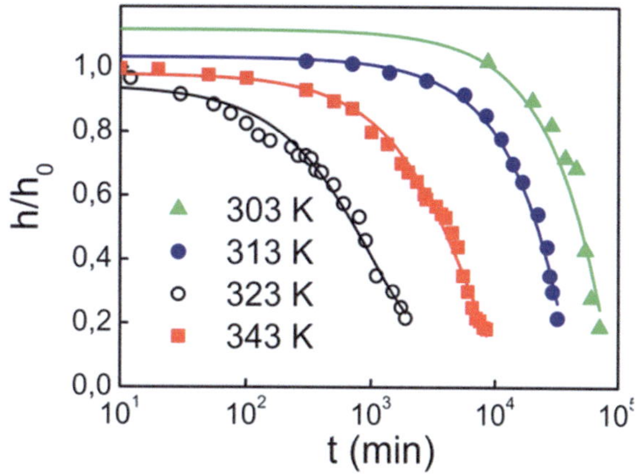

Fig. 11.6 Temporal evolution of the bump height for surfaces nanostructured at 25 °C aged at different temperatures. The continuous lines correspond to least-squares fits to exponential decay of the data at times longer than the decay time

distribution is observed. On the contrary, the size distribution increasingly broadens as the temperature of treatment is increased. The broad distribution is observed right after the treatment; it is not a consequence of the temporal evolution of the polymer surface. As can be observed in Fig. 11.5, the areal fraction occupied by large bumps increases with temperature of treatment, occupying most of the surface at 70 °C; however, zones of small bumps ($r < 30$ nm) can always be identified.

The subsequent evolution of the self-assembled structure is determined by the environmental conditions. The structure disappears after the structured surface is annealed at 95 °C for a few hours. Nevertheless, these patterns reappear if the surface is exposed again to degassed aqueous solutions. The patterns also relax under solvent annealing: after exposure to toluene vapor for 1 h most of the structuration is removed. Indeed, the surface of the film slowly relaxes back to the original smooth condition even when the films are conserved in air at room temperature. We observed a faster evolution of the films of PS of lower molecular weights. The temporal evolution for mean bump height at different temperatures below T_g is presented in Fig. 11.6a. For a number of samples structured under identical conditions at 25 °C, as can be expected, the relaxation time markedly decreases with increasing temperature. The relaxation process can be used to study the viscoelastic properties of the polymer surface, as has been described in studies of relaxation of grating-shaped surfaces [42] or nanoholes [33]. In agreement with these earlier studies, a significant reduction in T_g near the film surface is supported by our results.

11.5 IPN: Remote Control of Surface Structuration

IPN is due to partial reconstruction of the polymer surface. It is interesting to explore how deep in the film the surface modification is propagated. We have observed that the IPN surface pattern can be altered by modifying the substrate

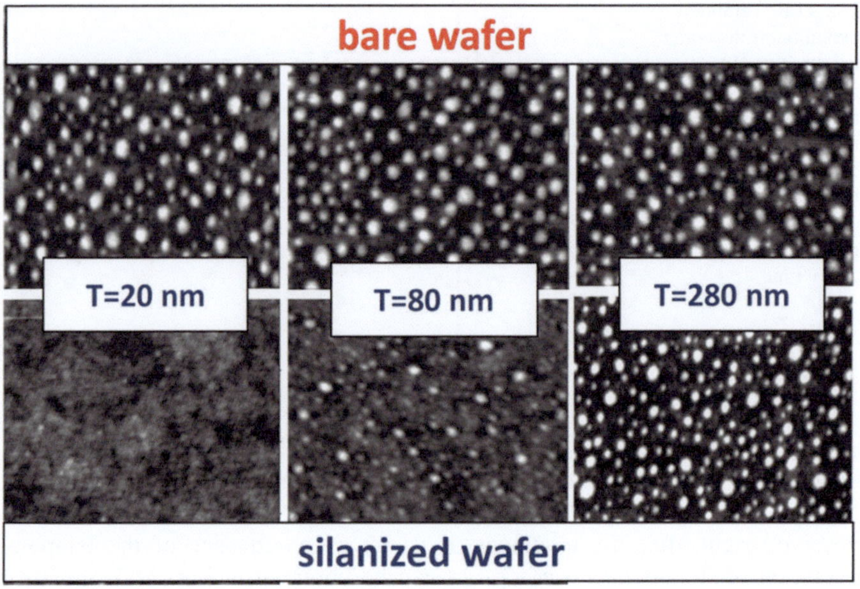

Fig. 11.7 Height AFM micrographs of films of 250 kDa polystyrene films measured in tapping mode in air. The films of indicated thickness were spun coated on bare and silanized silicon wafers (fractional OTS surface coverage 0.7) and then exposed to a degassed solution of nitric acid in double-distilled water at pH 1.5 at room temperature. The presence of asperities of regular nanometric size is clearly observed on some of the films exposed to the degassed solution during 5 min. The silanization of the wafer blocked the structuration of films thinner than the end-to-end radius of the polymer. All the films were featureless before treatment

supporting the film, as shown in Fig. 11.7. The typical radii and heights of the bumps formed in the surface are similar in oxidized Si wafers, and are independent of polymer film thickness down to 10 nm. On the contrary, the characteristic size of the bumps is substantially reduced—or they are not present—for films deposited on hydrophobized wafers for polymer films thinner than the end-to-end radius of the polymer. As mentioned above, the idea that polymer segments close to the surface are more mobile than bulk segments is now broadly accepted, although the properties of the mobile layer (thickness, variation with temperature, and molecular weight) are still debated. On the contrary, it is not obvious why this excess mobility should be impaired by the influence of the underlying substrate buried down several tens of nanometers below the polymer film surface.

The threshold thickness for surface structuration is determined by polymer size: the larger the molecular weight, the longer the range of the effect of the substrate on the structuration of the polymer layer. In all cases the observed effect has a range too large—up to 100 nm for the largest polymer investigated—to be interpreted in terms of the dispersion interaction between the solid substrate and water through the polymer film. The reason for the long-range effect of the substrate on the polymer structuration is related to the connectivity of the polymer chains. A complete study of this effect has been reported before [43].

11.6 IPN: An Easy Waterborne Process for Polymer Surface Structuration

The procedure described here may be of interest for producing controlled nanostructured structures on flat, non-planar, or hollow hydrophobic substrates in a single simple step allowing an easy and precise control of the nanopatterns produced. Some realizations of this idea are presented in Fig. 11.8: the nanostructuration of a monolayer of PS microspheres—a substrate that would be impossible to pattern with conventional lithographic techniques—or the selective patterning of a PS substrate partially protected by a PDMS mask were performed in a simple waterborne step. Selective patterning of the surface of a polymer film can be achieved if the film is deposited on a patterned substrate that selectively avoids IPN to proceed, taking advantage of the process described in the previous section, as illustrated in Fig. 11.9.

The obtained patterned polymer surfaces can also be replicated by metal thermal evaporation to produce nanostructured metallic films with holes or asperities of controlled size, as illustrated in Fig. 11.10. After deposition of a sufficiently thick metal layer, the polymer layer can be cleaved or dissolved away. This procedure allows an efficient and precise control of the metallic surface structure, with possible applications in materials science and photonics. The roughness of polydimethylsiloxane (PDMS) surfaces can be tuned by this technique if the PDMS is treated while cross-linking, which may be of interest for microfluidic applications. We have also observed that substrates of poly(methyl methacrylate) (PMMA), PS in the form of colloidal spheres and bulk, and semicrystalline films of polyethylene (PE) are prone to be structured by this technique, evidencing the versatility and potential for its widespread use. It may find applications in many different scientific and technological fields like nanolithography, microfluidics, or flexible electronics.

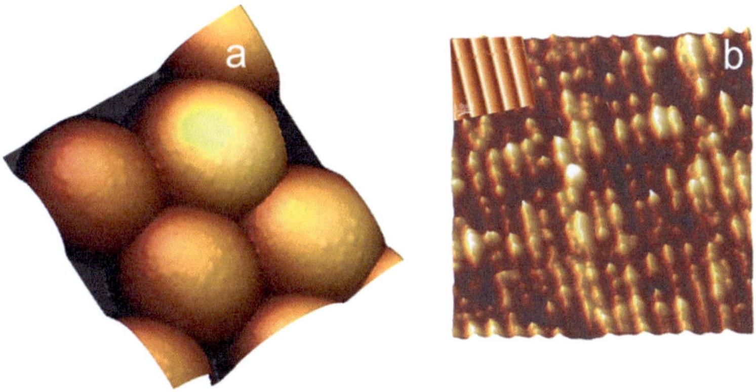

Fig. 11.8 AFM micrographs taken in air show that the surface nanostructuration was also observed in (**a**) a layer of PS latex (radii 0.2 μm) deposited by dip-coating from a concentrated aqueous solution (10 g/l) on top of a smooth mica surface, (**b**) in bulk PS partially protected by a sinusoidal micro-wrinkled PDMS stamp ($\lambda = 2$ μm; see *inset*). Adapted with permission from [21]. Copyright (2011) American Chemical Society

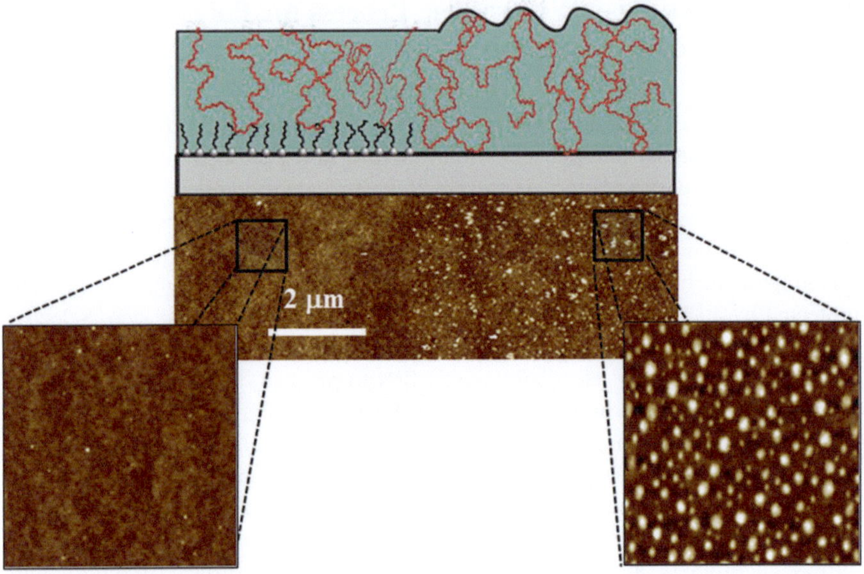

Fig. 11.9 Height tapping-mode AFM micrograph measured in air of 50 nm thick 250 kDa polystyrene films after exposure to a degassed solution of nitric acid in double-distilled water during 5 min. pH 1.5; T 25 °C. The film was spin coated on a half silanized silicon wafer. Scale bar corresponds to 2 µm. The presence of asperities of regular nanometric size bumps is clearly observed on the bare silicon wafer. On the contrary, no modification was detected in the region which was coated with OTS. Adapted with permission from (47). Copyright (2011) American Chemical Society

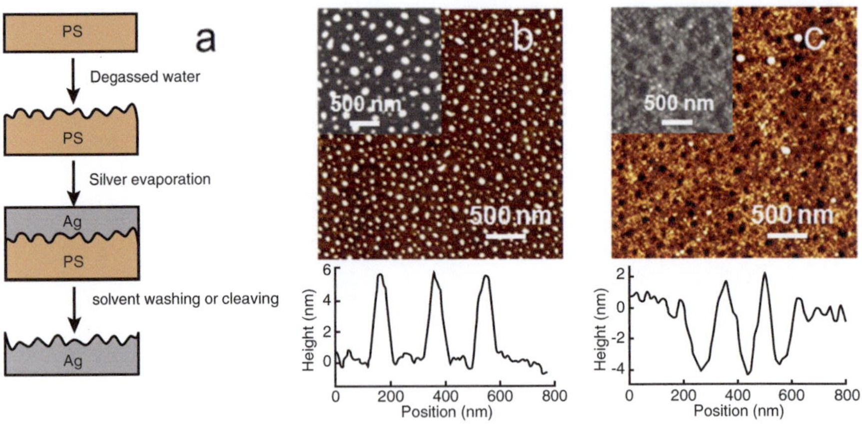

Fig. 11.10 The polymer-modified surfaces can also be used as a template to pattern metallic films as outlined in (**a**). Evaporating a 250 nm thick silver film on a structured PS film and removing the polymer film after gluing the stack onto a convenient substrate allow producing a metallic film of controlled surface structure with bumps (**b**) or holes (**c**). Adapted with permission from [21]. Copyright (2011) American Chemical Society

References

1. Parker, J.L., Claesson, P.M., Attard, P.: Bubbles, cavities, and the long-ranged attraction between hydrophobic surfaces. J. Phys. Chem. **98**, 8468–8480 (1994)
2. Zhang, X.H., Zhang, X., Sun, J., Zhang, Z., Li, G., Fang, H., Xiao, X., Zeng, X., Hu, J.: Detection of novel gaseous states at the highly oriented pyrolytic graphite-water interface. Langmuir **23**, 1778–1783 (2007)
3. Seddon, J.R.T., Bliznyuk, O., Kooij, E.S., Poelsema, B., Zandvliet, H.J.W., Lohse, D.: Dynamic dewetting through micropancake growth. Langmuir **26**, 9640–9644 (2010)
4. Attard, P.: The stability of nanobubbles. Eur. Phys. J. Spec. Top. 1–22 (2013). doi:10.1140/epjst/e2013-01817-0
5. Seddon, J.R.T., Lohse, D.: Nanobubbles and micropancakes: gaseous domains on immersed substrates. J. Phys. Condens. Matter **23**, 133001 (2011)
6. Zhang, X., Khan, A., Ducker, W.: A nanoscale gas state. Phys. Rev. Lett. **98**, 136101 (2007)
7. Martinez, J., Stroeve, P.: Transient behavior of the hydrophobic surface/water interface: from nanobubbles to organic layer. J. Phys. Chem. B **111**, 14069–14072 (2007)
8. Karpitschka, S., Dietrich, E., Seddon, J.R.T., Zandvliet, H.J.W., Lohse, D., Riegler, H.: Nonintrusive optical visualization of surface nanobubbles. Phys. Rev. Lett. **109**, 066102 (2012)
9. Switkes, M., Ruberti, J.W.: Rapid cryofixation/freeze fracture for the study of nanobubbles at solid–liquid interfaces. Appl. Phys. Lett. **84**, 4759 (2004)
10. Seddon, J.R.T., Kooij, E.S., Poelsema, B., Zandvliet, H.J.W., Lohse, D.: Surface bubble nucleation stability. Phys. Rev. Lett. **106**, 056101 (2011)
11. Ducker, W.A.: Contact angle and stability of interfacial nanobubbles. Langmuir **25**, 8907–8910 (2009)
12. Brenner, M., Lohse, D.: Dynamic equilibrium mechanism for surface nanobubble stabilization. Phys. Rev. Lett. **101**, 214505 (2008)
13. Mazumder, M., Bhushan, B.: Propensity and geometrical distribution of surface nanobubbles: effect of electrolyte, roughness, ph, and substrate bias. Soft Matter **7**, 9184 (2011)
14. Onsager, L., Samaras, N.N.T.: The surface tension of Debye-Hückel electrolytes. J. Chem. Phys. **2**, 528 (1934)
15. Noah-Vanhoucke, J., Geissler, P.L.: On the fluctuations that drive small ions toward, and away from, interfaces between polar liquids and their vapors. Proc. Natl. Acad. Sci. U. S. A **106**, 15125–15130 (2009)
16. Kudin, K.N., Car, R.: Why are water-hydrophobic interfaces charged? J. Am. Chem. Soc. **130**, 3915–3919 (2008)
17. Jungwirth, P., Tobias, D.J.: Molecular structure of salt solutions: a new view of the interface with implications for heterogeneous atmospheric chemistry. J. Phys. Chem. B **105**, 10468–10472 (2001)
18. Gray-Weale, A., Beattie, J.K.: An explanation for the charge on water's surface. Phys. Chem. Chem. Phys. **11**, 10994–11005 (2009)
19. Zangi, R., Engberts, J.B.F.N.: Physisorption of hydroxide ions from aqueous solution to a hydrophobic surface. J. Am. Chem. Soc. **127**, 2272–2276 (2005)
20. Hunter, R.J.: Foundations of colloid science, 2nd edn. Oxford University Press, Oxford (2001)
21. Siretanu, I., Chapel, J.P., Drummond, C.: Water-ions induced nanostructuration of hydrophobic polymer surfaces. ACS Nano **5**, 2939–2947 (2011)
22. Pashley, R.M.: Effect of degassing on the formation and stability of surfactant-free emulsions and fine teflon dispersions. J. Phys. Chem. B **107**, 1714–1720 (2003)
23. Beattie, J.K., Djerdjev, A.M.: The pristine oil/water interface: surfactant-free hydroxide-charged emulsions. Angew. Chem. Int. Ed. Engl. **43**, 3568–3571 (2004)
24. Maeda, N., Rosenberg, K.J., Israelachvili, J.N., Pashley, R.M.: Further studies on the effect of degassing on the dispersion and stability of surfactant-free emulsions. Langmuir **20**, 3129–3137 (2004)

25. Snoswell, D.R.E., Duan, J., Fornasiero, D., Ralston, J.: Colloid stability and the influence of dissolved gas. J. Phys. Chem. B **107**, 2986–2994 (2003)
26. Meyer, E.E., Lin, Q., Israelachvili, J.N.: Effects of dissolved gas on the hydrophobic attraction between surfactant-coated surfaces. Langmuir **21**, 256–259 (2005)
27. Pashley, R.M., Rzechowicz, M., Pashley, L.R., Francis, M.J.: De-gassed water is a better cleaning agent. J. Phys. Chem. B **109**, 1231–1238 (2005)
28. Dai, Z., Fornasiero, D., Ralston, J.: Influence of dissolved gas on bubble-particle heterocoagulation. J. Chem. Soc. Faraday Trans. **94**, 1983–1987 (1998)
29. Cottin-Bizonne, C., Cross, B., Steinberger, A., Charlaix, E.: Boundary slip on smooth hydrophobic surfaces: intrinsic effects and possible artifacts. Phys. Rev. Lett. **94**, 056102 (2005)
30. Wang, Y., Bhushan, B., Zhao, X.: Nanoindents produced by nanobubbles on ultrathin polystyrene films in water. Nanotechnology **20**, 045301 (2009)
31. Janda, P., Frank, O., Bastl, Z., Klementová, M., Tarábková, H., Kavan, L.: Nanobubble-assisted formation of carbon nanostructures on basal plane highly ordered pyrolytic graphite exposed to aqueous media. Nanotechnology **21**, 095707 (2010)
32. Tarábková, H., Janda, P.: Nanobubble assisted nanopatterning utilized for ex situ identification of surface nanobubbles. J. Phys. Condens. Matter **25**, 184001 (2013)
33. Fakhraai, Z., Forrest, J.A.: Measuring the surface dynamics of glassy polymers. Science **319**, 600–604 (2008)
34. Manciu, M., Ruckenstein, E.: On the interactions of ions with the air/water interface. Langmuir **21**, 11312–11319 (2005)
35. Levin, Y.: Polarizable ions at interfaces. Phys. Rev. Lett. **102**, 147803 (2009)
36. Parsegian, V.A.: Ion-membrane interactions as structural forces. Ann. N. Y. Acad. Sci. **264**, 161–174 (1975)
37. Horinek, D., Herz, A., Vrbka, L., Sedlmeier, F., Mamatkulov, S.I., Netz, R.R.: Specific ion adsorption at the air/water interface: the role of hydrophobic solvation. Chem. Phys. Lett. **479**, 173–183 (2009)
38. Shanahan, M.E., Carré, A.: Spreading and dynamics of liquid drops involving nanometric deformations on soft substrates. Colloids. Surf. A. Physicochem. Eng. Asp. **206**, 115–123 (2002)
39. Zimmermann, R., Freudenberg, U., Schweiß, R., Küttner, D., Werner, C.: Hydroxide and hydronium ion adsorption – a survey. Curr. Opin. Colloid Interface Sci. **15**, 196–202 (2010)
40. Jungwirth, P., Tobias, D.J.: Specific ion effects at the air/water interface. Chem. Rev. **106**, 1259–1281 (2006)
41. Siretanu, I., Chapel, J.-P., Bastos-González, D., Drummond, C.: Ions-induced nanostructuration: effect of specific ionic adsorption on hydrophobic polymer surfaces. J. Phys. Chem. B **117**, 6814–6822 (2013)
42. Buck, E., Petersen, K., Hund, M., Krausch, G., Johannsmann, D.: Decay kinetics of nanoscale corrugation gratings on polymer surface: evidence for polymer flow below the glass temperature. Macromolecules **37**, 8647–8652 (2004)
43. Siretanu, I., Chapel, J.P., Drummond, C.: Substrate remote control of polymer film surface mobility. Macromolecules **45**, 1001–1005 (2012)

Chapter 12
Nanobubble-Assisted Nanopatterning

Pavel Janda

In the previous chapter it was described how degassing aqueous solutions in contact with hydrophobic polymers open pathways for polymer surface patterning. In the absence of degassing, nanobubbles can nucleate on hydrophobic surfaces. In this chapter the structuring effect of nanobubbles on hydrophobic surfaces is discussed.

12.1 Introduction

Surface gaseous nanobubbles (NB) were first recognized in 1994 by Phil Attard et al. [1] as long-range (10^2 nm) attractive forces between two adjacent hydrophobic planar solids immersed in water. First nanobubble images made in situ by atomic force microscopy (AFM) in tapping mode were published by Phil Attard in 2001 [2]. However, even then nanobubbles were not widely accepted and were instead considered as AFM artifacts. Since then, gaseous nanobubbles have been recognized as a real, though somewhat peculiar phenomenon, existing at the solid/liquid interface. Their extensive study is resulting in exponentially grown number of publications dealing with aspects of their existence and various models in an attempt to explain their seeming disobedience of some thermodynamic rules such as Young-Laplace law [3]. Surface gaseous nanodomains—nanobubbles and nano/micro-pancakes appear frequently as rounded objects adhering to immersed solid surfaces with the coverage varying from few percent to more than 80 %. From that point of view gaseous nanostructures became the issue of growing importance even for some diverse fields like heterogeneous catalysis, liquid immersion lithography, electrochemistry, and interfacial physical chemistry.

P. Janda (✉)
Scanning Probe Microscopy Group, Department of Electrochemical Materials, J. Heyrovsky Institute of Physical Chemistry AS CR, v.v.i. Dolejskova 3, 182 23 Prague 8, Czech Republic
e-mail: pavel.janda@jh-inst.cas.cz

J. Rodríguez-Hernández, C. Drummond (eds.), *Polymer Surfaces in Motion*,
DOI 10.1007/978-3-319-17431-0_12

12.2 Imaging

Nowadays gaseous nanodomains are studied extensively in situ by AFM in semicontact (tapping) mode, which so far appears to be the most reliable method for nanobubble high-resolution imaging (Fig. 12.1).

Moreover AFM-derived methods, namely force spectroscopy and phase and lateral force imaging, allow investigating nanobubble mechanical properties.

During the last few years optical visualization techniques have also been introduced. Among them the total internal reflection fluorescence excitation (TIFR) microscopy [4] and optical interference-enhanced reflection microscopy [5] appear to be the most promising nonintrusive techniques. Their resolution, however, does not even approach the resolution of atomic force microscope and optical techniques may thus serve as an image survey of nanobubbles at 300 nm level (diameter) which is so far their resolution limit.

12.3 Interactions with Solid Polymer Surfaces

While nanobubble properties including gas exchange dynamics, mechanical properties, stability, composition, and thermodynamics have been extensively examined, substantially less attention has been paid to nanobubble interactions

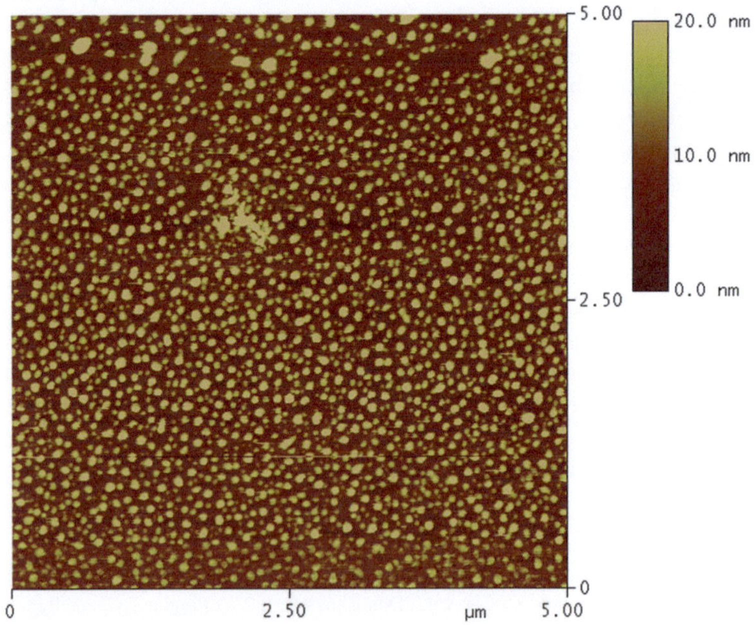

Fig. 12.1 In situ AFM (tapping) image showing the nanobubble population found on hydrophobic surface immersed in deionized water at room temperature

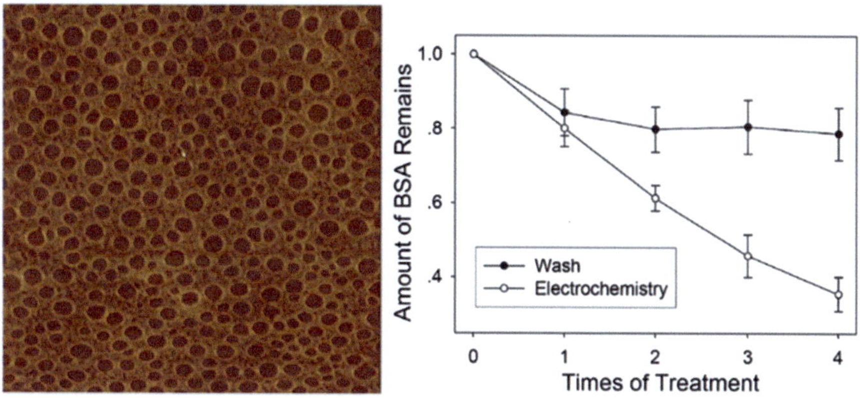

Fig. 12.2 5 × 5 μm^2 AFM height image of BSA protein adsorbed on HOPG with nanobubbles formed prior to protein adsorption (*left*, image scale: 5 × 5 μm, height scale: 10 nm) and BSA removal (surface fraction on axial axis) by electrochemically generated nanobubbles (*right*) (from Z.H. Wu et al. [7])

with solid surfaces and to forces exerted at the nanobubble ternary interface. Following is the overview on the progress in the field keeping its chronological order:

Zhihua Wu et al. [6, 7] have found that nanobubble interfacial forces can interact with surface deposits performing defouling and blocking the surface, respectively, preventing its contamination. The layer of biopolymer bovine serum albumin (BSA) deposited on basal plane highly ordered pyrolytic graphite (HOPG) was patterned by nanobubbles preventing BSA from adsorption, once nanobubbles were formed prior to BSA deposition (Fig. 12.2).

While surface areas preoccupied by nanobubbles can be prevented from adsorption, electrochemically produced nanobubbles can cause ultimate removal of deposits.

Similar observation was made by V. Kolivoška et al. [8] who correlated appearance of gaseous nanobubbles and nanopancakes on BSA-coated substrate with the pattern formed on the surface.

In 2009 Y. Wang, B. Bhushan, and X. Zhao [9] came with the interesting finding that nanobubbles may cause rearrangement of polymeric hydrophobic surfaces, which they adhere to. They correlated in situ AFM images of nanobubbles appearing on supported thin polystyrene film immersed in water with nanoindents gradually formed at nanobubble positions (Fig. 12.3).

Nanoindents with diameter ~20 nm and density 2×10^8 mm^{-2} were found to develop slowly within almost 3-h period. The authors presented a model ascribing the surface rearrangement to forces existing at three-phase contact line of nanobubble on polymer surface. Using the horizontal force component F_{h} determined by the circular diameter of nanobubble contact line D and surface tension of water $\gamma = 72$ mN m^{-1} the value of tensile stress σ at nanobubble ternary interface was found to be $\sigma = 24.8$ MPa (for polymer film thickness $h = 2.8$ nm). Though this value is below polystyrene bulk tensile stress (30–60 MPa), considering reduced

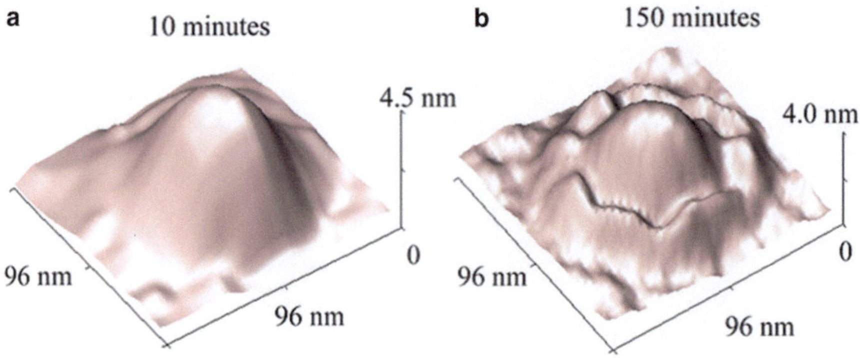

Fig. 12.3 3D AFM image (tapping) of nanobubble at the polystyrene surface immersed into deionized water. Time development after 10 min (**a**) and 150 min (**b**), respectively, shows rim formation around the nanobubble (from. Y. Wang et al. [9])

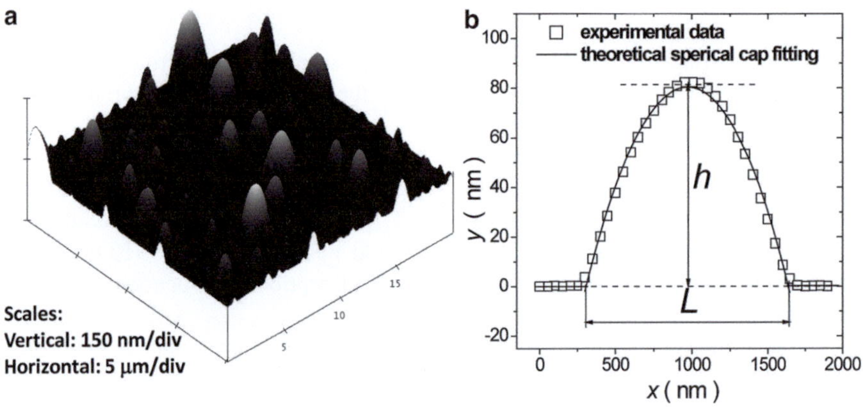

Fig. 12.4 AFM image of PMMA micro/nanobubbles after 186-min soaking in water (*left*) and the spherical cap model fitting (*right*). The film thickness is 47 nm (from B. Jing et al. [10])

entanglement density between polystyrene chains in thin layers can make the film rearrangement by NB interfacial forces feasible.

Morphological instability of polymethylmethacrylate (PMMA) film on silicone substrate was reported by B. Jing et al. [10] who found partial detachment of PMMA film from the supporting substrate as a result of formation of PMMA micro- and nanobubbles after immersion in water (Fig. 12.4).

The process is reversible and the radius of bubble curvature was found to be dependent on the thickness of PMMA film (Fig. 12.5) and on its Young modulus, while it was independent on thermal annealing treatment.

The equibiaxial compressive stress within the film, responsible for PMMA bubble formation, is developed by swelling the outer layer of the PMMA film, which is in contact with water (Fig. 12.6). The change in PMMA interfacial tension

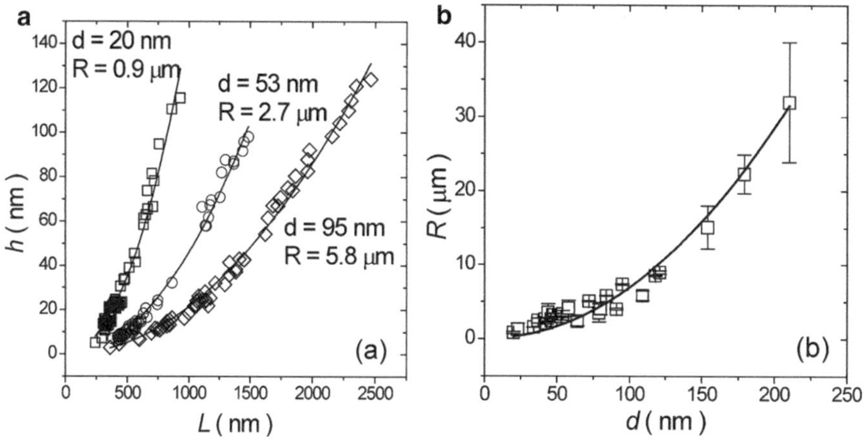

Fig. 12.5 Dimensions (h, L) of PMMA bubble at different film thickness (**a**). Radius of curvature (***R***) of the bubbles as a function of thicknesses of the PMMA film (**b**). The *solid line* is obtained from the theoretical model (from B. Jing et al. [10])

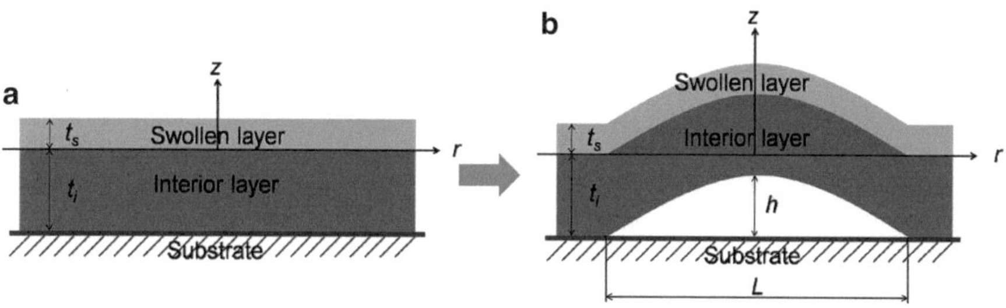

Fig. 12.6 Scheme of deformation of supported PMMA film upon water adsorption: Immediately after water swelling (**a**), layer deformation upon swelling for longer period of time (**b**) (from B. Jing et al. [10])

is expected to be negligible. The internal space of PMMA bubble is believed to be filled with ambient gas (air), for which thin PMMA film has high permeability.

This finding is in accordance with earlier observation of Tanaka et al. [11] who reported on nonuniform swelling of PMMA in water indicated by a neutron reflectivity measurement. They found that the thickness of the swelling layer was ~10 nm.

While polymeric blister and bubble formation both lead to the local lifting of polymer film from supporting substrate and impaired polymer-substrate adhesivity represents an important prerequisite, gaseous nanobubbles appear to act from the top on strictly localized area.

The existence of forces acting at the nanobubble ternary interface was further confirmed by the discovery of nanobubble-assisted exfoliation of basal planes of HOPG (Fig. 12.7), which was found to take place at room temperature in deionized water [12].

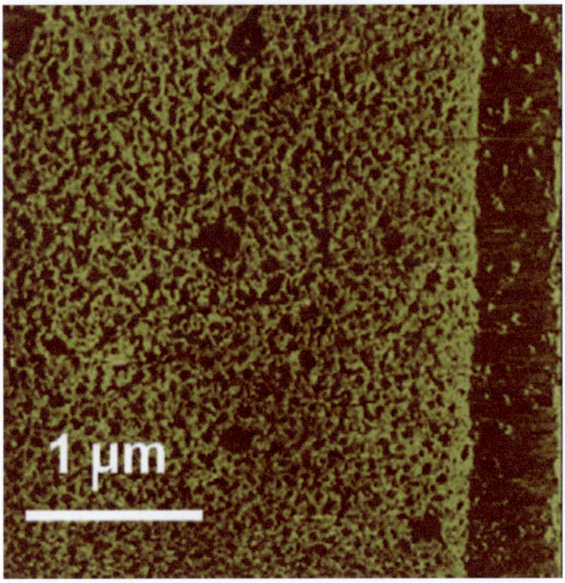

Fig. 12.7 In situ AFM (tapping) image of basal plane HOPG surface rearranged by nanobubbles acting in deionized water at room temperature (from P. Janda et al. [12])

Advantageously, the surface rearrangement of graphene layers leading to graphene-based nanoparticles was independently confirmed by Raman spectroscopy in situ showing increase of D-mode and also by diffraction techniques. Accordingly, transmission electron microscopy revealed graphene-based nanoscrolls and buckles formed by tearing graphene sheets and by subsequent scrolling.

Unlike nanobubble-oriented patterning models, Siretanu et al. [13] ascribed self-assembled rearrangement of hydrophobic polymer surface immersed in aqueous media to electro-hydrodynamic instability due to adsorption of hydronium and hydroxyl ions at the interface. Reportedly [13], both the instability and nanostructure are controlled by the charge of hydrophobic surface, and depend on the pH and inversely on the solvent saturation by gas. Authors describe the most pronounced nanostructuring process on 300 nm thick polystyrene film proceeding during immersion in degassed aqueous media for time period of minutes at high (11.3) and low (1.5) pH values as shown in Fig. 12.8. The phenomena was described in the previous chapter.

The authors [13] claim that interaction strength between hydrophobic substrate and aqueous phase is in fact negatively affected by the dissolved gas, which can protect the hydrophobic substrate from access of aqueous phase. They assume the existence of a mobile surface layer of polymer, which can be restructured under external influence. Observed relaxation of polymer nanostructures after a month's period allowed elucidating the model, according to which the enhanced mobility of the top of polymer film rather than its irreversible plastic deformation plays the role in the surface rearrangement.

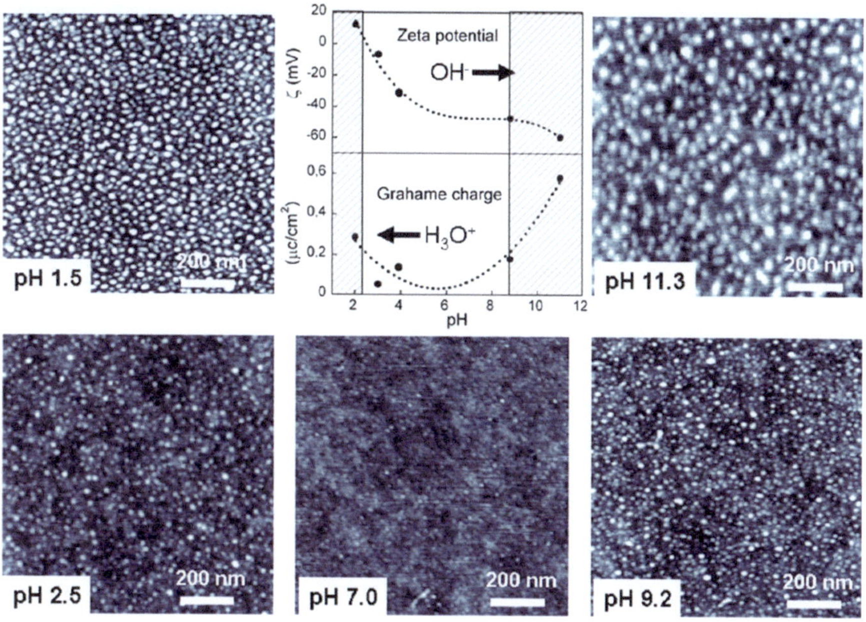

Fig. 12.8 Ex situ AFM images (tapping) of 300 nm polystyrene film treated with degassed aqueous solution at different pH: 1.5 (*top left*), 11.3 (*top right*), 2.5 (*bottom left*), 7.0 (*bottom, middle*), and 9.2 (*bottom right*), varied with HNO_3 and NaOH; zeta potential and charge density plots (*top, middle*) (from I. Siretanu et al. [13])

Away from the abovementioned electro-hydrodynamic instability concept, gaseous nanobubbles and forces at the ternary (gas/liquid/solid) interface in particular were recognized as having significant interface rearranging potential. A. Duisterwinkel and S. Yang [14] addressed again the surface defouling issue (Fig. 12.9), pointing to nanobubble cleaning capability.

Nanobubble cleaning was found to be highly effective on both plain and nanopatterned (trench/ridge) wafer without damaging the wafer surface.

The nanopatterning of polystyrene film was reported by H. Tarábková and P. Janda [15]. Supported polystyrene film was immersed in deionized water at room temperature and exposed to nanobubbles formed at the PS/water interface. It was assumed that the surface coverage by nanobubbles depends on flooding conditions and may reach up to 80 % (Fig. 12.1). Within the time period of seconds the surface nanopatterning was formed as imaged by AFM (Fig. 12.10).

Random nanobubbles and nanobubble aggregates were found to form random and netlike nanopatterning with higher degree of ordering, respectively, as shown in Fig. 12.10. The good agreement between statistical distribution functions of both nanobubbles and nanopattern indicates their mutual relation [15]. This work also points out the possibility of using decoration of surface nano/microbubble positions formed on hydrophobic polymer (PS) matrix, for examination of nano/microbubble appearance ex post by relatively simple ex situ AFM imaging.

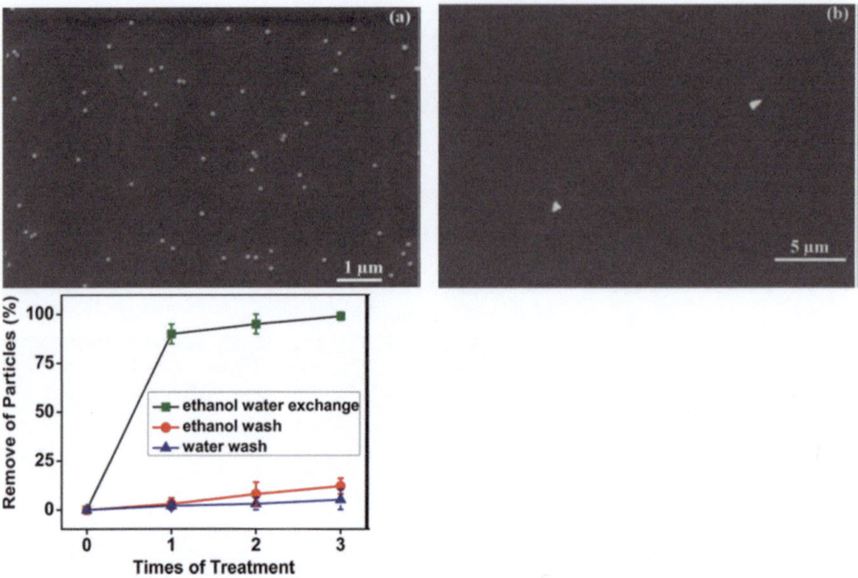

Fig. 12.9 The SEM image of plain wafer surface deposited with the 100 nm polystyrene particles (**a**). The same surface after treated with the ethanol-water exchange process, known to generate surface gaseous nanobubbles (**b**). Nanoparticles are removed with 100 % removal efficiency with no extra damage of the surface as compared with removal rate by ethanol and water rinsing itself (*bottom plot*) (from S. Yang and A. Duisterwinkel [14])

Further in-depth study of nanobubble-assisted nanopatterning of various polymeric surfaces is presented by H. Tarábková, Zdeněk Bastl, and Pavel Janda [16] (Fig. 12.11).

The nanopatterning appearance is regarded as the nanobubble decoration formed by the polymer rearranged by nanobubble interfacial forces, arising upon nanobubble expansion after mild pressure drop is applied on the interface.

Tensile stress imposed by shrinking nanobubble perimeter is expected to squeeze and push up the surface delimited by surface nanobubble foot radius (Fig. 12.12); hence the nanoprotrusions are formed. Nanomorphology, elastic properties, and thickness of the hydrophobic film are among properties influencing pattern, which may appear in a form of nanopinholes or nanoprotrusions [16].

Forces acting at nanobubble ternary interface originate in surface tension Γ_{LG} determined by the composition of the interface. For ambient gas atmosphere (*a*ir) and deionized *w*ater the surface tension then generates force with the lateral component

$$F_{L} = 2\pi r\ \Gamma_{aw} \cos(180 - \Theta) \quad (12.1)$$

where θ corresponds to interfacial (water-air) contact angle (measured from aqueous phase).

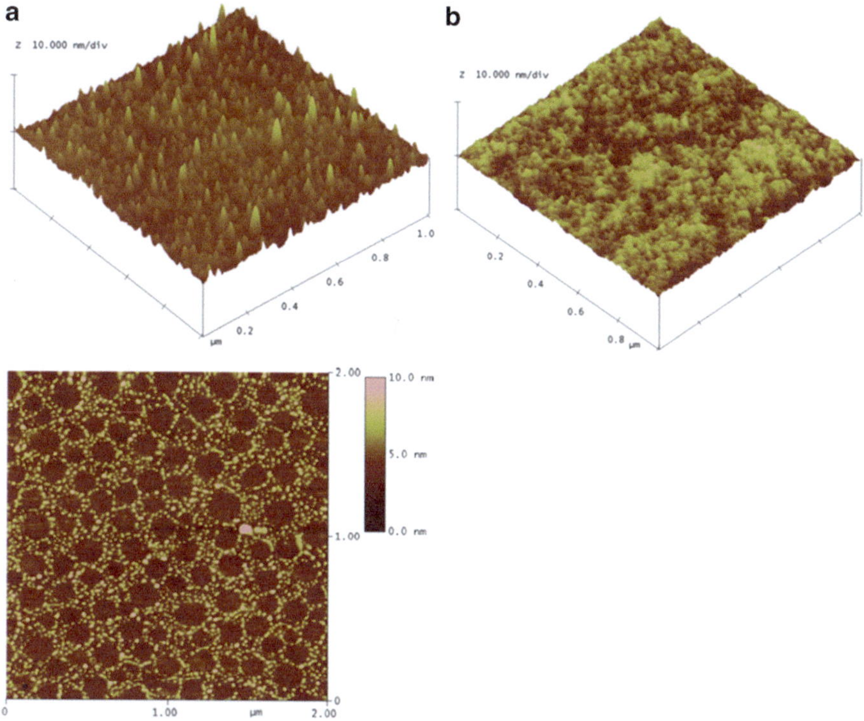

Fig. 12.10 Ex situ AFM image of random nanopatterning of the PS surface after its exposition to nanobubbles in water (**a**), as received PS surface after spin coating on the Si wafer as a reference (**b**), net-nanopatterning by nanobubble aggregates (*bottom image*) (from H. Tarábková and P. Janda [15])

Considering nanobubble interface pinning [17] on solid surface, the force F_L is transferred via pinned interface areas to the underlying solid, developing thus tension stress σ within the polymer surface layer. Its magnitude is given by the magnitude of projected lateral component F_L and by the thickness of the film h:

$$\sigma = F_L / 2\pi r h \tag{12.2}$$

Density functional theory (DFT) calculations [18] suggest the tangential pulling force F_T as a solid/liquid adhesion originated in long-range solid/liquid intermolecular attraction

$$F_T = \Gamma_{LG}(1 + \cos\theta) = \Gamma_{LG} + \Gamma_{SG} - \Gamma_{SL}, \tag{12.3}$$

with interfacial tension components *s*olid/*g*as, *l*iquid/*g*as, and *s*olid/*l*iquid, respectively.

For contact angle θ the lateral and normal components of resulting force (Fig. 12.13) are

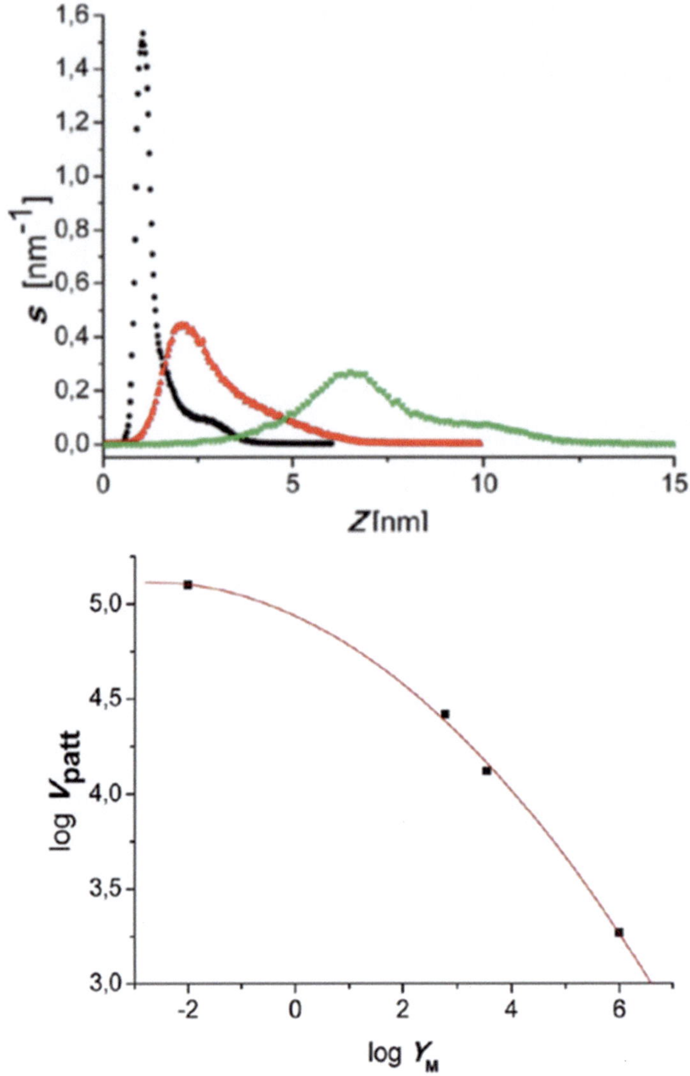

Fig. 12.11 The height density distribution (σ) of nanopattern formed by nanobubbles on different surfaces—HOPG (*black*), PS (*red*), and paraffin (*green*) (*top*) and the pattern volume (V_{patt}) dependence on material properties expressed by elasticity Young modulus (Y_{M}). V_{patt} decreases in the direction of increasing Young modulus in the sequence paraffin, PTFE, PS, and basal plane HOPG (*bottom*) (from H. Tarábková, P. Janda et al. [16])

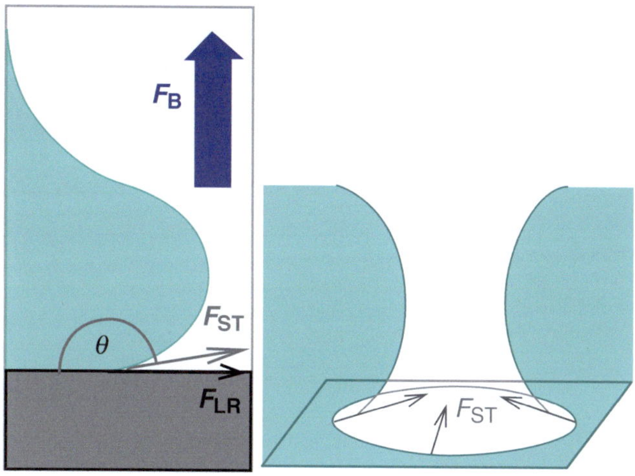

Fig. 12.12 Illustration of forces acting at pinned bubble ternary interface during neck phase of the expansion. Interfacial (surface tension) force F_{ST} is pointed inward the bubble contact perimeter (from H. Tarábková, P. Janda et al. [16])

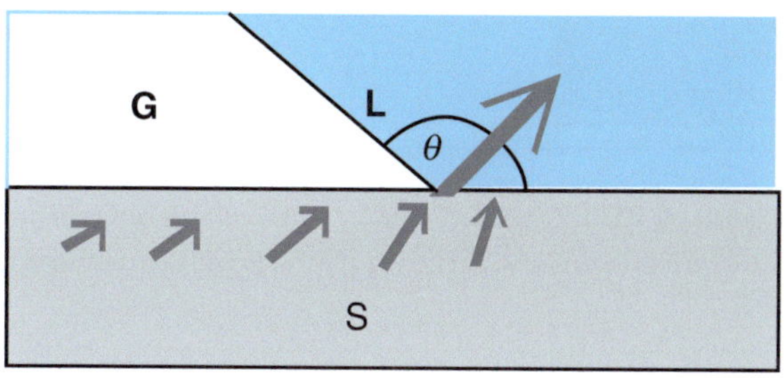

Fig. 12.13 Schematic drawing of forces acting on the *s*olid surface at the *g*as (*bubble*)/*l*iquid interface. Their distribution at the contact (interface) line (contact angle θ) is represented by a projection of solid/liquid adhesion pointed towards liquid phase. Accordingly, the resultant uncompensated pulling force is headed towards liquid phase, while in the distance from the interface the mutual *S–L* attraction is balanced [19]

$\Gamma_{LG}(1 + \cos\theta)$ and Γ_{LG} sinθ, respectively [19]. For nanobubbles with assumed pinned contact interface, when their expansion enters neck phase [20], the solid surface is pulled by the interface attempting to shrink its radius.

The aggregated effect of lateral force developed by the shrinking interface (12.1) together with pulling force (Fig. 12.13) on the solid surface is manifested (via rising tensile stress) by squeezing surface layer within shrinking bubble circular contact line. The magnitude of developed tensile stress is estimated to reach order of 10^2 MPa, which can overbalance the material elasticity point causing its irreversible plastic deformation. Thus, the nanoprotrusion is formed on thick layer, while on

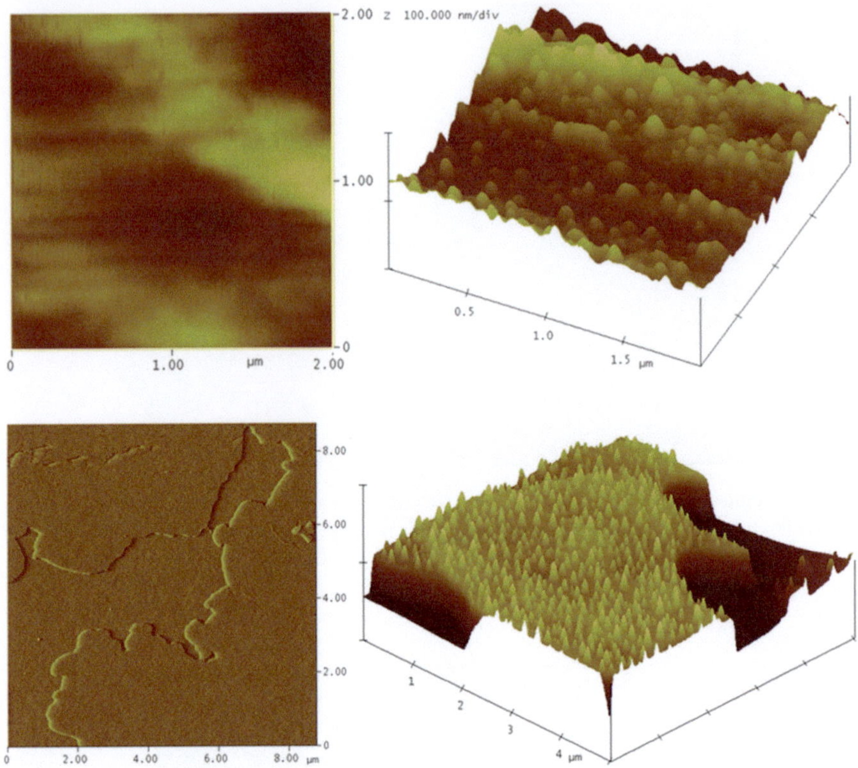

Fig. 12.14 AFM image of PTFE (*top*) and paraffin film (*bottom*) before (*left*) and after (*right*) nanopatterning by surface gaseous nanobubbles (ambient gas) in deionized water, at room temperature (pressure drop −10 kPa)

thin layers with lowered elasticity modulus it may lead to its ultimate removal, forming nanopinholes [16].

Figure 12.14 shows nanobubble-assisted nanopatterning of solid hydrophobic surfaces of different composition and material properties—PTFE (Teflon) and paraffin immersed in water, after mild pressure drop (~ −10 kPa) was applied on the interface.

Conclusively, according to indications gathered, nanobubbles should be considered as large-scale surface-nanopatterning elements. The solid surfaces, which elasticity point is transcended by tension stress exerted by surface nanobubble interfacial forces, are restructured by surface nanobubbles at mild conditions, in the absence of chemical reaction or processes requiring high-energy inputs.

References

1. Parker, J.L., Claesson, P.M., Attard, P.: Bubbles, cavities, and the long-ranged attraction between hydrophobic surfaces. J. Phys. Chem. **98**, 8468 (1994)
2. Tyrrell, J.W., Attard, P.: Images of nanobubbles on hydrophobic surfaces and their interactions. Phys. Rev. Lett. **87**, 176104 (2001)
3. Joost, H., Weijs, D.L.: Why surface nanobubbles live for hours. Phys. Rev. Lett. **110**, 054501 (2013)
4. Chan, C.U., Ohl, C.D.: TIRF microscopy for the study of nanobubble dynamics. Phys. Rev. Lett. **109**, 174501 (2012)
5. Karpitschka, S., Dietrich, E., Seddon, J.R.T., Zandvliet, H.J.W., Lohse, D., Riegler, H.: Nonintrusive optical visualization of surface nanobubbles. Phys. Rev. Lett. **109**, 066102 (2012)
6. Zhihua, W., Zhang, X., Zhang, X., Li, G., Sun, J., Zhang, Y., Li, M., Jun, H.: Nanobubbles influence on BSA adsorption on mica surface. Surf. Interface Anal. **38**, 990–995 (2006)
7. Wu, Z., Chen, H., Dong, Y., Mao, H., Sun, J., Chen, S., Craig, V.S.J., Hu, J.: Cleaning using nanobubbles: defouling by electrochemical generation of bubbles. J.Colloid.Interface.Sci **328**, 10–14 (2008)
8. Kolivoška, V., Gál, M., Hromadova, M., Lachmanová, Š., Tarábková, H., Janda, P., Pospíšil, L., Turoňová, A.M.: Bovine serum albumin film as a template for controlled nanopancake and nanobubble formation: In situ atomic force microscopy and nanolithography study. Colloids Surf. B: Biointerfaces **94**, 213–219 (2012)
9. Wang, Y., Bhushan, B., Zhao, X.: Nanoindents produced by nanobubbles on ultrathin polystyrene (PS) film in water. Nanotechnology **20**, 045301 (2009)
10. Jing, B., Zhao, J., Wang, Y., Yi, X., Duan, H.: Water-swelling-induced morphological instability of a supported polymethyl methacrylate thin film. Langmuir **26**, 7651–7655 (2010)
11. Tanaka, K., Fujii, Y., Atarashi, H., Akabori, K.-i., Hino, M., Nagamura, T.: Nonsolvents cause swelling at the interface with poly(methylmethacrylate) films. Langmuir **24**, 296–301 (2008)
12. Janda, P., Frank, O., Bastl, Z., Klementová, M., Tarábková, H., Kavan, L.: Nanobubble-assisted formation of carbon nanostructures on basal plane highly ordered pyrolytic graphite exposed to aqueous media. Nanotechnology **21**, 095707 (2010)
13. Siretanu, I., Chapel, J.P., Drummond, C.: Nanostructuration of hydrophobic polymer surfaces. ACS Nano **5**, 2939–2947 (2011)
14. Yang, S., Duisterwinkel, A.: Removal of nanoparticles from plain and patterned surfaces using nanobubbles. Langmuir **27**, 11430–11435 (2011)
15. Tarábková, H., Janda, P.: Nanobubble assisted nanopatterning utilized for ex situ identification of surface nanobubbles. J. Phys. Condens. Matter **25**, 184001 (2013)
16. Tarábková, H., Bastl, Z., Janda, P.: Surface rearrangement of water-immersed hydrophobic solids by gaseous nanobubbles. Langmuir **30**, 14522–14531 (2014)
17. Liu, Y., Zhang, X.: Nanobubble stability induced by contact line pinning. J. Chem. Phys. **138**, 014706 (2013)
18. Das, S., Marchand, A., Andreotti, B., Snoeijer, J.H.: Elastic deformation due to tangential capillary forces. Phys. Fluids **23**, 072006 (2011)
19. Marchand, A., Weijs, J.H., Snoeijer, J.H., Andreotti, B.: Why is surface tension a force parallel to the interface? Am. J. Phys. **79**, 999 (2011)
20. Bari, S.D., Robinson, A.J.: Experimental study of gas injected bubble growth from submerged orifices. Exp. Therm. Fluid. Sci **44**, 124 (2013)

Chapter 13
Future Trends of Unconventional Methods in Polymer Surface Patterning

Carlos Drummond and Juan Rodríguez-Hernández

13.1 Patterning Based on Surface Instabilities: State of the Art

The outstanding progress in synthetic polymer science during the last century stimulated the development of a large number of low-cost methods of polymer production, making a growing variety of complex materials available. During the last few decades important efforts have been devoted to developing strategies of polymer patterning, aiming to achieve new functionalities for these widespread materials, largely inspired by examples found in nature.

Instabilities, regardless of their origin, have demonstrated to be an interesting alternative to obtain different morphologies on polymeric surfaces, morphologies that would be difficult or even impossible to obtain by employing "traditional" patterning techniques. In contrast to the "classical" surface structuration approaches that have been usually grouped either in top-down or bottom-up methods, the intrinsic nature of the spontaneous processes occurring in the instability-based patterning methodologies permits the fabrication of a rich number of structures with intricate and complex patterns. The morphology and dynamics of the formed structures are a rich source of information related to the physical processes governing the instability; this knowledge makes it possible to control the output

C. Drummond (✉)
Centre de Recherche Paul Pascal, CNRS-Université Bordeaux 1,
Avenue Albert Schweitzer, 33600 Pessac, France
e-mail: drummond@crpp-bordeaux.cnrs.fr

J. Rodríguez-Hernández (✉)
Instituto de Ciencia y Tecnología de Polímeros, CSIC, Juan de la Cierva 3,
28006 Madrid, Spain
e-mail: jrodriguez@ictp.csic.es

J. Rodríguez-Hernández, C. Drummond (eds.), *Polymer Surfaces in Motion*,
DOI 10.1007/978-3-319-17431-0_13

(by tuning the relevant control variables or guiding the process dynamics) and to envisage practical applications of the process.

The methods described in this volume are typically easy to carry out, are relatively fast, and do not require expensive equipment. Thus, they offer unique possibilities to be transferred to the industrial or prototype level, since they are easily scalable. The tools to pattern polymer surfaces described through this book included approaches in which scientists have been interested during decades (e.g., surface-directed patterning or block copolymer self-assembly) with others that are not more than 5 years old (e.g., nanobubbles, coffee rings, or electrostatic patterning in aqueous media). Old and new methods share common features and allow the preparation of particular surface patterns with typical lengths from the micrometer down to the nanometer scale.

On the downside, there are complications associated to most of the processes described in this volume. The most apparent is probably the randomness accompanying the outcome of most of these methods, which generates a certain degree of uncertainty on the formed structures. As has been illustrated through the book, some control can be achieved by guiding the processes using patterned substrates, electrodes, or masks. While a complete control of the resulting morphology is necessary in many applications, some structural randomness is not harmful (and can even be beneficial) in many others. Adhesion or friction can be determined by the presence of a particular microstructure, regardless of its fine details. Thus, applications in enhanced and reusable adhesives, haptic technology, or drag reduction do not require detailed control of the surface mesostructure. Similarly, applications related to super-hydrophobicity or self-cleaning surfaces are independent of the fine details of the structure. For some optical applications (e.g., antireflective coatings, structural colors, specific visual properties, cosmetic applications) the existence of a characteristic size or a certain hierarchy in the surface structure may be enough to achieve a particular function.

Even though detailed control of the surface morphology may be out of reach for most of the techniques presented, in most of them the general patterns and characteristic length scales produced can be easily delimited. Some of the methods described make use of a control variable to master the process. Some others require a fine control of the process to adjust the formed structure. In the first group we found the methods that resort to external forces (electrical, mechanical, etc.) to induce the instability that finally results on a modification of the surface morphology. As a consequence, the control of the intensity of the external stimulus directs the extent in which the surface morphology can be altered and the dimensions of the final pattern. In the second group we found techniques based on the dynamic evolution of the systems in response to a changing environment. A fine control of the process is then necessary to master the resulting pattern.

13.2 Future Developments in a Young Field

As mentioned above, whereas some approaches and the mechanisms involved to obtain a particular pattern are already well known, some of these methodologies have been only recently reported. Thus, further advances towards the understanding of the fundamental mechanism directing the modification of the surface morphology are necessary. New pathways of surface morphology control should result of these studies.

The approaches covered in this book have been explored in separate basis in the different chapters. These individual strategies may be the foundation for further developments in which two or more patterning processes act simultaneously, to achieve several levels of structuration and greater complexity. As an example, the breath figures methodology employed to prepare micrometer-size porous films can be combined with the self-assembly process occurring in block copolymers and produce unprecedented hierarchically structured materials. Analogously, the combination of chemical and morphological patterning can be easily implemented by some of the methods presented. This multi-technique approach will open pathways for controlled, cost-effective production of complex, three-dimensional polymer materials in the near future, which may find applications in a wide range of fields, from tissue engineering and other biomedical applications to flexible and advanced microelectronics.